Middleware
for Network Eccentric and
Mobile Applications

Benoît Garbinato · Hugo Miranda ·
Luís Rodrigues
Editors

Middleware
for Network Eccentric and
Mobile Applications

 Springer

Benoît Garbinato
Université de Lausanne
Fac. Hautes Etudes
Commerciales (HEC)
Dorigny
1015 Lausanne
Switzerland
benoit.garbinato@unil.ch

Hugo Miranda
Universidade de Lisboa
Fac. Ciencias
Depto Informática
Campo Grande
1749-016 Lisboa
Bloco C6, Sala 6.3.12
Portugal
hmiranda@di.fc.ul.pt

Luís Rodrigues
INESC-ID
Instituto Superior Técnico
Rua Alves Redol 9
1000-029 Lisboa
Portugal
ler@ist.utl.pt

ISBN 978-3-642-10053-6   e-ISBN 978-3-540-89707-1

ACM Computing Classification (1998): C.2, D.2

Cover design: KuenkelLopka GmbH

Printed on acid-free paper

9 8 7 6 5 4 3 2 1

springer.com

To the memory of Kimmo Raatikainen.

Preface

During the past decade, we have observed impressive scientific, technological and experimental advances in the area of mobile ad hoc networks (MANETs). Although this emerging network technology is considered one of the main enabler for future mobile applications, today we are lacking appropriate middleware abstractions that adequately address the requirement of such challenging environments. Yet, middleware is a critical component needed to leverage the development of a wide range of applications for MANETS. Such applications range from so-called mobile and ubiquitous applications to mobile peer-to-peer applications.

The challenges of building middleware for mobile environments are considerable. In particular, middleware solutions must abandon the strong assumption that there exists some underlying networking infrastructure upon which they can rely for communication. In some sense, this new class of emerging applications, and thus its supporting middleware, are not only mobile but also *network eccentric*, meaning they no longer rely on a network infrastructure at their core. In this context, the diversity of networking constraints and application requirements promotes the development of specialized middleware solutions, which are then hard to re-use in different contexts. Furthermore, in order to build successful middleware for mobile ad hoc networks, one requires complementary expertise in the area of distributed systems, networks, algorithms, programming languages, software engineering, and application development.

The challenges we just sketched are exactly those addressed in the context of the ESF MiNEMA (Middleware for Network Eccentric and Mobile Applications) programme, which brings together European research groups from different communities working on middleware for mobile environments. Along that line, this book precisely emerged from the idea to have MINEMA experts compile recent results on middleware for ad hoc networks in a single and consistent volume. For this reason, the book provides a comprehensive introduction to the main fundamental problems, technologies, paradigms, and services that developers of middleware for mobile environments need to know to be successful.

Content

This book offers a coherent and comprehensive set of tutorials on the main topics of relevance for the construction of middleware for mobile environments. Contributions have been carefully selected and revised to provide a broad coverage of MANETs, from networking aspects to application-level aspects. The book structure has been heavily inspired by the contents of the very successfully thematic schools organized by the ESF MiNEMA programme.

Presentation

Each chapter is written in a self-contained manner and starts with the identification of the problem being addressed, followed by a brief discussion of the related work. Each chapter then proceeds to describe illustrative instances of how the issue has been addressed in the literature.

The book is planned to serve as a textbook for advanced courses in mobile ad hoc networking at the graduate level. It is assumed that the reader has some introductory knowledge on existing networking technology and principles and wants to learn more about the specific problems posed by mobile ad hoc networks and the current state of the art in the field.

Organisation

The book is organised in a bottom-up manner and is divided in four main parts, including basic networking and programming issues, communication models, middleware issues, and high-level issues that are specific to a certain class of mobile ad hoc applications. The content of each of these parts and of their respective chapters is sketched hereafter.

Networking and Programming Issues.   This part exposes the reader to a set of fundamental issues related to mobile ad hoc networks that the programmer of middleware and applications needs to be aware of. These include the need to make software that is energy-aware, the constraints that exist when programming for mobile devices, how the mobility patters of the devices can be modeled and, finally, what routing algorithms are available to ensure network connectivity. Namely:

- In Chap. 1, R. Mini and A. Loureiro address the need to consider energy consumption at every stage of the design and operation of mobile devices. They discuss the main energy-aware techniques that can be applied to the design of applications, algorithms and protocols for mobile devices, with emphasis on wireless sensor networks.

- Besides general purpose devices like PDAs and laptops, ad hoc networks are also expected to serve specialised devices, like sensors. These specialised devices impose severe restrictions on developers, making robust programming quite challenging. In Chap. 2, L. Lopes, F. Martins, and J. Barros, survey the current state of the art in programming languages and runtime systems for specialised devices that make up wireless sensor networks.
- Node motion may adversely impact the performance of mobile ad hoc networks by breaking routes and disconnecting nodes. In Chap. 3, M. Musolesi and C. Mascolo present a state-of-the-art survey on the mobility models commonly used in networking and systems research.
- Not surprisingly, packet routing was one of the first problems addressed in the scope of ad hoc routing. It is also one of the most well-studied issues in this domain. Chapter 4 by F. Araújo and H. Miranda surveys the main existing routing algorithms under different network and node assumptions, including the important case where nodes are aware of their own location and can use this information to route packets.

Communication Models. Depending on the way mobile devices are deployed and used, different interaction patterns are established among them. A communication model captures what is common to a representative set of interaction patterns. Thus, these models make the bridge between networking issues and middleware-specific issues. The second part of the book gathers chapters presenting key communication models used in ad hoc networks.

- In Chap. 5, J. Aspnes and E. Ruppert introduce a theoretical model for a collection (or population) of tiny mobile agents that interact with one another to carry out a computation. Their chapter surveys results that describe what can be computed in various versions of the population protocol model.
- Opportunistic networking relies on the idea of exploiting the mobility of nodes to provide connectivity in scenarios where the source and destination nodes might never be connected to the same network at the same time. Routing issues specific to opportunistic networking are discussed in Chap. 6 by M. Conti, J. Crowcroft, S. Giordano, J. Crowcroft, P. Hui, H. Nguyen, and A. Passarella.
- Chapter 7 describes an emerging network architecture, known as mesh network, that is characterised by the partial coverage of the network with infrastructure support. That is, a mesh network can be considered as a mix between a network-centric architecture and a network-eccentric architecture. This type of networks is discussed in Chap. 7 by J. Ishmael and N. Race.

Middleware Issues. The third part covers different problems that typically solved ate middleware level, such as efficient broadcast, structured message dissemination, event dissemination and subscription, distributed coordination, security, and adaptation. Namely:

- R. Friedman, A.-M. Kermarrec, H. Miranda and L. Rodrigues address in Chap. 8 the use of gossip-based protocols in mobile ad hoc networks. Gossip-based protocols have the advantage of requiring little or no structure to operate, making them particularly suitable for dynamic systems such as wireless self-organizing networks.

- In Chap. 9, M. Allani, B. Garbinato and F. Pedone address the problem of structured dissemination of data, which is usually solved via some application layer multicast. In particular, the authors show how it is possible to efficiently disseminate information without forcing nodes to have global knowledge of their networking environment, typically by relying on some form of overlay creation and maintenance.
- Chapter 10 by R. Baldoni, L. Querzoni, S. Tarkoma, and A. Virgillito addresses the issue of achieving scalable information dissemination using the publish/subscribe paradigm. This paradigm is of particular interest for mobile ad hoc networks, as it offers both anonymity and asynchrony to communicating nodes.
- Chapter 11 addresses the implementation of tuple spaces, an abstraction that supports data sharing and coordination among components of a distributed system. P. Costa, L. Mottola, A. Murphy, and G. Picco concisely present the state-of-the-art concerning middleware platforms based on the tuple space abstraction and expressly designed for wireless scenarios.
- Mobile device applications can be highly security- or privacy-sensitive, which is often difficult to ensure due to the inherent limitations of mobile device platforms. In Chap. 12, B. De Win, T. Goovaerts, W. Joosen, P. Philippaerts, F. Piessens, and Y. Younan elaborate on possible threats and applicable security solutions for mobile devices.
- A fundamental characteristic of mobile systems is the variability of their deployment and execution environments. In this context, dynamic adaptation is an essential technique when it comes to ensure that these systems continually provide the required level of service, in spite of continuous changes. In Chap. 13, P. Grace investigates the software techniques for performing adaptation, as well as the adaptive middleware technologies that have been used to develop dynamic mobile applications.

Applicative Issues.   The last part of the book focuses on aspects that are specific to some concrete application areas. In particular, we gather here chapters that discuss issues that are relevant to build context-aware applications, the problem of self-configuration, cyber foraging, and specific issues in the application area of vehicular networking.

- The part begins with Chap. 14, where P. Eugster, B. Garbinato, and A. Holzer address the issue of building context-aware applications, i.e., applications that can explicitly learn about a part of their deployment and execution context (individual or social), and act accordingly. The authors focus in particular on the problem of making context information available to the programmer and propose a categorization of middleware solutions to this problem.
- Entering an ad hoc network requires some negotiation, for example to ensure that each device will own a unique address. Ideally, the negotiation

should be performed without user intervention, in what is usually called self-configuration. The problems raised when trying to achieve self-configuration are addressed by J. Manner in Chap. 15.

- Chapter 16 addresses cyber foraging, the process of opportunistically using resources available in the surroundings. J. Porras, O. Riva, and M. Kristensen give an overview of the cyber foraging process, with a special focus on the challenges that arise in every step of the process, and offer an overview of some existing prototype systems that support cyber foraging.
- Finally, Chap. 17, by A. Senart, M. Bouroche, V. Cahill and S. Weber, discusses recent results on vehicular networks and applications. These networks pose a number of specific challenges, due to the high speed and predefined movement pattern, as well as to the real-time communication requirements that are essential to safety applications.

In summary, the present book provides a unique collection of tutorials, which will allow graduate students in computer science to get familiar with state-of-the-art researches in mobile ad hoc networks and applications, typically on their road to doctoral studies.

It is noteworthy that this book was made possible not only thanks to the dedication and expertise of our contributing authors, many from the ESF MiNEMA programme, but also thanks the excellent work done by our reviewers; we wish to thank them here. Their effort resulted in a book that is an invaluable tool for researchers and developers working this emerging domain.

September 2008 *B. Garbinato, H. Miranda, L. Rodrigues*

Acknowledgements

MiNEMA is a programme sponsored by the European Science Foundation that has gathered researchers from the most relevant European research institutions in the field, and many of the authors of the chapters in the book have actively participated in the programme activities.

The editors are grateful to the European Science Foundation for supporting the Middleware for Network Eccentric and Mobile Applications programme (MiNEMA), a forum that was fundamental for the preparation of this book. The editors also express their gratitude to all authors, for their enthusiastic cooperation in the preparation of this book. Finally, the editors are grateful to the reviewers of this book, for their excellent work and their insightful comments and suggestions.

Contents

Part II Communication Models

Part III Middleware Issues

Acronyms

AACS	Arithmetic Attribute Constraint Summary
ACM	Association for Computing Machinery
AEMD	Adaptive Algorithm for Efficient Message Diffusion
AODV	Ad Hoc On-Demand Distance Vector Routing Protocol
AOP	Aspect-Oriented Programming
API	Application Programming Interface
ARP	Address Resolution Protocols
ASLR	Address Space Layout Randomization
BTP	Banana Tree Protocol
CAR	Context Aware Routing
CCTV	Closed-Circuit Television
CLDC	Connected Limited Device Configuration
CORBA	Common Object Request Broker Architecture
CSMA	Carrier Sense Multiple Access
CUWiN	Champaign-Urbana Community Wireless Network
CWMN	Community Wireless Mesh Networking
DHCP	Dynamic Host Configuration Protocol
DHT	Distributed Hash Table
DNS	Domain Name Service
DSDV	Destination Sequence Distance Vector
DSR	Dynamic Source Routing
DT	Delaunay triangulation
DTM	Distributed Token Machine
DTN	Delay Tolerant Network
DYMO	Dynamic MANET On-demand Routing Protocol
GASR	Gambling Algorithm for Scalable Resource-Aware Streaming
GEDIR	GEographical DIstance Routing
GENA	Generic Event Notification Architecture
GPS	Global Positioning System
GPCR	Greedy Perimeter Coordinator Routing
GPRS	General Packet Radio Service

GPSR   Greedy Perimeter Stateless Routing
HiBOp   History Based Opportunistic Routing
HMTP   Host Multicast Tree Protocol
IEEE   Institute of Electrical and Electronics Engineers, Inc.
IETF   Internet Engineering Task Force
IIOP   Internet Inter-ORB Protocol
IP   Internet Protocol
ISO   International Organization for Standardization
LAN   Local Area Network
MAC   Medium Access Control
MAN   Metropolitan Area Network
MANET   Mobile Ad Hoc Network
MAODV   Multicast AODV
MIDP   Mobile Information Device Profile
MCL   Mesh Connectivity Layer
MFR   Most Forward within Radius
MIT   Massachusetts Institute of Technology
MMS   Multimedia Messaging Service
MTU   Maximum Transmission Unit
MUP   Multi-Radio Unification
NC   Nearest Closer Algorithm
ODMRP   On-Demand Multicast Routing Protocol
OLSR   Optimized Link State Routing Protocol
OWL   Web Ontology Language
OSPF   Open Shortest Path First
PADIS   Power-Aware data DIStribution algorithm
P2P   Peer to Peer
PDA   Personal Digital Assistant
Propicman   Probabilistic Routing Protocol for Intermittently Connected Mobile
            Ad hoc Network
PROPHET   Probabilistic ROuting Protocol using History of Encounters and Tran-
            sitivity
QoS   Quality of Service
RDG   Restricted Delaunay Graph
RF   Radio Frequency
RFID   Radio Frequency Identification
RFC   Request for Comments
RIP   Routing Information Protocol
RMI   Remote Method Invocation
RMX   Reliable Multicast proXy
RNG   Relative Neighborhood Graph
RPC   Remore Procedure Call
RTP   Real-Time Transport Protocol
RTT   Round-Trip Time
SACS   String Attribute Constraint Summary

SETI	Search for Extra-Terrestrial Intelligence
SDP	Session Description Protocol
SDP	Service Discovery Protocol
SIG	Special Interest Group
SLP	Service Location Protocol
SMS	Short Message Service
SNR	Signal-to-Noise Ratio
SOAP	Simple Object Access Protocol
SOC	Service-Oriented Computing
SPIN	Sensor Protocols for Information via Negotiation
SQL	Structured Query Language
SRE	Security-Related Events
SSDO	Simple Service Discovery Protocol
STAG	Subnet Topology-Aware Grouping
STRAW	Street Random Waypoint
SxC	Security-by-Contract
TAG	Topology-Aware Grouping
TBCP	Tree Building Control Protocol
TBF	Trajectory Based Forwarding
TEDD	Trajectory and Energy-based Data Dissemination
TCP	Transmission Control Protocol
TIFF	Tagged Image File Format
TOMA	Two-tier Overlay Multicast Architecture
UDG	Unit Disk Graph
UDP	User Datagram Protocol
UPnP	Universal Plug and Play
URL	Uniform Resource Locator
UUID	Universally Unique Identifiers
V2I	Vehicle-to-Infrastructure
V2V	Vehicle-to-Vehicle
VITP	Vehicle Information Transfer Protocol
VANET	Vehicular Ad Hoc Network
VCP	Vehicular Communication Platform
WAN	Wide Area Network
WLAN	Wireless Local Area Network
WCETT	Weighted Cumulative Expected Transmission Time
WEP	Wired Equivalent Privacy
WiMAX	Worldwide Interoperability for Microwave Access
WMN	Wireless Mesh Network
WPA	Wi-Fi Protected Access
WSDL	Web Services Description Language
WSN	Wireless Sensor Network
XML	eXtended Markup Language
XSDF	eXtensible Service Discovery Framework

Part I
Networking and Programming Issues

This part exposes the reader to a set of fundamental issues related to mobile ad hoc networks that the programmer of middleware and applications needs to be aware of. These include the need to make software that is energy-aware, the constraints that exist when programming for mobile devices, how the mobility patters of the devices can be modeled and, finally, what routing algorithms are available to ensure network connectivity. Namely:

- In Chap. 1, R. Mini and A. Loureiro address the need to consider energy consumption at every stage of the design and operation of mobile devices. They discuss the main energy-aware techniques that can be applied to the design of applications, algorithms and protocols for mobile devices, with emphasis on wireless sensor networks.
- Besides general purpose devices like PDAs and laptops, ad hoc networks are also expected to serve specialised devices, like sensors. These specialised devices impose severe restrictions on developers, making robust programming quite challenging. In Chap. 2, L. Lopes, F. Martins, and J. Barros, survey the current state of the art in programming languages and runtime systems for specialised devices that make up wireless sensor networks.
- Node motion may adversely impact the performance of mobile ad hoc networks by breaking routes and disconnecting nodes. In Chap. 3, M. Musolesi and C. Mascolo present a state-of-the-art survey on the mobility models commonly used in networking and systems research.
- Not surprisingly, packet routing was one of the first problems addressed in the scope of ad hoc routing. It is also one of the most well-studied issues in this domain. Chapter 4 by F. Araújo and H. Miranda surveys the main existing routing algorithms under different network and node assumptions, including the important case where nodes are aware of their own location and can use this information to route packets.

Chapter 1
Energy in Wireless Sensor Networks

Raquel A.F. Mini, *Pontifical Catholic University of Minas Gerais, Brazil*
Antonio A.F. Loureiro, *Federal University of Minas Gerais, Brazil*

1.1 Introduction

Wireless sensor networks (WSNs) are composed of cooperating sensor nodes that can perceive the environment to monitor physical phenomena and events of interest. WSNs are envisioned to be applied in different applications, including, among others, habitat, environmental and industrial monitoring, which have great potential benefits for the society as a whole. Each sensor node typically includes a sensing component responsible for data acquisition from the physical environment. The node also has a processing unit for local data processing and storage, a wireless communication interface for data communication between nodes, and a power source to supply the energy used by the node to perform the programmed task. The power source often consists of a battery with a limited energy capacity. In many applications, sensor nodes may not be easily accessible because of the locations where they are deployed or the large scale of such networks. In both cases, network maintenance for energy replenishment becomes impractical. Furthermore, in case a sensor battery needs to be frequently replaced the main advantages of a wireless sensor network are lost, i.e., its operational cost, freedom from wiring constraints, and possibly more important, many sensing applications may become impractical.

WSNs form a new kind of ad hoc network with a new set of characteristics and challenges. Unlike conventional mobile ad hoc networks (MANETs), a wireless sensor network potentially has hundreds to thousands of nodes. Sensors have to operate in noisy environments and higher densities are required to achieve a good sensing resolution. Therefore, in a sensor network, scalability is a crucial factor. Different from nodes of a customary mobile ad hoc network, sensors are generally stationary after deployment. Although nodes are static, these networks still have dynamic network topology. During periods of low activity, the network may enter a dormant state in which nodes go to sleep to conserve energy. Also, nodes go out of service when the energy of the battery runs out or when a destructive event takes place. Another characteristic of these networks is that sensors have limited resources, such as limited computing capability, memory and energy supplies, and they must balance these restricted resources to increase the lifetime of the network. In addition, sensors will be battery powered and it is often very difficult to change or recharge batteries for these nodes. Therefore, in sensor networks, the designer is interested in prolonging the lifetime of the network and thus the energy conservation is one of the most important aspects to be considered in the design of these networks.

Wireless sensor networks pose new research challenges related to the design of algorithms, network protocols, and software that will enable the development of applications based on sensor devices. WSN design often employs some approaches as energy-aware techniques such as in-network processing, multi-hop communication and density control techniques to extend the network lifetime. In addition, WSNs should be resilient to failures due to different reasons such as physical destruction of nodes or energy depletion. Fault tolerance mechanisms should take advantage of nodal redundancy and distributed task processing. Several challenges still need to be overcome to have ubiquitous deployment of sensor networks. These challenges include efficient energy management, dynamic topology, device heterogeneity, lack of quality of service, application support, manufacturing quality, and ecological issues.

Energy consumption is a very important design consideration for protocols and algorithms in sensor networks. Energy management in WSNs involves not only reducing the energy consumption of a single sensor node but also maximizing the lifetime of the entire network. The network lifetime can be maximized only by incorporating energy awareness into every stage of the wireless sensor network design and operation, thus empowering the system with the ability to make dynamic tradeoffs between energy consumption, system performance, and operational fidelity [685].

The primary goal of WSNs is to transport the sensed information from sensors to the monitoring node, also referred to as the sink node. Thus, data communication is of paramount importance in all WSN design. Furthermore, communication spends more energy than the other network activities. In particular, power efficient communication paradigms for a given application should consider both routing and media access algorithms. The routing algorithms must be tailored for efficient network communication while maintaining connectivity when required to relay packets. In this case, the research challenge of the routing problem is to find a power efficient method for scheduling the nodes such that a multi-hop path may be used to relay the data. But, when it is considered the particular aspects of the monitoring application, it is possible to apply, for instance, information fusion and density control algorithms to reduce, respectively, the amount of data packets to be relayed and sensor nodes that need to be active, and, thus, saving energy.

1.2 Energy

In science, energy is a concept that relates the capacity of matter to perform work as the result of its motion or its position in relation to forces acting on it. The term "energy" is widely used in various spheres of life and many meanings are often ascribed to it. In general terms, the word is useful to explain changes that characterize

Fig. 1.1 Typical elements of
a battery

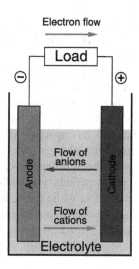

natural processes and phenomena. In this chapter, it is studied the role of energy in
the design of wireless sensor networks.

Energy sources can be classified into primary and secondary [184]. A primary
energy source is one that already exists in nature and can be used directly, converted
or re-directed into a form of energy that satisfies some particular needs. Examples of
primary energy sources are solar energy, fossil fuels (e.g., coal, oil and gas), nuclear
fuels (e.g., radioactive isotopes), wind energy, geothermal heat, and tidal energy. A
secondary energy source, such as electricity contained in a battery, does not exist in
nature but can be obtained from a primary energy source. A secondary source plays
an important role in society because it is frequently easier to use it than a primary
source.

A battery or electric cell is a device that produces electricity from a chemical reaction. Strictly speaking, a battery consists of two or more cells connected in series
or parallel, but the term is generally used for a single cell. It consists of negative
and positive electrodes, called anode and cathode, respectively, and an electrolyte,
which conducts ions and may be aqueous (composed of water) or non aqueous (not
composed of water), in liquid, paste, or solid form. Figure 1.1 depicts the typical elements of a battery. When a battery is connected to a device to be powered, which is
called load, the negative electrode supplies a current of electrons that flow through
the load and are accepted by the positive electrode. When the external load is removed the reaction ceases and the current of electrons stops.

As the intensity of the discharge current increases, an insoluble component develops between the inner and outer surfaces of the cathode of the batteries. The
inner surface becomes inaccessible as a result of the phenomenon, rendering the
cell unusable even while a sizable amount of active materials still exists. This effect
depends on the actual capacity of the cell and the discharge current.

A battery is a perishable product that starts deteriorating from the moment it
leaves the factory. Its physical operation depends on different factors such as di-

mensions, type of electrode material used, depth of discharge, diffusion rate of the active materials in the electrolyte, environmental conditions, charge methods if possible, and maintenance procedures. Each battery chemistry behaves differently in terms of aging and wear through normal use, which means that it needs a proper management.

Battery management is concerned with problems that lie in the selection of battery technologies, finding the optimal battery capacity, and scheduling a battery to increase its capacity. A battery presents a recovery effect that is the capacity of recovering charges under idle conditions. The lifetime of a sensor node can be extended by draining power for short time intervals followed by idle periods. During the idle periods, also called relaxation time, the battery can partially recover the charge lost while delivering the current impulse. This is particularly important for sensor nodes that perform sensing, processing and communication activities, each one for a different period of time with a different energy cost.

1.3 Sensor Energy Consumption

The energy consumption of the sensing element depends on the specific sensor device. It can be either low compared with the energy consumed by the processing element or even greater than the energy needed for data transmission. Experimental measurements published in the literature [685] using commercial sensor nodes have shown that data transmission is an expensive operation in terms of energy consumption, whereas data processing consumes significantly less. Furthermore, the energy consumption of transmitting a single bit is approximately the same of processing a thousand operations in a typical sensor node [681]. This observation will be probably valid in the years to come.

As the sensor network starts to operate, it may be necessary to adjust the functionality of individual nodes. This refinement can take several different forms. Scalar parameters, like duty cycle or sampling rates, may be adjusted using self-configuration and self-organization algorithms. This process may occur in different ways along the operation of the network lifetime.

A promising solution to extend the network lifetime is energy harvesting [854]. In some cases it is possible to scavenge energy from the external environment such as electromagnetic fields, Radio Frequency (RF) signals, solar energy, thermal gradients, radioactivity, and mechanical movements. The challenges are how to harvest enough energy to keep the operation of sensor nodes and how to have an efficient conversion process. Investigating these directions is very important because energy harvested from the environment is pollution free, which is highly desirable, and it is a renewable solution that potentially allows sensor nodes to run unattended for virtually unlimited time.

Energy harvesting for sensor nodes is still in its early stages. Furthermore, external power supply sources often exhibit a non-continuous behavior so that a battery is needed as well. In the long term, energy harvesting techniques represent a promising solution. Given the current state-of-the-art of battery development, energetic resources available at sensor nodes must be used very sparingly. Hence, energy harvesting and energy conservation are two key principles around which sensor networks and systems should be designed [320]. In the next sections, it will be discussed some of the main techniques to reduce energy consumption in a wireless sensor network, thus prolonging its lifetime.

1.4 Wireless Networking Considering Energy

In this section, it is discussed how protocols for WSNs are dealing with energy restriction. Protocols for WSNs try to adapt to this new kind of environment by reducing the amount of spent energy. The goal of this section is to present a taxonomy of the techniques used by protocols to save energy. In Sect. 1.4.1, this taxonomy is proposed and, from Sects. 1.4.2 to 1.4.5, the components of the taxonomy are presented.

1.4.1 Taxonomy of Energy Saving Techniques

As the conservation of energy is one of the main issues in the design of WSNs, a natural question that arises is how protocols are dealing with energy restriction. To answer this question, it is presented a taxonomy of energy saving techniques used by protocols for WSNs. Figure 1.2 illustrates the proposed taxonomy. It is important to point out that this taxonomy classifies techniques used by protocols and not the protocols themselves. A protocol can use more than one of these techniques in order to adjust to the energy restriction of WSNs. For instance, a protocol can use both the "local energy information" and "turn off parts of sensor nodes" techniques so nodes with low battery turn off their radio. On the other hand, a technique can be used at different layers of the protocol stack to incorporate energy awareness in the WSN design. As an example, the "local energy information" technique can be used by MAC and routing protocols. In fact, the presented techniques provide a basic building block out of which a variety of solutions for WSNs can be implemented. In the following sections the techniques described in this taxonomy will be presented with a few examples of their usage to protocols.

The first class of energy saving techniques used by protocols is named "no energy information". In this class, all techniques, whose goal is to save energy without taking into consideration the amount of available energy in sensor nodes, are put together. The use of "multi-hop communication", "turn off parts of sensor nodes" and performing "in-network processing" are examples of techniques that do not use any energy information. Section 1.4.2 presents the techniques classified in this category.

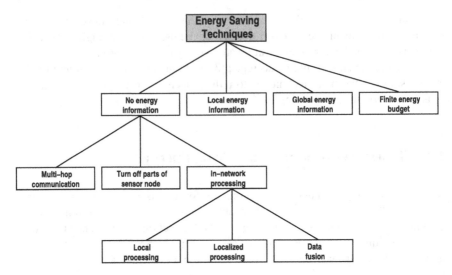

Fig. 1.2 Taxonomy of energy saving techniques

The second class is comprised of techniques that use the amount of available energy in the node to make a decision. The basic operation of these protocols consists in using a given energy threshold to define two different behaviors for a sensor node, each one for energy values higher or lower than the threshold. The use of this basic scheme allows sensors to cut back on certain activities (e.g., forwarding third-party data) when their energy is low. Section 1.4.3 presents the "local energy information" technique.

The third class uses the information about the remaining available energy in all nodes of the network. The global information about the amount of available energy in each part of the network can help prolonging the lifetime of the WSN [575]. In Sect. 1.4.4, the "global energy information" technique is presented.

The last technique is the "finite energy budget" paradigm. In this paradigm, it is specified the maximum amount of energy each network activity can spend, and ask this activity to achieve its best performance spending only the specified budget. Due to the paramount importance of energy conservation in these networks, the idea of defining energy budget to each network activity is a very promising paradigm [576]. The finite energy budget technique is presented in Sect. 1.4.5.

1.4.2 No Energy Information

In this section, it is presented three techniques used to save energy: multi-hop communication, turn off parts of sensor node and in-network processing. These are the most basic schemes used to save energy and most of the protocols for WSNs use at least one of them. The common characteristic of these techniques is that they perform a given task with the goal of saving energy without considering the amount of available energy in the network.

Multi-hop Communication

Communication is the network activity that consumes more energy, especially over long distances. If all sensor nodes communicate directly with the monitoring node using only one hop, the spent energy with this communication would fast draw all energy reserves of the network. Thus, communication strategies have a large influence on energy efficiency and network lifetime. Several factors affect the power consumption characteristics of a radio, including the modulation scheme, data rate, transmit power (determined by the transmission distance), and the operational duty cycle [685].

The minimum output power required to transmit a signal is d^α, where d is the transmission distance and α the path loss exponent ($2 \leq \alpha < 4$) [748]. As an example, considering $\alpha = 3$, the minimum output power required to transmit over one-hop communication with $d = 20$ m is $d^\alpha = 20^3 = 8000$ J. On the other hand, if there is a four-hop communication with 5 m per hop, the minimum output power is $4 \times d^\alpha = 4 \times 5^3 = 500$ J. Then, as stated in [748], depending on radio parameters, it can be more energy economic to transmit many short-distance messages than one-long distance message. Therefore, in WSNs, the large number of nodes can take advantage of short-range, multi-hop communication (instead of long-range communication) to conserve energy [681].

As long distance wireless communication is expensive in terms of energy consumption, instead of using single hop communication, information is routed from monitoring node to sensor node and vice-versa though multi-hop communication (Fig. 1.3(a)). Therefore, WSNs use a multi-hop infrastructureless architecture where all nodes cooperatively maintain network connectivity [31, 681] (Fig. 1.3(b)).[1]

Turn off Parts of Sensor Node

The idea of the second technique, classified as no energy information, is to turn off parts of sensor node. As stated in [370], the best way to save energy is to make unused components inactive whenever possible. In WSNs, nodes or parts of nodes that are not in use should be turned off to conserve energy. Thus, a WSN should embrace the philosophy of getting the work done as quickly as possible and going to sleep. During periods of low activity, the network should enter a dormant state in which nodes go to sleep to save energy. This can be achieved in a framework in which the nodes have various operation modes with different levels of activation, and, consequently, with different levels of energy consumption [577]. As the radio is the component of sensor node that consumes more energy, the turn off parts of

---

[1] Multi-hop communication in MANETs is addressed in Chap. 4.

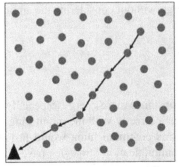 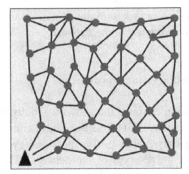

(a) Multi-hop communication from sensor nodes to monitoring node

(b) WSN forms a multi-hop network

Fig. 1.3 WSN uses multi-hop communications

sensor node technique is used by most MAC protocols for WSNs in such a way that they turn off the radio of sensor nodes in order to save energy [678, 834, 881, 882].

When MAC protocols turn off the radio of sensor nodes, they are saving energy from one of the major sources of energy waste, named *idle listening*, that happens when nodes keep listening in order to receive possible data that does not arrive. Many measurements have shown that idle listening consumes 50–100% of the energy required for receiving data [881]. Most energy in traditional MAC protocols is wasted by idle listening, since a node does not know when it will receive a message from a neighbour, it must keep its radio in receive mode at all times [834]. The waste of energy of idle listening is especially critical in many sensor network applications. If nothing is sensed, nodes are in idle mode for most of the time [881]. As the main focus of MAC layer for WSNs is to minimize the energy consumption, diminishing the waste of energy from idle listening is one of the major goals of these protocols.

Sensor MAC (S-MAC) [881] is a MAC protocol explicitly designed for wireless sensor networks. One of the main components of S-MAC is called *periodic listen and sleep* and, using the taxonomy of energy saving techniques proposed in this work, this component can be classified as turn off parts of sensor node. In the periodic listen and sleep component of S-MAC, the listen time is reduced by letting a node go into periodic sleep mode. The basic scheme is shown in Fig. 1.4. Each node goes to sleep for some time, and then wakes up and listens to check whether another node wants to transmit to it. During the sleep period, a node turns off its radio, and sets a timer to awake itself later. Using this scheme, the duty cycle is reduced and the latency is increased, since a sender must wait for the receiver to wake up before it can send out data. In order to minimize the additional latency, S-MAC uses synchronization to form virtual clusters in a way that neighbouring nodes use the same sleep schedule.

Using the same idea of the periodic listen and sleep component of S-MAC, other MAC protocols for WSNs, such as T-MAC (Timeout MAC) [834], also reduce the energy waste of idle listening by turning off the radio of sensor nodes. T-MAC

Fig. 1.4 Periodic listen and sleep [881]

dynamically determines the length of the listen period considering that it depends mainly on the message rate. It uses an adaptive duty cycle in such a way that the messages are transmitted in bursts of variable length. Its listen period is ended when nothing is heard.

In-network Processing

A WSN can be viewed as a collection of autonomous computing devices interconnected by a communication technology. "Interconnected" means that sensors are able to exchange information and, "autonomous computing devices" means that sensors are able to perform some kind of computation. In-network processing defines a set of techniques employed when sensor nodes use their processing abilities to diminish the amount of information sent to the monitoring node. The main goal of all in-network processing techniques is to exchange communication by computation. As communication is the major consumer of energy, the use of in-network processing techniques can help to save the scarce energy resources [881].

As depicted in Fig. 1.2, there are basically three kinds of in-network processing techniques: "local processing", "localized processing" and "data fusion". Local in-network processing technique happens when a computation is performed locally in a sensor node. As an example, consider that a sensor node, after performing a computation, decides between sending or dropping the sensed data. Using this idea, instead of sending the raw data to the monitoring node, sensor nodes use their processing abilities to locally carry out simple computations and transmit only the required and partially processed data, saving energy of the network [12].

The second in-network processing technique is called localized processing. In WSN, different sensor nodes often detect the same phenomenon, and, thus, there will be some redundancy in the transmission of the sensed data to a monitoring node [478]. To diminish the amount of data transmitted, neighbouring nodes exchange messages to ensure that only useful information is transferred to the monitoring node. This paradigm moves from the common *address-centric* approaches used in traditional computer networks (finding routes between pairs of nodes) to a *data-centric* approach (finding a particular data object stored on an unknown subset of nodes) [478]. In the data-centric approach, data generated by sensor nodes is named and it can be requested by sending a query containing the named data. Only data matching the interest is replied, saving the scarce energy resources [405]. Localized in-network processing technique is also known as cooperation or negotiation and it is used by some routing protocols [362, 478, 483].

In [362, 483], it is presented a family of adaptive protocols, called SPIN (Sensor Protocols for Information via Negotiation), that uses a localized in-network processing technique. SPIN disseminates information among sensors in an energy-constrained wireless sensor network. The SPIN family of protocols has two key components: data negotiation and resource management. Data negotiation of SPIN is a kind of localized in-network processing technique in which nodes negotiate with each other before transmitting data. This strategy helps ensure that only useful information will be transferred and also that nodes will never waste energy on useless transmissions. The second component of SPIN, called resource management, is implemented by making nodes aware of local energy resources. This scheme allows sensors to cut back on activities whenever their energy resources are low to increase their longevity. According to the taxonomy presented in this work, resource management of SPIN is a local energy information technique, and it will be presented in Sect. 1.4.3.

The third in-network processing technique is called data fusion or aggregation. This technique consists of combining data that comes from different sensors with the goal of eliminating redundancy, minimizing the number of transmissions and the number of collisions and, consequently, saving energy. In data fusion, intermediate nodes, in the path from source to monitoring node, aggregate various events into a single event in order to reduce the number of transmissions and the amount of data sent to the monitoring node, saving energy [610]. Data fusion requires store-and-forward processing of messages and, like localized in-network processing techniques, it also requires a data-centric approach in which the data generated by sensor nodes are named.

Directed Diffusion [405] is a routing protocol that uses data fusion in order to save energy. In this protocol, when a monitoring node wants to perform a dissemination throughout the network, it sends an interest. An interest message is a query or an interrogation which specifies what a user wants. This dissemination sets up gradients within the network designed to draw the data that matches the interest (events). Events start flowing toward the originators of interests along multiple gradient paths. The sensor network reinforces one or a small number of these paths that will be used to draw events from source to destination. As the data goes along the reinforced path, intermediate nodes might aggregate the data by combining reports from several sensors to create a more accurate information about the sensed event. Figure 1.5 illustrates the fusion of packets "a" and "b" into a single packet "ab" along the reinforced path of the Directed Diffusion.

The main goal of in-network processing techniques is to reduce the amount of information that must be transmitted to the monitoring node. The main advantage of these techniques is that the number of transmissions is decreased and, consequently, the energy is saved. On the other hand, the main disadvantage is the increase in the latency. In [478], the authors present a study that analyzes the trade-off between the energy savings due to aggregation in terms of the number of transmissions, and the aggregation latency, and also how this trade-off is affected by factors such as source-sink placements and the density of the network. Results showed that significant

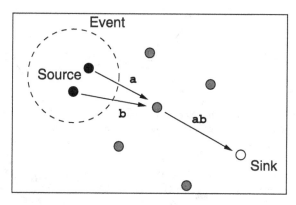

Fig. 1.5 Data fusion in Directed Diffusion (adapted from [405])

Algorithm 1.1: Local energy information

---

if *myEnergy < thresholdEnergy* **then**
    instructions related to low energy level
else
    instructions related to high energy level
end

---

energy gains are possible when the number of sources is large, and when the sources are located relatively close to each other and far from the monitoring node.

1.4.3 Local Energy Information

To save energy, some protocols use the amount of available energy in a sensor node to make a decision. Most of the protocols that use this energy saving technique perform a kind of computation similar to the one presented in Algorithm 1.1. The behavior of these protocols depends on the amount of available energy in the sensor node.

As stated above, the SPIN family of protocols has two main components: data negotiation and resource management. Data negotiation is a localized in-network processing technique and it was presented in Sect. 1.4.2. Resource management is a local energy information technique and it will be presented in the following.

Before presenting the resource management component of SPIN, it is necessary to see how sensor nodes communicate in this protocol. A SPIN node uses three types of messages to communicate:

- ADV (new data advertisement): when a SPIN node has data to share, it can advertise this fact by transmitting ADV messages containing meta-data.
- REQ (request for data): a SPIN node sends an REQ message when it wishes to receive some data.

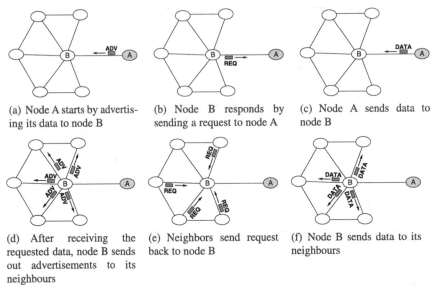

(a) Node A starts by advertising its data to node B

(b) Node B responds by sending a request to node A

(c) Node A sends data to node B

(d) After receiving the requested data, node B sends out advertisements to its neighbours

(e) Neighbors send request back to node B

(f) Node B sends data to its neighbours

Fig. 1.6 Basic operation of SPIN protocols [483]

- DATA (data message): DATA messages contain sensor data with a meta-data header.

Figure 1.6 shows an example of communication in this protocol. Upon receiving an ADV packet from node A, node B checks whether it possesses all of the advertised data (Fig. 1.6(a)). If not, node B sends an REQ message back to A (Fig. 1.6(b)), listing all of the data that it would like to acquire. When node A receives the REQ packet, it retrieves the requested data and sends it back to node B as a DATA message (Fig. 1.6(c)). Node B, in turn, sends ADV messages advertising the new data it received from node A to all of its neighbours (Fig. 1.6(d)). It does not send an advertisement back to node A, because it knows that node A already has the data. These nodes then send a request back to node B (Fig. 1.6(e)). Node B, in turn, sends the data packet to all of its neighbours that requested the new data (Fig. 1.6(f)), and the protocol continues.

The resource management of SPIN uses the local energy information technique in a way that a node will only participate in a stage if it believes that it can complete all the other stages of the protocol without going below the low-energy threshold. As an example, if a node receives an advertisement (ADV message), it does not send out a request if it does not have enough energy to transmit the request and receive the corresponding data. Similarly, if a node receives some new data, it only initiates the three-stage protocol if it believes it has enough energy to participate in the full protocol with all of its neighbours. This approach does not prevent a node from receiving, and therefore expending energy on ADV or REQ messages below its low-energy threshold. It does, however, prevent the node from ever handling a DATA message below this threshold.

The local energy information technique is used to save energy of nodes whose energy is below a given threshold. Its main goal is to use just the current amount of available energy in the node to make a processing decision. Protocols that use this scheme allow sensors to not execute certain activities such as forwarding third-party data when their energy is below a given threshold.

1.4.4 Global Energy Information

In many cases, to only look at the amount of the available energy in a node may either be insufficient or lead to an undesirable solution. In this scenario, it would be interesting to evaluate whether a global energy information could provide a better solution. In this section, it is presented protocols that use the information about the remaining available energy in all nodes of the network. This energy information is maintained at the monitoring node and it is called *Energy Map*. In this section, it is discussed the usefulness and the construction of the energy map and also the *Trajectory Based Forwarding* (TBF), a routing protocol for WSNs based on a trajectory (curve). TBF does not use the energy map, but it is presented to illustrate the basic idea of routing on a curve that is used by another routing protocol, also presented here, named *Trajectory and Energy-based Data Dissemination* (TEDD). The key idea of TEDD is to combine concepts presented in TBF with the information provided by the energy map to determine routes in a dynamic fashion.

The Energy Map

The information about the available energy at each part of the network is called the energy map and it is maintained at the monitoring node. The energy map of a WSN can be represented as a gray level image, in which light shaded areas represent regions with more remaining energy, and regions short of energy are represented by dark shaded areas, as illustrated in Fig. 1.7. The energy map can be constructed using a naive approach, in which each node sends periodically only its available energy to the monitoring node. However, this approach would spend so much energy due to communication, that the usefulness of the energy information would not compensate the amount of spent energy in this process. The work proposed in [908] obtains the energy map of sensor networks by using an aggregation based approach. A sensor node only needs to report its local energy information when there is a significant energy level drop compared to the last time the node reported it. Furthermore, energy information of neighbouring nodes with similar available energy are aggregated to decrease the number of packets in the network.

The energy map can also be constructed using a prediction-based approach [575]. The key idea is that there are situations in which the node can predict its energy

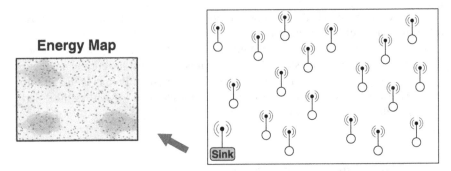

Fig. 1.7 Example of an energy map of a WSN

consumption based on its own past history. If a sensor can predict efficiently the amount of energy it will dissipate in the future, it is not necessary to frequently transmit its available energy. This node can just send one message with its available energy and the parameters of the model that describes its energy dissipation. Using these parameters, the monitoring node can update its local information about the available energy of this node.

The energy map can help prolonging the lifetime of the network. Using the energy map, a user may be able to determine if any part of the network is about to suffer system failures in near future due to depleted energy [908]. The knowledge of low-energy areas can aid in incremental deployment of sensors because additional sensors can be placed selectively on those regions short of resources. The choice of the best location for the monitoring node can also be made based on the energy map. Probably nodes near the monitoring node will spend more energy because they are used more frequently to relay packets to the monitoring node. Therefore, if the monitoring node is moved to areas with more remaining energy, the lifetime of the network can be prolonged. Other possible applications of the energy map are reconfiguration algorithms, query processing and data fusion.

Routing protocols can also take advantage of the available energy information at each part of the network. A routing algorithm can make a better use of the energy reserves if it selectively chooses routes that use nodes with more remaining energy, so that parts of the network with small reserves can be preserved. This protocol can also form a virtual backbone connecting high energy islands. In WSNs, routing protocols can be divided into three categories [323], as depicted in Fig. 1.8: from sensors to a monitoring node (data collection), among neighbouring sensors (cooperation), and from a monitoring node to sensors (data dissemination). Data collection is used to send the sensed data to the monitoring node. Cooperation often happens when some kind of communication among nodes is needed to make a decision about a sensing event. Data dissemination is normally used to disseminate a piece of information that is important to sensor nodes. Reliable data dissemination is crucial to WSN since a monitoring node has to perform some specific activities, such as to change the operational mode of part or the entire WSN, broadcast a new

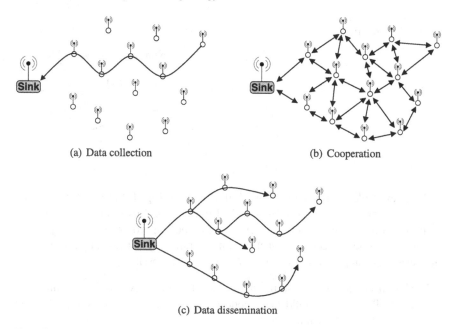

(a) Data collection

(b) Cooperation

(c) Data dissemination

Fig. 1.8 Data communication schemes in wireless sensor networks

interest to the network, activate/deactivate one or more sensors, and send queries to the network.

As the energy map is constructed at the monitoring node, data dissemination protocols can use this map to choose routes that use nodes with more remaining energy. Next it is presented the TBF, a data dissemination protocol whose basic idea is used by another data dissemination protocol named TEDD. The energy map is used by TEDD in a way that energy efficient routes are determined in a dynamic fashion.

Trajectory Based Forwarding (TBF)

In [629], Niculescu and Nath propose the *Trajectory Based Forwarding* (TBF) algorithm to disseminate messages in dense wireless networks. The key idea is to embed a curve (trajectory) in the packet to be disseminated from a monitoring node to sensor nodes, and then let the intermediate nodes forward it in a unicast manner to those nodes that lie close to the curve. In TBF, the curve can be represented by possibly a few equation parameters as opposed to a source routing protocol in which the source node inserts all nodes of the path into the packet as a discrete set of points, generating a considerable overhead and making its use in WSNs impracticable. Two main advantages of TBF are compact representation of a route and node independence, since no particular node address is specified in the trajectory.

Fig. 1.9 Routing on a curve
(adapted from [629])

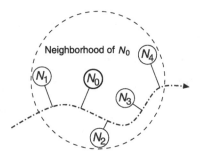

As an example, consider the network illustrated in Fig. 1.9 in which the neighbourhood of node N_0 is shown by a dashed line. In the forwarding technique, the main question is how to choose a next hop that best approximates the trajectory. Assume that node N_0 receives a packet with the trajectory indicated by the curve as illustrated, several policies of choosing a next hop are possible [629]:

- Minimum deviation: choose the node closest to the curve, with the minimum residual. This policy would favor node N_2 and would tend to produce a lower deviation from the ideal trajectory.
- Most forwarding within radius, choosing N_4. This policy should also be controlled by a threshold of a maximum acceptable residual, to limit the drifting of the achieved trajectory. It would produce paths with fewer hops than the previous policy, but with higher deviation from the ideal trajectory.
- Centroid of the feasible set, favoring N_3: the centroid is a way to uniformly designate clusters along the trajectory, and a quick way to determine the state of highly dense networks.
- The node with most battery left.
- Randomly choose between best three: useful when node positions are imperfect, or when it may be necessary to route around obstacles.

TBF is a sender-based algorithm since the current node systematically chooses the next hop of the route. This forwarding decision is based on the curve and a neighbouring table. To update this table, nodes exchange beacon packets periodically.

Despite its advantages, TBF has two main drawbacks. Firstly, the overhead required to update the neighbour tables increases the number of transmitted packets and, consequently, the total spent energy. In dynamic topology environments, such as WSNs in which nodes frequently enter a sleep mode to save energy, mechanisms for neighbour table maintenance have a prohibitive cost. Secondly, TBF is not fault tolerant in scenarios in which topological changes are faster than the neighbour table updates. In this case, broken trajectories always happen when the selected node is unavailable (e.g., the node is sleeping). Therefore, there is a trade-off between the neighbour table update overhead and the protocol efficiency.

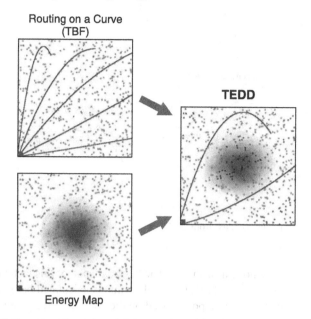

Fig. 1.10 Basic operation of TEDD

Trajectory and Energy-based Data Dissemination (TEDD)

TEDD is an energy-efficient data dissemination protocol that uses the global energy information technique. The key idea of TEDD, as illustrated in Fig. 1.10, is to combine concepts presented in TBF with the information provided by the energy map to determine routes in a dynamic fashion. TEDD is comprised of two main components: dynamic trajectory generation and forwarding policies.

Dynamic trajectory generation is a method for specifying the trajectories dynamically based on the energy map of the network. The main idea is to select a set of nodes in the network that are most suitable for forwarding the packets sent by the monitoring node and to find the best set of curves passing through or near these selected points. The choice of the best set of curves can be based on different criteria, such as the amount of energy available at the forwarding nodes, the percentage of nodes the information disseminated by the monitoring node is supposed to reach, or the area at which the dissemination is aimed [323].

The second component of TEDD is a new packet forwarding mechanism which is a receiver-based approach, i.e., each node upon receiving a packet decides whether to relay it or not, as opposed to TBF which is a sender-based data dissemination algorithm. In TEDD, the decision to forward a packet or not is based on the node geographical location and the packet information. It is important to point out that both TBF and TEDD assume that each node knows its geographical location. The forwarding decision process uses a temporization policy: before relaying a packet, the current node waits a small time interval. After this time, if no neighbour has

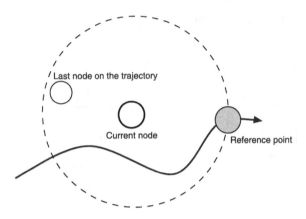

Fig. 1.11 Reference points (adapted from [537])

relayed the packet, the node transmits it. The key challenge of this technique is how
to estimate the delay time. In TEDD, the delay time is calculated based on the distance from the current node to a point ahead on the curve called *reference point*. In
particular, the reference point is the location where the trajectory leaves the neighbourhood of the current node as illustrated in Fig. 1.11. In each relay, the selection
of the reference point is determined by the previous hop of the trajectory. Each node
that receives a packet adjusts its delay time using its distance to the reference point
sent in the packet. The packet forwarding mechanism of TEDD introduces two improvements to the trajectory based forwarding process. Firstly, it eliminates the need
of neighbour table maintenance, which is very expensive in terms of radio transmissions. Secondly, it presents a more robust behavior in dynamic topology scenarios,
such as WSNs.

TEDD can be used to perform energy efficient data broadcasting and multicasting in the network. Figures 1.12(a) and 1.12(b) illustrate two sets of broadcasting curves, selected for two different energy maps. Figures 1.12(c) and 1.12(d)
illustrate two sets of curves generated to perform multicasting in which the squares
represent the multicasting area. It is possible to see that by using the energy map,
the curves generated by TEDD try to avoid low energy areas, extending the lifetime
of the network.

The main goal of the global energy information technique is to design WSN solutions that use the amount of available energy in the network to extend the lifetime
of the sensor network. It is important to point out that the cost of constructing the
energy map can be distributed (amortized) among algorithms and applications that
would benefit from it. In fact, it is difficult to think of an application and/or an algorithm that does not benefit from the use of the energy map.

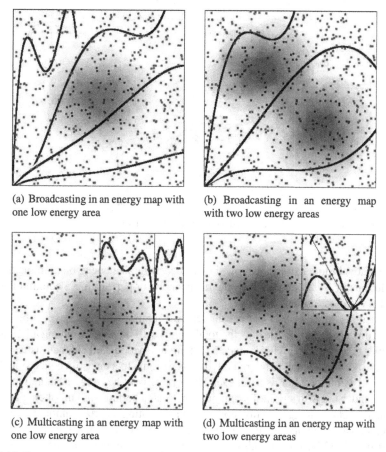

(a) Broadcasting in an energy map with one low energy area

(b) Broadcasting in an energy map with two low energy areas

(c) Multicasting in an energy map with one low energy area

(d) Multicasting in an energy map with two low energy areas

Fig. 1.12 Curves to perform broadcasting and multicasting [537]

1.4.5 Finite Energy Budget

The large use of WSNs depends on the design that considers the energy conservation a fundamental issue and devises mechanisms for extending the network lifetime. Furthermore, the total amount of energy available in the network is finite. This scenario leads us naturally to the following problem: what is the best performance a protocol can achieve given that it can spend only a finite amount of energy? Using this idea, it is possible to associate a finite energy budget for each network activity, and ask this activity to achieve its best performance using only its budget. This technique is named finite energy budget and it is the last energy saving scheme presented in this chapter.

Finite energy budget is a new way of dealing with network related problems, and it should be considered a new paradigm to design algorithms for networks that are battery powered, especially for WSNs. In [576], the authors present the finite

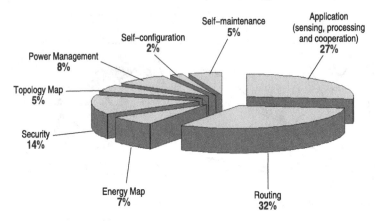

Fig. 1.13 Associating a finite energy budget to each network activity

energy budget model and show how this new paradigm can be applied in the design of solutions of problems in which the decision about performing a communication or not is the most important one.

To use the finite energy budget model, two main challenges have to be faced. Firstly, it is needed to decide how the total energy will be divided among all network activities. This decision will be made by taking into account all characteristics of the network. Secondly, each network activity has to come to a decision concerning how it will use its budget. In this case, each network activity has to be designed to make its best under the restriction of using no more than a certain amount of energy.

The first challenging situation consists in associating a certain amount of energy to each network activity. As an example, it is possible to associate the energy budget as illustrated in Fig. 1.13. In this case, all routing performed in the network, including data and management information packets, should spend only 32% of the total network energy. Based on this budget, the routing algorithm should make some decisions such as when and how to perform fusion and compression activities. What is important is that the routing design will be done under a finite energy budget. In the example of Fig. 1.13, all sensing, processing and cooperation performed by the application should spend no more than 27% of the energy. The energy map construction, security, power management, self-configuration and self-maintenance should spent no more than 7%, 14%, 5%, 8%, 2% and 5%, respectively.

The decision of how the energy budget should be distributed among the network activities depends on application features such as the deployment strategy, communication models, data delivery model, topology, density and dynamic behavior of sensors, phenomena and observers. As an example, a WSN can have a random or planned deployment strategy. When sensor nodes are deployed in a random topology, there is no a priori knowledge of node location. In this case, nodes will need to determine their relative positions and self-organize themselves. On the other hand, when nodes are organized in a planned topology, the exact position of each node is

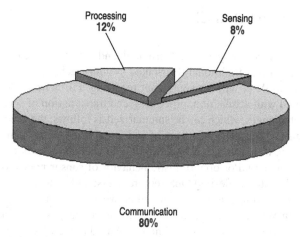

Processing
12%

Sensing
8%

Communication
80%

Fig. 1.14 Associating a finite energy budget to each node activity

carefully studied and known a priori. In this case, the self-configuration budget of a planned network will be smaller than the one of a random network.

The second problem that has to be faced when using the finite energy budget is how each network activity should use the received budget. It is necessary to plan the node behavior in order to achieve the best performance. To this end, each network activity will have to decide, for each sensor node, the percentage of time it will be sensing, processing and communicating. Figure 1.14 shows an example of this association. In this case, suppose the node can spend 80% of this total energy doing communication, 12% doing computation, and 8% sensing. This kind of budget association can be different depending on the location of the node or on its role in the network.

Using the finite energy budget model, all network activities will have to be designed to work under an energy restriction. The network characteristics will guide this new design. As an example, in the continuous data delivery model (sensors communicate their data continuously at a pre-specified rate) [807], some management data can be sent using piggybacking and, consequently, the communication budget can be smaller than in an observer-initiated data delivery model network (sensors only report their results in response to an explicit request from the observer, either directly, or indirectly through other sensors), in which management data has to be sent in a separate packet.

Another interesting issue that needs to be further investigated is to negotiate dynamically the remaining budget of the network. In some situations, it would be necessary to transfer energy budget from one activity to another that needs more energy. This problem is similar to the dynamic channel allocation in cellular network. If a cell has more channels in relation to its demand for communication in its coverage area, it can yield some channels to another cell that is with a debit on its demand. The same idea can be used to support dynamic budget association in WSNs.

1.5 Summary and Outlook

The power-constrained, lossy, noisy, distributed, and remote nature of wireless sensor networks mean that traditional distributed systems techniques often cannot be applied directly to them without significant recasting. Furthermore, the new challenges associated with acquisition, processing and transmission of sensor data pose new research challenges, which can be summarized as follows: the need to cope with the autonomous and spontaneous nature of nodes that leads to a dynamic and unpredictable topology, which in turn may cause link failures and network partitions; the need for cooperative and/or opportunistic behavior of sensor nodes to achieve the application goals with the desired QoS requirements; and battery limitations which imply constraints on transmission power and network functioning.

Adaptation in WSNs is motivated by those new challenges but in particular because of energy constraints of sensor nodes. An adaptive algorithm or protocol makes decisions for the future based solely on past and current information. These decisions must take into account the remaining energy available in the network, and the estimated energy consumption of the new activities so the network can accomplish its goals.

In this chapter it is discussed the main energy techniques that can be applied to the design of applications, algorithms and protocols for wireless sensor networks. These energy strategies should be further investigated to be effectively used in future solutions for WSNs. Energy awareness should be incorporated into every stage of the wireless sensor network design and operation, thus empowering the system with the ability to make dynamic tradeoffs between energy consumption, system performance, and operational fidelity.

Chapter 2
Programming Wireless Sensor Networks

Luís Lopes, *CRACS/DCC-FCUP, Universidade do Porto, Portugal*
Francisco Martins, *LASIGE/DI-FCUL, Universidade de Lisboa, Portugal*
João Barros, *Instituto de Telecomunicações, Universidade do Porto, Portugal*

2.1 Introduction

Sensor networks can be viewed as a collection of tiny, low-cost devices programmed to sense the physical world and that communicate over radio links [12]. The devices are commonly called *motes* or *smart dust* [676], in allusion to their computational and sensing capabilities, as well as their increasingly small size.

Sensor networks are significantly different from other wireless networks in that: (a) the design of a sensor network is strongly driven by its particular application; (b) sensor nodes are highly constrained in terms of power consumption, CPU speed, and memory availability; and (c) large-scale sensor applications require self-configuration and distributed software updates without human intervention.

The potential applications of wireless sensor networks are vast: (a) in biology—environmental and habitat monitoring; (b) medicine—medical diagnosis and health monitorization; (c) city management—traffic control and inventory tracking; (d) high-energy physics–particle detectors, etc. Indeed, this variety of applications and proliferation of hardware platforms with distinct computational capabilities is one of the reasons behind the difficulties in programming and deploying these networks.

Moreover, while there has been a considerable amount of work on wireless sensor networks from the viewpoint of communication-oriented models, in which the sensor nodes are assumed to store and to process data, coordinate their transmissions, organize the routing of messages within the network, and relay the data to a remote receiver (see *e.g.* [731, 384, 65] and references therein), work on programming models and languages for these computing platforms is far more recent and comparatively scarce.

Traditionally, sensor networks have been programmed at a very low-level of abstraction, typically at operating system level. This state of affairs is well demonstrated by the large number of sensor network applications developed with nesC [308] and coupled with its sibling operating system TinyOS [805].

In recent years, however, there is a growing trend towards the development of high-level programming models and abstractions for wireless sensor networks. From a conceptual point of view, these models explore new *views* of sensor networks such as: (a) *streams*—modeling the network as data streams with no perception of the underlying hardware by the programmer; (b) *regions*—allowing the programmer to reason in terms of groups of sensors organized according to some membership criteria, either topological or based on sensor attributes; (c) *databases*—perceiving wireless sensor networks as dynamic data repositories that may be targeted by declarative languages such as SQL; and (d) *computing*—enabling the programmer to use the sensor network as a computational platform that may perform online computation, despite the nodes scarce resources, *e.g.* by hosting autonomous mobile agents.

2.2 Hardware and Operating Systems

The last decades of computing have shown that software development cannot keep up with the exponential evolution of hardware platforms. The situation is no different for sensing devices and wireless technology. In fact, while there exist already several different sensor networking prototypes available to the user, there is only a rather limited number of wide-spread real-world applications.

The current state-of-the-art in sensing devices is well represented by a class of multi-purpose sensor nodes called *motes* [206], which were originally developed at UC Berkeley and are being deployed and tested by several research groups and start-up companies. Typical motes, such as MICA, MICAZ, or TELOS-B, consist of a combination of different modules, namely a data acquisition card, a mote processor, and a wireless interface. Examples of sensor platforms that are already available in the market include ambient light, barometric pressure, GPS , magnetic field, sound, photo-sensitive light, photo resistor, humidity, and temperature. The mote processor performs all the information processing tasks, from analog-to-digital conversion up to coding and upper layer protocol functions. For computation purposes, motes rely on a dedicated processor and 512 KB of non-volatile memory. The wireless module operates at radio frequencies within the available ISM bands, which can vary from slightly over 300 MHz to almost 2500 MHz. Motes may come in different sizes, depending mostly on the chosen type of energy supply, since all other components have already reached an impressive degree of miniaturization.

Another example of prototype sensor nodes was developed by the Free University of Berlin. The so called Embedded Sensor Board (ESB) includes the same kind of sensing capabilities, but offers very low energy consumption. In idle mode and in active mode an ESB requires 8μA and about 10 mA, respectively, which, assuming ordinary AAA batteries and average transmission rates of 0.8 bytes per second, results in a network lifetime of 5 to 17 years.

More recently, Sun Microsystems introduced the Sun SPOT (Sun Small Programmable Object Technology) motes. Featuring IEEE 802.15.4 wireless networking, USB interface, 512 KB of RAM, 4 MB of flash memory and a 32-bit RISC ARM processor, these are resourceful sensors that can run the Squawk Java Virtual Machine developed for embedded systems. Another recent trend is the use of Zig-Bee, a new simple and low-cost networking standard, aimed at sensor networks that require low data rates, long battery lifes for nodes, and secure networking.

These platforms are characterized by the scarcity of resources they provide to the programmer or that they have available for self-maintenance. In general, the main resource bottlenecks are:

Battery Life: standalone sensing devices use batteries to power their hardware components. Recharging batteries is often not an option and this places stringent constraints on the amount of computing and communication the devices can engage into;

Microprocessor: sensors usually have very limited processing power due to battery restrictions and also to miniaturization and cost issues. This limits the online computing capabilities of a given sensor network;

Memory: memory in sensing devices is scarce for much the same reasons that the processing power is limited. This has important implications on the complexity and size of the software that may be run on the devices, as well as the amount of information it can store;

Communication Range: devices in a sensor network communicate through wireless channels and thus the signal decreases very rapidly (with the square of the distance to the device) and thus the range is rather limited. Increasing the communication range implies ramping up the power of the wireless transmitter, something that is undesirable given the power limitations of the devices.

Running software on these platforms requires a delicate balance between the application requirements and the hardware available resources. In this context, the introduction of operating system support for these devices presents several interesting problems. Naturally, the operating system must provide a very low overhead interface to the hardware. The current systems are usually composed of several modules of which only a part may be present at any time in a sensor node. In other words, the selection of the operating system modules and their deployment onto sensor nodes is driven by the application needs. As expected, communication, memory, and CPU management are the main tasks for such micro-kernels.

Most current sensor networks run on top of the TinyOS [805] operating system and its sibling programming language nesC [308]. TinyOS provides a very simple, event-based, single-threaded execution-model with non-preemptive tasks. The system is loaded onto the sensor nodes as a set of modules to be used by a target application.

Several alternative operating systems have been proposed. Despite adhering to the same event-driven model of TinyOS, Contiki [241] differs from the latter in that it supports multi-threaded execution of applications, based on very lightweight Protothreads [242], and the dynamic loading of program modules. SOS [357] is also event-driven and its main characteristic is the way it is constructed from smaller modules and its ability to dynamically load them, aided by a clever memory management scheme. MANTIS [85], Nano-RK [258], and BTnut[1] diverge from the above systems in that they support preemptive multi-threading, in other words, the operating system, not the application programmer, manages the CPU. Furthermore Nano-RK is designed with real-time sensor network applications in mind, providing fine grained mechanisms to control access to hardware resources. Finally, systems like the Squawk Java Virtual Machine [757], for the SunSPOT devices, run *on the bare metal*, without any external OS support.

2.3 Programming Languages and Systems

The discussion on sensor network hardware and operating systems from the previous section introduces the challenging aspects of these architectures when viewed as computing platforms. In fact, developing adequate programming models for sensor networking applications is a difficult problem for which multiple solutions may be justified, depending on the resource constraints of the nodes and the target application.

This section presents an overview of the state-of-the-art in languages and systems for programming sensor networks whilst establishing a classification based on the hardware, network, and data abstractions they support. The analyzed criteria are summarized in the following table.

Table 2.1: Proposed classification criteria

Hardware interaction	Network perception	Data acquisition
λ High-level	Sensor-based	C/λ Mobile agent
GLUE Middleware	C/λ Macroprogramming	Database
✿ Virtual machine	—	Stream
C Low-level	—	Message

In what concerns the level of abstraction from the hardware four categories are proposed: low-level, virtual machines, middleware, and high-level programming languages. A low-level programming language has full access to hardware and operating system resources. In contrast, a high-level programming language offers constructs to the programmer that hide almost all sensor specific information. As for network abstraction, systems can be broadly divided into those that promote either

---

[1] Available at http://www.ethernut.de/en/software/index.html.

sensor-based programming or macroprogramming, whether the system aims at programming each sensor *per se*, or considers the network as clusters of sensors or even as a whole, respectively. Last, but not least, the data abstraction view categorizes programming languages as they interpret the information gathered by the sensor network. Here, four categories can be identified: messages, streams, databases, and mobile agents. Messages represent the lowest level of data abstraction. Data collected by the sensors is gathered into amorphous, protocol dependent, packets and transmitted over the network. At a higher level, sensor networks may be seen as information flows organized as time varying streams of data. Another approach is to view a sensor network as a distributed database that may be queried to obtain and process information. Yet a different approach is to gather information from the network by sending software agents that travel thought the sensors, collect the information, and carry it to the base.

2.3.1 Level of Abstraction from the Hardware

Current systems and programming languages can be divided into four different categories depending on their level of abstraction from the underlying sensor network: low-level, virtual machines, middleware, and high-level programming languages.

Low-Level

Systems in this category promote system-level programming of the sensor networks. The abstraction level is very low and most programming is done in languages such as **nesC** [308], making direct calls to the appropriate modules of the supporting operating system, *e.g.* TinyOS [805]. In nesC, programs are nested collections of components, some of which may be provided by TinyOS itself. An application is built from several components that are linked into an executable that is run by the underlying TinyOS engine as a non-preemptive tasks. Besides running tasks, TinyOS also captures events and directs them to some handling code. Support for network re-programming is very limited in TinyOS and is provided by the XNP protocol [417]. XNP allows a limited form of network re-programming: only complete executable files may be uploaded in single-hop configurations with all nodes within bidirectional communication range from the source. Some systems, like **Protothreads** [242], try to provide more sophisticated computational models. These are very lightweight threads developed for the Contiki operating system to support multi-threaded applications with minimal resource consumption. In fact, despite tasks can be multi-threaded they are still non-preemptive as in TinyOS, thus avoiding the scheduling overheads otherwise required for their management by the operating system. At the extreme range of resource availability one finds systems like **Pushpin** [519]. This is an extremely lightweight system as there is not even an abstraction layer for programming. Pieces of native code called *pfrags* are transferred and executed in the nodes (called *pushpins*). The operating system provides a shared memory address space for communication between pfrags and a few basic system calls.

The main advantage of working at this abstraction level is the very fine control the programmer has on the resources of the sensor, allowing the fine-tuning of applications in a more effective way. However, from the point of view of programming and deploying large-scale sensor network applications, this approach promotes node-by-node programming and introduces considerable debugging and deployment problems. Systems like **Deluge** [388] try to overcome the difficulties in deployment of applications by providing reliable dissemination protocols for propagating large amounts of data over a multi-hop sensor network. Deluge uses an *epidemic* approach for robustness and to account for lossy communication channels. Retransmission of broadcasts (flooding) has long been known to produce the *broadcast storm problem* (redundancy, contention, and collision impairments to performance). Several approaches have been proposed to minimize re-transmissions (*e.g.* gossip and scoped-flooding) but network re-programming requires that all data must reach the nodes, hence the need for a reliable protocol. Deluge has similarities with other protocols such as SPIN-RL [483] and Trickle [510] in that it aims for low-level, link-layer protocols for efficient, reliable broadcasting. Other protocols include MOAP [772] that implements a multi-hop solution in which program fragments are advertised and passed one-hop at a time, and, Difference Based Protocols [700], that try to minimize the impact of transmissions by sending only the changes to programs, a kind of patching, that is also more conservative from the point of view of power consumption.

Virtual Machines

These systems try to provide a software infrastructure that allows programmers to abstract away from the sensor specific hardware and operating system. The programming level is sufficiently low for the programmer to retain some fine grained control of the applications whilst being insulated from more complex aspects of the network. Perhaps the best known example of this approach for sensor networks is **Maté** [508] and its associated language TinyScript. Maté is a compact virtual machine implemented on top of TinyOS. Programs, called *capsules*, may be injected in the network at any time to perform specific tasks. They are written in a very simple assembly language and have the capability to move between sensor nodes, a form of code mobility. Maté also provides a higher level programming language for the virtual machine, called **Mottle** [509], that supports more powerful primitive data structures and first class functions. Mottle is based on Scheme and is likewise dynamically typed.

Another example of a virtual machine developed for sensor networks is the **Distributed Token Machine** (DTM) [624] used in the Regiment programming language [623]. The execution and communication model associated with these machines is based on the processing of *tokens* by event handlers. Each token is a typed message with some data or code that triggers a specific handler upon reception. The execution is thus data-flow oriented. Associated with the virtual machine is an intermediate language, appropriately called Token Machine Language (TML), that can be used as a target for compilers of higher level programming languages for sensor networks.

Recently, Sun Microsystems introduced the **Squawk** virtual machine to support applications for their SunSPOT devices [757]. Squawk is a very compact Java virtual machine (as opposed to mainstream implementations like Sun's own HotSpot) with a simplified byte-code layout that runs without an underlying operating system on wireless devices. It has the particularity of being itself implemented (mostly) in Java, following the tradition of other object virtual machines such as Smalltalk's Squeak [403] and the Klein VM [823]. The architecture is based on a *split-vm* design in which the class loading and the execution of the byte-code are decoupled. Class loading and verification is assumed to take place at base stations with less resource restrictions. The objects representing loaded classes are serialized to files called *suites* and can be sent over a radio link to devices in the network. On arrival to a remote device, the suite is deserialized and eventually interpreted. Squawk supports a restricted form of multithreading using *green threads* and provides an API to manage the execution of applications (called *isolates*). This API, when coupled with the serialization mechanisms and a wireless communication API provides the means to support not only suite migration between devices but also of complete applications.

The use of virtual machines for supporting sensor network applications is particularly interesting given the prospects of portability of the application code and the possibility of seamlessly harnessing huge heterogeneous networks. It is also relevant from the point of view of security, allowing for example *proof carrying code* [614] to be developed more easily. However, the approach is not without problems. Indeed, virtual machines introduce computational overhead and memory demands that may be incompatible with the most restrictive sensor devices.

Middleware

Another approach championed by some systems is that of programming sensor networks using an API provided by an underlying middleware that hides the details of the sensor network from the programmer.

An example of such a system is **EnviroTrack** [2], which is based on an object-based distributed middleware. It exposes a new address space to the sensor network where the physical events in the environment are addressable. The idea is to label events and to address them directly from the application, *e.g.* assign a computation to the perception of a given event. This effectively decouples the application from the physical topology of the network. This view of the sensor network is implemented with a library that hides the complexity of object mobility, tracking, communication, and state. Applications are compiled to nesC and executed on top of TinyOS. Another system, **Impala** [527], is a middleware framework that targets application deployment and dynamic update. Here, native code is transferred to sensor nodes and dynamically linked without affecting existing applications. Internally, nodes toggle between multiple pieces of code to execute tasks. This mechanism is controlled by an underlying state machine that responds to a fixed set of standard events.

Although not specifically oriented for sensor networks, a lot of research has been targeted at studying middleware for mobile networks [555] in the past few years. These studies may be grouped, according to [219], into four middleware categories: RPC-based, data sharing-oriented, publish/subscribe, and tuple space based. RPC-based middleware addresses mainly volatile connections via temporary queuing of RPCs [433], via rebinding [733], or by providing ORBs suitable for lightweight devices, like Alice [351] and DOLMEN [704]. Data sharing-oriented middleware uses weak replica management in order to optimize the autonomy of disconnected mobile devices (cf. [433, 802, 895]). The publish/subscribe middleware builds on top of the publish/subscribe paradigm [265] to handle the specificities of mobile computing [135, 200, 267, 293]. Finally, tuple space based middleware [600, 550] exchange information via tuple spaces [310], which offer a kind of distributed shared memory where processes may asynchronously write or query tuples based upon a pattern matching scheme.

High-Level Programming Languages

Finally, some languages and systems provide programmers with very high-level views of the network by hiding all networking and communication details. The programmer implements a distributed application that usually is not targeted at a specific sensor network architecture or configuration. A specialized compiler takes this high-level view of the application and produces the node specific behavior for each sensor as required for the deployment of the application, without the intervention of the programmer. A good example of such a language is **Regiment** [623], a macroprogramming language that uses network *regions* and data *streams* as the basic programming abstractions. The run-time for the language is based on a distributed version of a token machine (DTM). Nodes perform sensing and computation in response to tokens received from the network or to tokens generated internally.

Another example, **Kairos** [342], provides language independent compile- and run-time tools that can be used to extend existing programming languages with primitives specific to sensor networks (the concrete implementation extends Python). Kairos adheres to macroprogramming as a single, centralized, application is developed and transparently tasked over a target sensor network. The sensor nodes interact through three basic primitives that the authors argue are intuitive and expressive: reading and writing of variables at nodes, one-hop communication and, addressing an arbitrary node in the network. A compiler transforms the Kairos macroprograms into node-specific versions. These are executed by the Kairos run-time system, present in every sensor node.

SNACK [330] is a system that provides the application programmer with a component composition language and a library of predefined components and services.

The programmer selects items in the library and specifies how to plug them with the composition language so that some relevant information flows through the network. The system is implemented on top of nesC but the language presents a very high-level view of the network since the network specific, low-level details, are insulated in the system library.

Other systems, whilst providing very high-level programming abstractions, require the programmer to compose the applications by programming the individual sensor behaviors and connecting them in some meaningful way. An example of this approach is **Agilla** [287] that introduces mobile-agents into sensor networks. Here, programs, or autonomous agents, move between nodes in a sensor network. These are dedicated virtual machines. The communication model is based on a distributed Linda tuple-space. Network re-programming is allowed by injecting new agents into the network and killing existing ones. One of the goals of Agilla is to transform more resourceful sensor networks into general purpose computing platforms. Another similar approach is taken by **SensorWare** [112], a system that given its size aims at more resourceful sensor nodes and networks. Programs appear as mobile scripts and a sensor node is seen as a dynamically changing entity where new scripts may be installed on the fly. The scripts may be injected in the network at any time. The language is TCL-based with primitives for timer services, acquisition and sensing data, mobility of scripts, and for a location discovery protocol. Another system, **Smart Messages** [771], is based on Java and allows messages between sensor nodes to carry code, data, and execution state. It extends the Java APIs with extra functionality such as support for *spatial programming*, i.e. allowing the programmer to select nodes based on topological or state information. **AmbientTalk** [220] is an object-oriented, prototype-based (classless), concurrent programming language featuring *active objects* as the main computational abstraction. Each active object is composed of a running thread, updateable state, a set of methods and a queue to process incoming messages. Active objects are deployed onto wireless devices and communicate via asynchronous message passing. The language also defines *passive objects* as objects without an associated thread that may be manipulated by active ones according to *containment* rules that avoid *race conditions* to occur. Finally, **SpatialViews** [627], takes the view that location and time, besides network node resources, are fundamental abstractions for programming sensor networks. The main concept is that of *virtual network*, that describes the services provided, and the locations of a subset of the nodes in a physical network. Many virtual networks may co-exist in a physical one. Each virtual network has an associated computation that migrates from node to node as it is being executed, and the computation may be replicated and executed in parallel. Virtual networks, called *spatial views*, are specified with a set of services (required by the computation to be performed), a *space*, and an iterator. The latter discovers the nodes within the spatial view that provides requested services by the application and controls the movement of the computation within that space. The discovery process is bounded in time.

2.3.2 Network Abstraction

In terms of network abstraction, current systems can be broadly classified into two categories, some focusing on sensor-based programming while others adhering to macroprogramming. In a sensor-based programming model applications are built by composing individual software components for each of the sensors in a network. In this way, the mapping between the software components and the physical sensors is made explicit to the programmer. Macroprogramming systems, on the other hand, allow a programmer to develop an application for sensor networks while abstract-ing away from network architecture details. In these systems, the mapping between the application constructs and the actual physical infrastructure is performed by ad-vanced compilers and supported by the associated runtime systems.

Macroprogramming Systems

These systems take the view that sensor network applications should be developed as typical distributed applications without requiring the developer to specify the be-havior of each computing node individually. A compiler or a set of run-time libraries should, based on this high-level specification of the application, provide the specific nodal behaviors required by the target network and the run-time should take care of their distribution, without any further intervention from the programmer. This ap-proach allows the programmer to focus on the application and abstract away from the low-level details of communication and network architecture.

Still under the macroprogramming philosophy, systems that emphasize on *global behavior* express applications in such a way that their implementation includes all nodes in the network, without using hierarchical partitioning of the network into subunits for the specification of the computation. Other systems use the notion of *local behavior*, focusing on the implementation of applications for sensor networks that use some form of network partitioning into regions to specify a computation. Such regions may be defined using the sensor nodes' geographic position, their distance measured in terms of number of hops, radio connectivity or received power, or in terms of the properties of the data gathered by the sensor nodes.

Region abstractions also have the important advantage of introducing program-ming constructs that help the developer to deal with the cost of communication in a sensor network. For example, data aggregation techniques that require every sensor in a network to report a value to a sink node are usually very inefficient and incom-patible with the resource limitations of such platforms. Each region in a network is capable of identifying neighboring nodes, of sharing data, and of performing in-ternal data aggregation and processing, thus reducing substantially the complexity of data processing and the bandwidth requirements of the network, with the added bonus of preserving the battery life of sensor nodes.

On the other hand, the use of region abstractions implies a trade-off between the accuracy of the data being gathered by the network (the individual sensor readings and other data are lost in the process of aggregating data over a region) and the relaxation of the physical limitations of the sensor network (the required network bandwidth is reduced and battery life is improved).

Several systems have been proposed that implement some form of region abstraction as described. **Hood** [864] implements a model of regions that supports the discovery of neighborhood nodes and the sharing of data among nodes and regions. No primitives for data aggregation nor for fine-tuning the trade-off between accuracy and resource consumption are given. These primitives are available in another system, **Abstract Regions** [862], implemented on top of TinyOS and partially inspired in Hood. Regiment [623], Kairos [342], and SNACK [330] are examples of languages and systems, described elsewhere in this chapter, that also provide programming abstractions based on network regions.

Cosmos [37], on the other hand, is an architecture for macroprogramming sensor networks that focuses on distributed data processing. It is composed of a basic operating system, mOS, and an associated programming language, mPL. Applications are built from reusable functional components, plugged together using an interaction primitive to specify the data-flow through the components. Each functional component features a type declaration and a block of C code that implements its functionality. The operating system provides the runtime system for the applications and supports dynamic re-programming. Interaction between remote components uses asynchronous message passing.

Other protocols and systems, more concerned with the lower level communication aspects of operating a sensor network, also include region abstractions usually for optimizing network resources when disseminating or aggregating data. Rather than using typical data routing protocols for disseminating information in a network, they rely on data-centric or storage-centric routing techniques to disseminate and aggregate data.

Data-centric routing means that a programmer does not request a data item from the network by specifying a node or collection of nodes to inquire, but rather by stating the type of data it is interested in by means of some high-level specification language. An underlying system must then map the abstract request presented by the developer into low-level interactions between nodes. Examples of such systems are **Directed Diffusion** [405] and **TAG** [538] that use region abstractions, namely sensor state or interest in data to generate dynamic pathways in the network, yielding basic statistic views of the network (*e.g.* medians, averages, maximum values). **SPIN** [483] (Sensor Protocols for Information via Negotiation) goes one step further by annotating the data, and using the annotations to eliminate redundant data transmissions and to adapt the communication and routing strategies to the available network resources.

Data-centric storage avoids the flooding of the network with requests for data, typical of the previous approach, by mapping all data producing events in a sensor network to individual nodes and redirecting any incoming data request to that restricted set of nodes. There is no need to look for the data, rather it is only required to compute a path from the sink node to the nodes that are associated with it. Examples of such systems are **GHT** [691], which replicates the synchronization points for the data producing events, by hashing them using geographic information, and retrieves the data from the synchronization point closest to the node were detection takes place. Queries in GHT are rather simple, indicating only whether or not a given condition has occurred. **DIFS** [329] extends the type of queries available to range conditions on sensor attributes and tries to balance the load of the query over the network.

Sensor-Based Programming

Here, it is the developer's responsibility to explicitly implement the behavior of each sensor in the network, to compile it, and to deploy it for execution. This approach requires the programmer to be aware of the architecture of the sensor network and in cases even some details of the hardware that makes up the sensor nodes. Low level systems such as nesC, Pushpin, and Deluge are based on this approach to programming sensor networks, as well as high-level programming languages such as Agilla [287], SensorWare [112], and Smart Messages [771].

2.3.3 Data Abstraction

Sensor networks can be viewed from distinct perspectives according to the application in mind. One can view the networks as communicating systems exchanging messages, *e.g.* if the goals are to work at the system level and to develop energy efficient communication protocols and middleware—a *communication-centric* approach. Another possibility is to look at a far higher level of abstraction and think of the sensor networks as time varying data streams or data repositories—a *data-centric* approach. One can also view them as collections of collaborative mobile computations (the sensors being the computing platform)—a *computation-centric* approach.

Communication-Centric

At the lowest level of data abstraction, all that exists for the sensor network programmer is the data generated by the sensor nodes packed in messages and transmitted over the network. This view is usually associated with communication-centric programming. Most software developed at this level of abstraction corresponds to implementations of message routing and of data and code dissemination protocols (*e.g.* XNP [417], Deluge [388], and Pushpin [519]).

Data-Centric

Sensor networks may be viewed as streams of values originating in sensor nodes. This view abstracts away from the underlying sensor network hiding lower level aspects from the programmer such as the architecture, the hardware, and the communication protocols. In this approach, programs manipulate streams of user defined data. The operations on streams are typically very simple such as *fold* (aggregate data) and *map* (apply some processing function to a stream).

A concrete instance of such an approach coupled with topological abstractions is that of *region streams*, defined as spatially distributed, time-varying collections of node state, *e.g.* the collection of values from the sensor nodes or its aggregate value, within a region. Regiment [623] and its sibling, **Flask** [547], are canonical examples of such programming languages. Other similar languages and systems include SNACK [330] and Kairos [342].

Still in the data-centric approaches, some systems view sensor networks as data repositories upon which standard database processing primitives may be used. This is another data-centric view in which information generated in the sensors can be the object of queries and processing. Such systems allow for example an interface with web-services permitting users in a conventional network to connect and extract knowledge from a sensor network. One example is **TinyDB** [539], a macroprogramming system that allows a programmer to reason about a sensor network as a database. Accordingly, the user may write declarative style queries to request a particular *view* of the data being generated by the network. This is done at a very high-level of abstraction without the user having any notion of the underlying network. A sophisticated compiler decomposes the user queries into low-level, sensor specific operations based on primitives like sampling, application of filters to data, data aggregation, and data broadcast. A similar approach is taken in **Cougar** [298], which allows users to specify high-level queries for data views to be extracted from a sensor network. The system then analyses the queries and decomposes them into a sequence of network operations optimized to minimize resource consumption. Other projects focus on the interface between the Internet and the sensor networks viewed as Web resources. In this line, **IrisNet** [313] is a Java based system that allows programmers to specify distributed database services through XML documents and provides a set of high-level APIs to compose data queries to sensor networks and to process the resulting data.

Computation-Centric

Finally, some systems view sensor networks as computational infrastructures in which applications composed of multiple, interacting, mobile agents evolve. Using mobile agents to program sensor networks is not trivial. In fact, agent traffic in the network poses serious autonomy problems for sensors given the large cost of broadcasts. Moreover, an underlying middleware must be provided so that agents may move between nodes in the network. This requires processor and memory resources not commonly available in sensors. Such systems are thus more appropriate for applications that use more resourceful sensor nodes. Mobile agents introduce a very natural way of solving one of the standing problems in programming sensor networks, that of in-network re-programming or code update. Examples of such systems are SensorWare [112], based on TCL, and Smart Messages [771], based on Java, both previously described in this chapter.

2.4 Discussion

This chapter surveys different proposals for programming wireless sensor networks with respect to three abstraction perspectives: hardware exposure/interaction, network perception, and data acquisition.

Table 2.2 summarizes the classification of the programming languages and systems presented in the survey, according to the proposed criteria. The evolution of the research in this field seems to show a clear tendency towards high-level programming languages based on hardware independent virtual machines. Many of these systems enjoy properties that promote a robust programming discipline such as strong typing and type safeness.

Table 2.2: Classification grid for programming languages and systems

Proposals/Abstractions	Hardware interaction	Network perception	Data acquisition
Abstract Regions [862]	C	(icon) C/A	M
Agilla [298]	λ	(icon)	C/λ
AmbientTalk [220]	λ	(icon)	M
Cosmos [37]	λ	(icon) C/A	M
Cougar [298]	λ	(icon) C/A	(cylinder)
Deluge [388]	C	(icon)	M
DIFS [329]	λ	(icon) C/A	(cylinder)
Directed Diffusion [405]	λ	(icon) C/A	(cylinder)
Distributed Token Machine [624]	(gear)	(icon) C/A	M

Table 2.2: Classification grid (continuation)

Proposals/Abstractions	Hardware interaction	Network perception	Data acquisition
EnviroTrack [2]	GLUE		M
Flask [547]	λ	C/λ	
GHT [691]	λ	C/λ	
Hood [864]	C	C/λ	M
Impala [527]	GLUE		M
IrisNet [313]	λ	C/λ	
Kairos [342]	λ	C/λ	
Maté [508]	✲		M
Mottle [509]	λ		M
nesC [308]	C		M
Protothreads [242]	C		M
Pushpin [519]	C		M
Regiment [623]	λ	C/λ	
Smart Messages [771]	λ		M
SensorWare [112]	λ		C/λ
SpacialViews [627]	λ		C/λ
SPIN [483]	λ	C/λ	
Squawk [757]	✲		M

Table 2.2: Classification grid (continuation)

Proposals/Abstractions	Hardware interaction	Network perception	Data acquisition
SNACK [330]	λ	🔍 C/λ	✌
TAG [538]	λ	🔍 C/λ	⬡
TinyDB [539]	λ	🔍 C/λ	⬡

However, some important issues still remain. One problem stems from the *semantic gap* between the high-level programming language abstractions or constructs and their concrete, low-level, implementations. For example, languages such as Regiment [623] (and Flask [547]) or TinyDB [539], are mapped at compile time into intermediate representations such as Distributed Token Machines [624] and nesC [308]. This mapping is encoded in the language compiler and therefore it is rather difficult to ensure that the semantics of the high-level application is equivalent to the semantics of its low-level compiled version.

The low-level programming languages (like nesC or Hood) lack a formal specification, making it difficult to reason about program properties and composition and making the development of applications difficult and error-prone. This situation emerges from the absence of an adequate hardware abstraction, *e.g.* a virtual machine, that allows a complete formal specification for the semantics of sensor network applications. Some important work has been made in this direction, most notably Maté [508]. The Squawk Java Virtual Machine also represents an important breakthrough as the type-safety of an important sub-set of the language has been established [239].

A formal model for such a core language has been proposed in CSN (Calculus of Sensor Networks) [532]. In this approach, the *communication* and *concurrency* aspects of sensor networks is modeled with process calculi [574, 375, 111]. This approach is justified by the fact that most high-level languages, even those that fully abstract from the networking aspects and view sensor networks as time varying data streams or data repositories, ultimately map their high-level constructs into a lower level communication-centric language and run-time system. The authors show that the programming model is robust in the sense that it is type safe [554]. This work contributes to a more thorough understanding of the semantics of the sensor networks, something that is fundamental to provide a rigorous background for programming systems.

2.5 Summary and Outlook

Robust programming of large-scale sensor networks is a difficult problem. The severe restrictions imposed by the underlying hardware platform and the choice of an appropriate set of abstractions to model the processes occurring in these networks makes the task quite formidable.

Despite these difficulties, the potential pay-off is immense with applications in many areas from mundane (*e.g.* domotics) to state-of-the-art science (*e.g.* high energy physics). It is therefore not surprising that a considerable amount of research has been devoted to the subject and that some solutions have emerged with various degrees of success.

This chapter provides an overview of the state-of-the-art of programming models, programming languages, and systems for sensor networks. The main goal was to classify current approaches subject to a minimum set of criteria that are arguably relevant for the design of such systems. Also briefly sketched is the current state-of-the-art in hardware and operating systems for sensor networks, since their capabilities ultimately influence much of the work in programming languages.

Despite the advances presented in this survey, current programming languages and systems for sensor networks appear to suffer from a sustainability problem in the sense that the gap between the high-level presentation of the language, including its abstractions, and the underlying implementation is large. This leads to systems for which implementing provably correct and robust macroprograms is difficult or impossible.

Recent developments however introduce techniques that promise to change this picture such as high-level strongly typed languages, hardware independent and verifiable virtual machine architectures, and the appearance of formal models that allow a rigorous understanding of sensor network semantics.

Acknowledgements

The authors are partially supported by project CALLAS of the Fundação para a Ciência e Tecnologia (contract PTDC/EIA/71462/2006).

Chapter 3
Mobility Models for Systems Evaluation

Mirco Musolesi, *Dartmouth College, USA*
Cecilia Mascolo, *University of Cambridge, UK*

3.1 Introduction

Mobility models are used to simulate and evaluate the performance of mobile wireless systems and the algorithms and protocols at the basis of them. The definition of realistic mobility models is one of the most critical and, at the same time, difficult aspects of the simulation of applications and systems designed for mobile environments. There are essentially two possible types of mobility patterns that can be used to evaluate mobile network protocols and algorithms by means of simulations: traces and synthetic models [130]. Traces are obtained by means of measurements of deployed systems and usually consist of logs of connectivity or location information, whereas synthetic models are mathematical models, such as sets of equations, which try to capture the movement of the devices.

Currently, there are very few and very recent public data repository of traces capturing movement of people. Examples are GPS traces and Bluetooth connectivity traces (i.e., traces containing the Bluetooth identifiers of the devices that have been in radio range of a device). For instance, researchers at the Intel Research Laboratory and the University of Cambridge distributed Bluetooth devices to people, in order to collect data about human movements and study the characteristics of the colocation patterns among people. These experiments were firstly conducted among students and researchers in Cambridge [153] and then among the participants of INFOCOM 2005 [391]. Examples of similar projects are the Wireless Topology Discovery project at UCSD [559] and the campus-wide WiFi traffic measurements that have been carried out at Dartmouth College [367]. At this institution, a project with the aim of creating a repository of publicly available traces for the mobile networking community has also been started [472].

In general, synthetic models have been largely preferred [130]. The reasons of this choice are many. First of all, as mentioned, the publicly available traces are limited. Telecommunication companies usually collect and analyze large sets of data but these are kept secret since they may represent a source of competitive advantage, for example, for investments and marketing choices. Secondly, these traces are related to very specific scenarios (such as campus environments) and it is currently difficult to generalize their validity. However, it is important to note that these data show surprising common statistical characteristics, such as the same distribution of the duration of the contacts and inter-contacts intervals.[1] Thirdly, the available traces do not allow for sensitivity analysis of the performance of algorithms, since the values of the parameters that characterize the simulation scenarios, such as the distribution of the speed or the density of the hosts, cannot be varied. Finally, in some cases, it may be important to have a mathematical model underlying the movement of the hosts in simulations, in order to formally analyze its impact on the design of protocols and systems.

For these reasons, many mobility models for the generation of synthetic traces have been presented [130]. The most widely used of such models are based on random individual movements; the simplest, the Random Walk mobility model (equivalent to Brownian motion), is used to represent pure random movements of the entities of a system [249]. Another widely adopted random model is the Random Waypoint mobility model [429], in which pauses are introduced between changes in direction or speed. More recently, a large number of more sophisticated random mobility models for ad hoc network research have been presented [492, 413, 540].

However, all synthetic models are suspect because it is quite difficult to assess to what extent they map reality. It is not hard to see, even only with empirical observations, that the random mobility models generate behavior that is most unhuman-like. This analysis is confirmed by the examination of the available real traces [472]. As we will discuss later in this chapter, mobility models based on random mechanisms generate traces that show properties (such as distributions of the duration of the contacts between the mobile nodes and the inter-contacts time between two subsequent connections) that are different from those observed in real scenarios.[2]

An alternative approach to the problem of modeling human mobility is designing *synthetic* models starting from *real* traces. The challenge is to capture and model the key statistical properties of the traces in order to be able to reproduce and, possibly, to generalize them providing sets of realistic input data for simulators. The first examples of this kind of models are [380, 820], in which the authors considered, respectively, the movement traces collected from a campus scenario and direct

---

[1] We define *contact duration* as the time interval during which two devices are in radio range. We define *inter-contacts time* as the time interval between two contacts. These indicators are particularly important in ad hoc networking and, in particular, in delay tolerant mobile ad hoc networks [605, 359], since inter-contacts times define the frequency and the probability of being in contact with the recipient of a message or a potential message carrier in a given time period.

[2] However, as we will discuss in Sect. 3.4, Karagiannis et alii in [442] demonstrate that the inter-contacts time distributions generated by means of classic random mobility models such as the Random Waypoint model show properties that can also be observed in real traces such as power-law behavior in a certain range of values and an exponential tail after a characteristic time. Power-law distributions are characterized by the following form:

$$P(x) = x^{-k}$$

with $k \geq 0$. A power-law distribution is also called scale-free since it remains unchanged to within a multiplicative factor under a re-scaling of the independent variable x [620].

empirical observations of pedestrians in downtown Osaka as a basis of the design of their models. Many refined models have been presented in the last years such as [410, 505, 890, 459]. A key research area is the analysis and mathematical characterization of the available traces. The goal is to derive the fundamental properties of human mobility and connectivity. In fact, connectivity models derived from the analysis of the traces have also been proposed [128, 901]. Finally, another promising approach is the application of social network theory results to the design of mobility models, since mobile devices are carried by humans and, therefore, the resulting mobility and connectivity patterns are strongly influenced by human relationships [603].

3.2 Purely Synthetic Models

We firstly consider the class of purely random synthetic mobility models. We outline the main characteristics of these models and the most recent relevant results about the analytical characterization of such models. In [130] Camp, Boleng and Davids provide an excellent survey of the most relevant and popular random synthetic mobility models used in ad hoc network research.

3.2.1 Random Walk Mobility Model

The simplest mobility model is the Random Walk mobility model [249, 609], also called Brownian motion; it is a widely used model to represent purely random movements of the entities of a system in various disciplines from physics to meteorology. However, it cannot be considered as a suitable model to simulate wireless environments, since human movements do not present the continuous changes of direction that characterize this mobility model.

3.2.2 Random Waypoint Mobility Model

Another example of random mobility model is the Random Waypoint mobility model [429]. This can be considered as an extension of the Random Walk mobility model, with the addition of pauses between changes in direction or speed. However, also in this case, the realism of the model in terms of geographical movement is far from being realistic. First of all, the initial placement of the nodes in the network does not mirror any real-world situation.[3] The model also suffers from the fact that the nodes concentrate in the middle of the area if we consider a bounded area.

---

[3] As discussed in the introduction, this position has been disputed in [442]. We will present more details about this current discussion in the community in Sect. 3.4.

A possible solution is to assume spherical or toroidal surfaces, but clearly these geometrical abstractions are utterly unrealistic. An additional problem is related to the stationarity of the model (i.e., the variance of the characteristics of the model over time). This model suffers from the fact that the transient (i.e., non-stationary) regime may last for a very long time. One method for avoiding such a bias is to remove the initial part of the simulations in order to avoid the transient regime. However, this does not guarantee that the simulation has reached a stationary regime, since the time that is necessary to reach a stationary regime may be longer than the duration of the simulation itself. Finally, it has also been shown that the model also exhibits speed decay over time [888]. A partial solution to this problem have been proposed in [889].

In [492, 493], the authors present a generalization of the Random Walk and Random Waypoint mobility models that they call Random Trip model. The authors introduce a technique to sample the initial simulation state from the stationary regime (a methodology that is usually called *perfect simulation*) based on Palm Calculus [491] in order to solve the problem of reaching time-stationarity. Perfect simulation for the Random Waypoint model was originally proposed by Navidi and Camp in [613].

The analytical properties of the Random Waypoint model have been analyzed in several works from different perspectives such as the stationary distribution of nodes [81, 80], the node spatial distribution [702] and the evolution of the distribution of the nodes by means of partial differential equations [306].

3.2.3 Other Synthetic Single Node Mobility Models

Starting from the Random Walk and Random Waypoint models, many variations have been proposed. The common characteristic of this class of models is that the movements of the nodes are independent from each other and that the movements are based on random distributions. Notable examples include the Random Direction mobility model [609], the Gauss-Markov mobility model [516] and the Smooth Random mobility model [79]. The choice of these mobility models is usually driven by the need of using a model that is easily mathematically tractable.

Other random mobility models were designed with the goal of reproducing movements in a urban space. The movements of the nodes are constrained by the topology of streets and their associated maximum speed. Examples of this class of models are the City section [130], the Freeway and the Manhattan models [44]. These models are particularly useful for applications of ad hoc mobile networking technologies to vehicular settings.

3.2.4 Synthetic Group Mobility Models

These models and similar existing ones are used to represent the movements of single mobile nodes, however, in some situations the behavior of mobile hosts that move together, such as platoons of soldiers, group of students or colleagues and so on need to be modeled. For these reasons, group mobility models have been devised such as the Reference Group mobility model [377], the Reference Velocity Group mobility model [849] and the Structured Group mobility model [94]. These models are based on a set of equations that link the movements of a node to the positions of a subset of the other nodes of the network. These models are useful to reproduce scenarios characterized by the presence of clusters of people, however, the generated movements do not map those observed in the real worlds since the groups move randomly in the simulation space. The membership mechanisms are also usually hard-wired and single nodes cannot join other groups during the simulation time. Recently, Piorkowski et alii propose a synthetic model called Heterogeneous Random Walk [675] that is able to reproduce the presence of clusters that are observed in real-world traces. The goal of this model is to have a mathematically tractable model to study and explain the emergence of clustered networks.

3.2.5 Modeling Obstacles

Another key issue is the modeling of obstacles (such as buildings and walls) in simulation scenarios. This problem is highly intertwined with the definition of realistic radio propagation models [726]. This is an open research area and very few solutions have been proposed. The most remarkable solution is probably [413], where the authors present a technique for the creation of more realistic mobility models that include the presence of obstacles. The specification of obstacles is based on the use of Voronoi graphs in order to derive the possible pathways in the simulation space. The approach proposed by the authors is general and can be applied to other mobility models.

3.3 Trace-Based Mobility Models

In recent years, many researchers have tried to refine existing models in order to make them more realistic by exploiting the available mobility traces [472]. The key underlying idea of these models is the exploitation of available measurements such as connectivity logs to generate synthetic traces that are characterized by the same statistical properties of the real ones.

Various pioneering measurement studies have been conducted both in infrastructure-based and infrastructure-less environments since the first wireless networks have been deployed. Extensive measurements about the usage of the early deployed Wireless Local Area Networks (WLANs) have been conducted, for instance, in [794], in [46], and in [51]. A detailed analysis of the usage of the WLAN of the Dartmouth College campus is presented in [367].

The first examples of mobility models are based on traces of WLAN campus usage. In [820] a mobility model based on real data from the campus WLAN at ETH in Zurich is presented. The authors use a simulation area divided into squares and derive the probability of transitions between adjacent squares from the data of the access points. Also in this case, the session duration data follow a power-law distribution. This approach can be considered as a refined version of the Weighted Waypoint mobility model [84, 380]. The authors of this model represent the probability of user movements between different areas of the USC campus by means of a Markov model. The model is extracted from data collected from user surveys (i.e., the users were asked to keep a diary of their movements for one month).

The Model T and it evolution, the Model T++, proposed in [410] and [505] generate traces also mirroring the spatial registration patterns of user movements inside a campus WLAN (i.e., the connections to the access points spread in the campus area). The authors define the concept of popularity gradient between different access points and its influence on users' movements. This model is evaluated using traces from Dartmouth College. In [890] another model extracted from real traces based on the study of probability of transitions between different locations is presented. The evaluation of the model is essentially based on the matching of the geographical movements and density of users, rather than on the analysis of the patterns of connectivity among them.

A mobility model based on the extraction of user mobility characteristics from the wireless network traces of the Dartmouth College WLAN is presented in [459]. The authors define popular regions in the campus and characterize the transitions among these areas by means of a Markovian model. Another key finding of the authors is the fact that pause time and speed follow log-normal distributions. These models only represent the transitions between five and sixth locations respectively. The data present characteristics, similar to [459], that evidently differ from those generated by means of classic synthetic random mobility models. In [703] Resta and Santi present a model of user movement between access points driven by the quality of service perceived by the users themselves. This approach is very generic and it is composed of different models that allow for the simulation of user mobility, network traffic, underlying wireless technology and quality of service.

Another interesting model representing the movement inside downtown Osaka is discussed in [540]: the authors reproduce the movements of pedestrians by analyzing the characteristics of the crowd in subsequent instants of time and maps of the city using an empirical methodology, without relying on any wireless measurements.

With respect to mobility models for vehicular networks, a large amount of traces mapping the movements of vehicles in cities and in highways are collected by the traffic authorities but they are not publicly available also for security reasons. Starting from these traces and from empirical observations, several models have been recently presented. Examples include the model proposed by Saha and Johnson [722] extracted from the TIGER traces [831], GrooveSim [551] and STRAW [177].

Finally, a model for the generation of the inter-contacts time duration between buses derived from the log traces of the DieselNet is presented in [901]. We note that this is not a mobility model, but a connectivity model, i.e., it is used to represent topological and not geographical information over time. We will discuss these models in detail in Sect. 3.6.

3.4 Characterization of Human Connectivity

A number of pioneering works [794, 46, 51, 367] have been focussed on traces in order to gain insight about human mobility patterns. A key study in this area is the work on connectivity patterns presented by Chaintreau et al. in [154] which illustrates the fundamental insight that contacts duration and inter-contacts time between individuals can be represented by means of power-law distributions and that these patterns may be used to develop more efficient opportunistic protocols.

The work confirms the results of other studies conducted at Dartmouth [367], UCSD [559] and University of Toronto [777]. At the same time, it is interesting to note that these observed connectivity patterns are at odds with those that can be extracted from random mobility models that show an exponential decay of inter-contacts time intervals [747]. In a previous work [391], similar connectivity patterns have also been observed among the participants of INFOCOM'05.

Recently, Karagiannis, Le Boudec and Vojnovic in [442] offered a novel perspective to the problem of the approximation of these distributions. The authors consider 6 sets of traces and derive several analytical results that can be summarized as follows. First of all, the authors verify the power-law decay of inter-contacts time CCDF between mobile devices. Secondly, they demonstrate that beyond a characteristic time of about 12 hours the CCDF exhibits exponential decay. This is the major novel contribution, especially with respect to the findings presented in [154]. Thirdly, they present an analytical framework demonstrating that mobility models such as the Random Waypoint model should not be abandoned since they are able to represent power-law decay of inter-contact time with an exponential tail after this characteristic time. Finally, they show that the return time of mobile nodes to the same location can be modeled by means of a function composed of a scale-free distribution for a certain range between 0 and a characteristic time with an exponential tail.

Connectivity patterns have been studied by the authors of the aforementioned Model T [410] and Model T++ [505]. The main result of these studies is that user registration patterns exhibit a distinct hierarchy, and that WLAN access points (APs) can be clustered based on registration patterns. Cluster size distributions, intra-cluster transition probabilities and trace lengths are highly skewed and can be modeled by a heavy-tailed Weibull distribution with a good degree of approximation.

The fraction of popular APs in a cluster, as a function of cluster size, can be modeled by exponential distributions.

The spatio-temporal correlation in the user registration patterns has also been investigated in [496]. The mobility patterns are modeled using a semi-Markov process by means of the transition probability matrix. The authors estimate the long-term wireless network usage among different access points. By comparing the steady-state distributions of semi-Markov models based on trace data collected at different time scales, they characterize the degree of correlation in time and location. The analysis is founded on the logs from Dartmouth College [473]. An analysis of the periodic properties of the movements between access points using Fourier transforms is presented in [458].

Rhee et alii proposed a possible modelization of human movement by means of Lévy flights [705] but the analysis show that this approximation is valid only considering a coarse-grained geographical scale. Recently, Gonzalez et al. [321] present the analysis of the movements of 100.000 mobile phone users by analyzing their registration patterns. According to this study human trajectories show a high degree of temporal and spatial regularity; each user usually move between a few highly frequented locations. They also disprove the theory that a pure Lévy flights model can be used to represent human trajectories, since random jumps typical of this model are not observed in their traces.

We would also like to mention briefly the considerable amount of work done by mathematical biologists in modeling animal movements [821]. One of the most interesting studies is that about animal foraging behavior. It was believed that the movement of animals for foraging can be modeled by means of Lévy flights; many different species have been studied including albatrosses [841], deer [842] and grey seals [35]. Lévy flights are random walks characterized by step lengths extracted from probability distributions with heavy tails: the result is that sequences of short steps are followed by rare long steps. However, in a study published on Nature in 2007 [246], by reanalyzing the data about albatrosses, the authors conclude that the movement can be modeled with gamma distributions with an exponential decay and not by means of a Lévy flights model.

3.5 Social Network Based Mobility Models

In this section we discuss a recent development of mobility modeling [603], i.e., the introduction of social networking concepts as a basis of the representation of people movements. These models are usually trace based, i.e., they are generally founded or evaluated by means of real traces. The modeling of these relationships and their implications to human mobility is of paramount importance to test protocols and systems that exploit the underlying social structure, such as socially-aware delay tolerant forwarding protocols [209, 192].

Social network mobility models are based on a simple observation. In mobile networks, devices are usually carried by humans, so the movement of such devices is necessarily based on human decisions and social behavior. A key characteristic is the presence of clusters that are usually dependent on the relationships among the members of the social group. In order to capture this type of behavior, mobility models dependent on the structure of the relationships among people carrying the devices have been defined. However, existing group mobility models fail to capture this social dimension [130].[4]

3.5.1 Social Network Models

In recent years, social networks have been investigated in considerable detail, both in sociology [855] and in other areas, most notably mathematics and physics [14]. Various types of networks (such as the Internet, the World Wide Web and biological networks) have been studied by researchers especially in the statistical physics community. Theoretical models have been developed to reproduce the properties of these networks, such as the so-called small worlds model proposed in [857] or various scale-free models [618, 840]. Excellent reviews of the recent advances in complex and social networks analysis can be found in [14] and [618].

As discussed in [622], social networks appear to be fundamentally different from other types of networked systems. In particular, even if social networks present typical small-worlds behavior in terms of the average distance between pairs of individuals (the so-called *average path length*), they show a greater level of clustering. In [622] the authors observe that the level of clustering seen in many non-social systems is no greater than in those generated using pure random models. Instead in social networks, clustering appears to be far greater than in networks based on stochastic models. The authors suggest that this is strictly related to the fact that humans usually organize themselves into *communities*. Examples of social networks used for these studies are rather diverse and include, for instance, networks of coauthorships of scientists [617] and the actors in films with Kevin Bacon [857].

3.5.2 The Community Based Mobility Model

In [602] the authors propose the Community based mobility model, founded on social network theory.[5] A key input of the mobility model is the social network that links the individuals carrying the mobile devices in order to generate realistic syn-

---

[4] These mobility models can also be used to test other types of networks. Within the emerging field of sensor networks, mobile hosts are not necessarily carried directly by humans. However, sensor networks are usually embedded in artifacts and vehicles (such as cars, planes or clothing) or are spread across a geographical area (such as environmental sensors). In the former case, the movements of the sensors embedded in cars or in airplanes, for instance, are not random but are dependent on the movements of the carriers; in the latter, movement is not generally a major issue.

[5] This model can be considered an evolution of the basic model initially proposed in [604].

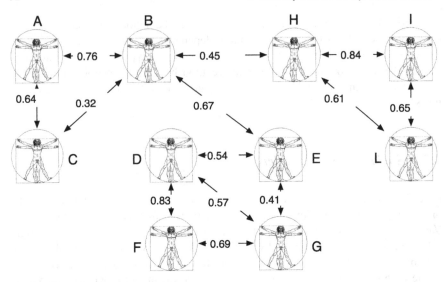

Fig. 3.1 Example of social network

thetic network structures [857]. The model allows collections of hosts to be grouped together in a way that is based on social relationships among the individuals. This grouping is only then mapped to a topographical space, with topography biased by the strength of social ties. The movements of the hosts are also driven by the social relationships among them. The model also allows for the definition of different types of relationships during a certain period of time (i.e., a day or a week). For instance, it might be important to be able to describe that in the morning and in the afternoon of weekdays, relationships at the workplace are more important than friendships and family one, whereas the opposite is true during the evenings and weekends.

The model is evaluated by means of real mobility traces provided by the Intel Research Laboratory [737]; the authors show that the model provides a good approximation of real movements in terms of some fundamental parameters, such as the distribution of the contacts duration and inter-contacts time. In particular, the data show that an approximate power law holds over a large range of values for the inter-contacts time. Instead, contacts duration distribution follows a power law for a more limited range of values. These statistical characteristics are also very similar to those observed by the researchers at the University of California at San Diego and Dartmouth College [153].

We now describe in more details the key aspects of the Community based mobility model, starting from the representation of the social graph. One of the classic ways of representing social networks is using *weighted graphs*. An example of social network is represented in Fig. 3.1. Each node represents one person. The weights associated to each edge of the network are used to model the strength of the interactions between individuals [736]. It is our explicit assumption that these weights, which are expressed as a measure of the strength of social ties, can also be

$$
\mathbf{M} = \begin{bmatrix}
1 & 0.76 & 0.64 & 0.11 & 0.05 & 0 & 0 & 0.12 & 0.15 & 0 \\
0.76 & 1 & 0.32 & 0 & 0.67 & 0.13 & 0.23 & 0.43 & 0 & 0.05 \\
0.64 & 0.32 & 1 & 0.13 & 0.24 & 0 & 0 & 0.15 & 0 & 0 \\
0.11 & 0 & 0.13 & 1 & 0.54 & 0.83 & 0.57 & 0 & 0 & 0 \\
0.05 & 0.67 & 0.24 & 0.54 & 1 & 0.2 & 0.41 & 0.2 & 0.23 & 0 \\
0 & 0.13 & 0 & 0.83 & 0.2 & 1 & 0.69 & 0.15 & 0 & 0 \\
0 & 0.23 & 0 & 0.57 & 0.41 & 0.69 & 1 & 0.18 & 0 & 0.12 \\
0.12 & 0.43 & 0.15 & 0 & 0.2 & 0.15 & 0.18 & 1 & 0.84 & 0.61 \\
0.15 & 0 & 0 & 0 & 0.23 & 0 & 0 & 0.84 & 1 & 0.65 \\
0 & 0.05 & 0 & 0 & 0 & 0 & 0.12 & 0.61 & 0.65 & 1
\end{bmatrix}
$$

Fig. 3.2 Example of an Interaction Matrix representing a simple social network

$$
\mathbf{C} = \begin{bmatrix}
1 & 1 & 1 & 0 & 0 & 0 & 0 & 0 & 0 & 0 \\
1 & 1 & 1 & 0 & 1 & 0 & 0 & 1 & 0 & 0 \\
1 & 1 & 1 & 0 & 0 & 0 & 0 & 0 & 0 & 0 \\
0 & 0 & 0 & 1 & 1 & 1 & 1 & 0 & 0 & 0 \\
0 & 1 & 0 & 1 & 1 & 0 & 1 & 0 & 0 & 0 \\
0 & 0 & 0 & 1 & 0 & 1 & 1 & 0 & 0 & 0 \\
0 & 0 & 0 & 1 & 1 & 1 & 1 & 0 & 0 & 0 \\
0 & 1 & 0 & 0 & 0 & 0 & 0 & 1 & 1 & 1 \\
0 & 0 & 0 & 0 & 0 & 0 & 0 & 1 & 1 & 1 \\
0 & 0 & 0 & 0 & 0 & 0 & 0 & 1 & 1 & 1
\end{bmatrix}
$$

Fig. 3.3 Example of a Connectivity Matrix representing a simple social network

read as a measure of the likelihood of geographic colocation. We model the degree of social interaction between two people using a value in the range [0, 1]. 0 indicates no interaction; 1 indicates a strong social interaction. Different social networks can be valid for different parts of a day or of a week.

As a consequence, the network in Fig. 3.1 can be represented by the 10×10 symmetric matrix **M** showed in Fig. 3.2, where the names of nodes correspond to both rows and columns and are ordered alphabetically. We refer to the matrix representing the social relationships as *Interaction Matrix*.

The generic element $m_{i,j}$ represents the interaction between two individuals i and j. We refer to the elements of the matrix as the *interaction indicators*. The diagonal elements represent the relationships that an individual has with himself and are set, conventionally, to 1. In Fig. 3.1, we have represented only the links associated to a weight equal to or higher than 0.25. A key issue of this model is the definition of this Interaction Matrix. This is clearly a simplified model of human relationships. The definition of these weights is an open research area also in sociology [855].

The Interaction Matrix is also used to generate a *Connectivity Matrix*. From matrix **M** we generate a binary matrix **C** where a 1 is placed as an entry c_{ij} if and only if $m_{i,j}$ is greater than a specific threshold t (i.e., 0.25). The Connectivity Matrix extracted by the Interaction Matrix in Fig. 3.2 is showed in Fig. 3.3. The idea behind this is that we have an *interaction threshold* above which we say that two people are interacting as they have a strong relationship.

The Interaction Matrix (and, consequently, the Connectivity Matrix) can be derived by available data (for example, from a sociological investigation) or using mathematical models that are able to reproduce characteristics of real social networks. The default implementation of the model uses the so-called Caveman model [857] for the generation of synthetic social networks with realistic characteristics (i.e., high clustering and low average path length). However, this is a customizable aspect and, if there are insights on the type of scenarios to be tested, a user-defined matrix can be used as input.

The simulation scenario is established by mapping groups of hosts to certain areas in the geographical space. After the definition of the social graph described above, groups, i.e., the highly connected set of nodes in the graph, need to be isolated. The authors use the algorithm proposed in [621] to detect the presence of community structures in social networks represented by matrices, like the Connectivity Matrix that we have defined in the previous section. This algorithm is based on the calculation of the so-called *betweenness* of edges. This provides a measure of the centrality of nodes.

In order to illustrate this process, let us now consider the social network in Fig. 3.1. Three communities (that can be represented by sets of hosts) are detected by running the algorithm: $C_1 = \{A, B, C\}, C_2 = \{D, E, F, G\}$ and $C_3 = \{H, I, L\}$. Now that the communities are identified given the matrix, they need to be associated with locations.

After the communities are identified, each of them is randomly associated to a specific location (i.e., a square) on a grid.[6] We use the symbol $S_{p,q}$ to indicate a square in position p,q. The number of rows and columns are inputs of the mobility model.

Going back to the example, in Fig. 3.4 we show how the communities we have identified can be placed on a 3×4 Grid (the dimension of the grid is configurable by the user and influences the density of the nodes in each square). The three communities C_1, C_2, C_3 are placed respectively in the grid in the squares $S_{a,2}, S_{c,2}$ and $S_{b,4}$. Each node of a certain community is placed in randomly selected positions inside the assigned square.

As described in the previous section, a host is initially positioned in a certain square in the grid. Then, in order to drive movement, a goal is assigned to the host. More formally, we say that a host i is associated to a square $S_{p,q}$ if its goal is inside $S_{p,q}$. Note that host i is not necessarily always positioned inside the square $S_{p,q}$, despite this association (see below).

The goal is simply a point on the grid which acts as *final destination* of movement like in the Random Waypoint model, with the exception that the selection of the goal is not as random. When the model is initially established, the goal of each host is randomly chosen inside the square associated to its community (i.e., the first goals of all the hosts of the community C_1 will be chosen inside the square $S_{a,2}$).

---

[6] A non random association to the particular areas of the simulation area can be devised, for example by deciding pre-defined *areas of interest* corresponding for instance to real geographical space. However, this aspect is orthogonal to the mechanisms at the basis of this model.

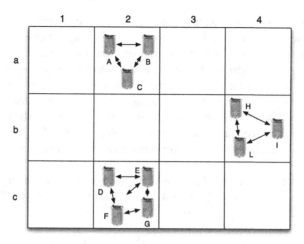

Fig. 3.4 Example of initial simulation configuration

When a goal is reached, the new goal is chosen according to the following mechanism. A certain number of hosts (zero or more) are associated to each square $S_{p,q}$ at time t. Each square (i.e., place) exerts a certain *social attractivity* to a certain host. The social attractivity of a square is a measure of its importance in terms of the social relationships for the host taken into consideration. The social importance is calculated by evaluating the strength of the relationships with the hosts that are moving towards that particular square (i.e., with the hosts that have a current goal inside that particular square). More formally, given $C_{S_{p,q}}$ (i.e., the set of the hosts associated to square $S_{p,q}$), we define *social attractivity* of that square towards the host i SA_{p,q_i}, as follows:

$$SA_{p,q_i} = \frac{\sum_{j \in C_{S_{p,q}}} m_{i,j}}{w} \tag{3.1}$$

where w is the cardinality of $C_{S_{p,q}}$ (i.e., the number of hosts associated to the square $S_{p,q}$). In other words, the social attractivity of a square in position (p, q) towards a host i is defined as the sum of the interaction indicators that represent the relationships between i and the other hosts that belong to that particular square, normalized by the total number of hosts associated to that square. If $w = 0$ (i.e., the square is empty), the value of SA_{p,q_i} is set to 0.

The mobility model allows for two alternative mechanisms for the selection of the next goal that are described, a *deterministic* one based on the selection of the square that exerts the highest attractivity and a *probabilistic* one based on probability of selection of a goal in a certain square proportional to their attractivities. Using the first one, the goals are chosen only inside the squares associated to the community, whereas with the second, the hosts may also randomly select their goals in other squares of the simulation area, with a certain non zero probability. In other words, the second mechanism allows for the selection of the destinations not only based on social relationships adding more realism to the model. According to this mechanism, the new goal is randomly chosen inside the square characterized by the highest social attractivity; it may be again inside the same square or in a different one. New goals are chosen inside the same area when the input social network is composed by loosely connected communities (in this case, hosts associated with different communities have, in average, weak relationships between each others). On the other hand, a host may be attracted to a different square, when it has strong relationships with both communities. From a graph theory point of view, this means that the host is located between two (or more) clusters of nodes in the social network.[7]

An alternative mechanism is based on a selection of the next goal proportional to the attractivity of each square. In other words, we assign a probability $P(s = S_{p,q_i})$ of selecting the square S_{p,q_i} as follows:

$$P(s = S_{p,q_i}) = \frac{SA_{p,q_i} + d}{\sum_{j=1}^{p \times q}(SA_{p,q_j} + d)} \qquad (3.2)$$

where d is a random value greater than 1 in order to ensure that the probability of selecting a goal in a square is always non zero.[8]

The parameter d can be used to increase the randomness of the model in the process of selection of the new goal. This may be exploited to increase the realism of the generated scenario, since in real situations, humans also move to areas without people or for reasons not related to their social sphere.

3.5.3 Other Social Network Based Mobility Models

Another notable example of mobility model founded on the social relationships between the individuals carrying the mobile devices is presented in [368]. This work is based on assumptions similar to [602], but it is considerably more limited in scope. Hosts are statically assigned to a particular group during the initial configuration process, whereas [602] accounts for movement between groups. Moreover, the authors claim that mobile ad hoc networks are scale-free, but the typical properties of scale-free networks are not considered in the design of the model presented by the authors. The scale-free distribution of mobile ad hoc networks is still not proven in general, since very limited measurements are available and it is worth noting that the scale-free properties are strictly dependent on the movements of hosts and, therefore, they are dependent on the actual application scenarios [317]. The idea of using communities to represent group movements in an infrastructured WiFi network has

---

[7] This is usually the case of hosts characterized by a relatively high betweenness that, by definition, means that they are located *between* two (or more) communities.

[8] The role of d is similar to the *damping factor* used in the calculation of the Google PageRank [118]. In fact, the transitions between squares can also be similarly represented using a Markov Chain model with $P(s = S_{p,q_i})$ as probability of transitions between states (squares).

also been exploited in [768] and in its time-variant extension presented in [381]. More specifically, this model preserves two fundamental characteristics, the skewed location visiting preferences and the periodical re-appearance of nodes in the same location. Recently, Ekman et alii propose a model based on the daily activities of the users (and group of users) and their movements between place of interests in a city map [251].

3.6 From Mobility to Connectivity Models

Another class of models for mobile networking research is that of connectivity models, that focusses on the evolution of the emergent connectivity graph that is changing over time as nodes move. Topological properties are fundamental for analyzing, for example, the performance of protocols and systems where (intermittent) connectivity plays an essential role such as protocols for delay tolerant networks or solutions for bandwidth provision in WLANs. This is a very open area and very few models have been proposed. Most of them are based on the analysis of the available connectivity traces, i.e. from logs of Bluetooth contacts or WLANs registration patterns.

In [128] the authors propose the Connectivity Trace Generator (CTG). This work differs from previous approaches in that probability distributions describing the patterns of colocation of mobile users (in terms of contact duration and inter-contacts time) are exploited for the first time as *direct inputs* of a synthetic traces generation tool. More specifically, the input parameters of this component are the relevant parameters of the connectivity model, namely: number of nodes, the contacts duration (i.e., the time interval in which two devices are in radio range) and inter-contact time (i.e., the time interval between two contacts), and node degree (i.e., number of neighbors) distributions.

All these distributions can be extracted by measurements of connectivity on real traces. The key steps of the proposed simulation framework are depicted in Fig. 3.5.

The input of the CTG is a set of real traces. These are processed by a trace analyzer to generate the parameters describing user connectivity required by the tool. These are essentially the coefficients of the curves used to approximate the distributions of the inter-contacts times, contacts durations and link degrees characterizing the social graph of the contacts among the users. This is a graph of the potential contacts between pair of nodes, i.e., an edge between two nodes exists if there is a probability different from zero that these two nodes will meet in a specified time interval equal to the simulation duration. This graph is also built from the traces. Additionally, a range of variations for the parameters is provided in input.

The process of generation is based on the selection of the desired number of hosts and on the construction of a connectivity graph of all the potential contacts of each host. In other words, we map each host to a node of the graph and we link a pair of nodes with an edge if the two hosts can potentially get in contact. The connectivity graph is then used to unfold a number of connection links between

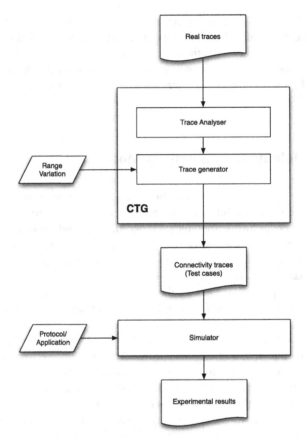

Fig. 3.5 Connectivity Trace Generator

users for each time instant. In other words, we use the connectivity graph as a basis for a *time-varying graph of instant connectivity* for each instant t. In these time-varying graphs (one for each time instant), each link is either active if the two hosts are colocated, or is not present if the two are not.

Each link is activated and de-activated according to the distributions of the contacts duration and inter-contacts time. Let us consider an example. Each pair at the beginning is initially disconnected. Let us consider the connectivity pattern between two hosts A and B. At the beginning, the model generates an initial disconnection time (an "inter-contacts time") sampled from the inter-contacts time distribution. Let us assume that this is equal to 15 seconds. Then after 15 seconds, the model has to generate a colocation time from the distribution of the contacts duration. Let us assume that this is equal to 30 seconds. An edge between A and B is activated for 30 seconds to represent the colocation of the hosts A and B between the instants 15 s and 45 s. In other words, the edge will be present in the graph for the next 30 seconds and then removed. Then the model generates another value, this time from the distribution of the inter-contacts time, for example 4 minutes. After 4 minutes, the link between *A* and *B* is re-activated for a duration time that is again generated from the distribution of the contacts duration interval and so on.

This process is completely automated and implemented by the trace generator component that produces traces containing the events of connections and disconnections for each pair of nodes of the simulation scenario and the time of each event. These traces can be used as test cases for the testing of opportunistic mobile systems [273].

As a concrete case study, the authors used the log session traces of the campus WLAN of Dartmouth College [473], to obtain empirical distributions for residence time, colocation and degree distribution of the nodes. These traces were used in conjunction with an original model developed by the authors that aims at representing the properties of the *colocation* of two users as a function of the probability for a user of being in a specific place for a given period of time. Two hosts were considered colocated if they were registered to the same access point.

As we said, the design of connectivity models is still an open research area; to the best of our knowledge, the other existing proposal is the position paper by Nykvist and Phanse [634]. With respect to vehicular networks modeling, the only existing example of connectivity models is that of the buses of the DieselNet project [901] discussed in Sect. 3.3. Another recent work analyzing the connectivity properties of a bus transportation system is [371].

3.7 Testing Tools and Mobility Modeling

The first step of any performance evaluation exercise based on simulations is the choice of the simulator tool. Various network simulators are available for the evaluation of protocols and systems of mobile networks; the most popular are ns–2 [803] with the so-called Monarch extension [428] (and the upcoming new version ns–3 [804]), Glomosim [896] and Opnet [644]. Another class of tools for simulation of generic complex systems (not only computer systems, but also economic, biological, industrial, etc.) are the so-called *discrete-event simulators*. These tools only provide primitives for the concurrent execution of multiple entities and communication among them usually by means of message passing based paradigms. OMNeT++ [839] and Parsec [43] are examples of this class of simulators.

These tools generally receive in input traces with different formats usually in the form of a series of triplets that specify when the change of direction has to take place, the next goal (that defines the direction of the host) and the node speed. Unfortunately, there is no standard format for this kind of traces. More in general, no standards have been defined also for measurement traces both of movements represented by means of geographical positions or connectivity traces (such as those collected by means of Bluetooth or ZigBee radio devices).

The results of simulations performed by means of different simulators may show significant differences; this fact may be explained by the various modeling techniques and assumptions and by the different levels of details offered by these simulators. In [148] the authors show and discuss the divergent results obtained by using OPNET Modeler, ns–2 and Glomosim. Other problems can be related to the methodology followed by the researchers and, unfortunately, this has caused a decreasing confidence in simulation results to evaluate the performance of protocols and systems: this is motivated by the apparent issues in terms of scientific standards of some of the existing published papers [485]. With respect to mobility modeling, the use of unrealistic mobility models or the absence of a meaningful number of runs to achieve a sufficient statistical validity of the results has contributed to this lack of confidence. It is interesting to note that there is a clear problem of achieving statistical validity when a limited set of traces is used to evaluate an algorithm or a protocol. More specifically, in presence of a limited set of nodes and/or of a limited duration of the traces, there is a critical issue of generality. For this reason we believe that tools like the CTG that allows the researchers to vary the parameters describing the mobility patterns distributions in order to explore their impact are needed.

There is also a growing interest in approaches for testing mobile systems and applications (see for instance [762]). Most of these approaches, however, concentrate on testing aspects related to context awareness (see, for example, [853]). Mobility and connectivity can be considered as context elements; however, these tools do not provide specific support for modeling these essential aspects of this class of systems.

The CTG presented in Sect. 3.6 provides automatic generation of connectivity test cases in order to evaluate the performance of communication protocols and applications in opportunistic mobile systems. The approach allows flexible performance testing of new protocols and applications. Indeed, when a system is being prototyped, usage patterns logs could be collected through a small scale trial. The connectivity traces could then be analyzed and, using the methodology proposed by the authors of the CTG, a simulation on a larger scale could be carried, using larger synthetic traces by a higher number of hosts or different colocation or inter-contacts time distributions.

A tool for the generation of traces for vehicular networking simulations is presented in [69]. The model allows for the generation of traces that reproduce steady-state random trips on real road topology from the Swiss Geographic Information System (GIS).

However, there are no comprehensive solutions for the verification of mobile systems; for example the CTG lacks a metric for coverage criteria of the generated test cases. An investigation along these lines for a similar problem has been presented in [720]: the authors of the CTG leave the issue of evaluating coverage conditions open for future work.

3.8 Summary and Outlook

In this survey, we have presented a description of the state of the art in mobility modeling, considering different classes of synthetic and trace-based models. We have also discussed the analytical models that have been developed to understand human movements. Finally, we have presented the concepts at the basis of the design of mobility models based on social networks. We now present a summary of the open research challenges in this area, outlining a research agenda for the mobile networking and systems community in this area. The research challenges can be summarized as follows:

Specificity of Available Models   The available traces describe very specific situations like campuses or conference environments and, for this reason, it is difficult to generalize the results obtained using the traces directly or the mobility models derived from the analysis of these traces. With high probability, different types of mobility patterns characterize specific application scenarios, both in terms of contacts distribution and scale of movements in the geographical space. The main research challenge resides in the identification of the common features of human mobility and the characterization of the specificity of a set of deployment scenarios. This problem will be tackled more and more effectively with the increasing availability of mobility traces extracted from heterogeneous environments.

Mobility Models vs Connectivity Models   In this survey, we have introduced and discussed the concept of connectivity models. This kind of models are not *alternative* but *complementary* to the existing models. In fact, mobility models (i.e., containing information about the locations of the nodes) are necessary for testing several classes of protocols and applications such as geocasting protocols [423] or location-aware applications [361]. An open problem is how to integrate the use of connectivity and mobility models in an effective way to characterize human mobility. Connectivity models can be derived by mobility models but the former represent a much more powerful tool for the statistical characterization of colocation patterns. These models are very useful for designing and evaluating protocols and systems where these aspects are fundamental such as in the case of performance evaluation of delay tolerant protocols or wireless peer-to-peer systems (for example to evaluate the available transmission bandwidth among a set of hosts). As for mobility models, further investigations are needed to characterize common properties of human connectivity and distinct features of specific application environments. Another open issue is the characterization of the interaction between human movement and the surrounding environment: more specifically, the influence of the geographical features of the simulation spaces such as the presence of obstacles (e.g., buildings, hills, green areas) on human connectivity and mobility patterns has not been studied yet.

Benchmarks for Protocol and System Evaluation   Unfortunately, the choice of values for parameters of simulations for mobile (in particular, ad hoc) networks research is extremely variable. In fact, the ad hoc and delay tolerant research communities lack of consistent scenarios to validate and to benchmark the different solutions. As cited previously, in [485] Kurkoswski, Camp and Colagrosso reported an analysis of the performance evaluation of papers published at Mobi-Hoc from 2000 to 2005, showing evident flaws of a large number of works from a scientific point of view in terms of simulation methodology. The community should define a common set of mobility traces that should be used to verify the performance of protocols. A possible idea is to define a series of sets of traces for different classes of application scenarios such as dense networks, urban environments and sparse networks, for instance for the evaluation of delay tolerant networking protocols or Bluetooth based systems. This can be seen as a medium term goal, given also the limited amount of available traces. However, following the introduction of more powerful and, at the same time, affordable devices such as phones equipped with GPS units and Bluetooth, we believe that the amount of available information will increase hugely in the next few years.

Tools   There is a very limited number of available tools, in particular open source and free, for academic and industrial testing of mobile applications. With respect to mobility modeling, there is a concrete need of network emulators that are able to simulate connectivity based on an underlying mobility model (or directly on traces). An interesting example of this kind of systems is [633]. Another very useful class of systems for performance evaluation studies are emulators based on virtualization techniques on a single machine for testing multiple instances of mobile applications by means of virtual communication interfaces (such as Bluetooth or ZigBee) and infrastructure-based (such as based on access points or GPRS), also providing radio propagation models.

Standardization of the Trace Formats   Unfortunately, the available traces (see, for example, those stored in the CRAWDAD repository [472]) do not follow a common standard and scripts are needed to convert them to the various formats in order to be processed by the different simulators. The mobile networking and systems community should allow for common standards in order to promote an easy data exchange among researchers for cross-comparisons, also for the establishment of benchmarks for the community, as it happens in other fields of computer science.

Chapter 4
Ad Hoc Routing

Filipe Araujo, *University of Coimbra, Portugal*
Hugo Miranda, *University of Lisbon, Portugal*

4.1 Introduction

Chapter 1 presented multi-hop communication as one of the opportunities to reduce the energy consumption of the devices, thus extending their lifetime. The development of efficient multi-hop routing protocols is therefore a fundamental aspect for the successful deployment of wireless ad hoc networks. However, routing in wireless ad hoc networks is fundamentally different from the apparently similar problem in wired networks, because bandwidth is scarce and topology changes frequently. This raises a double challenge for multi-hop routing in wireless ad hoc networks: nodes are less powerful and routing is more difficult. Given this fact, it is not surprising to observe that multi-hop routing has been one of the most active fields of research in MANETs.

This chapter reviews two major types of routing for wireless ad hoc networks: topology-based and position-based routing. In general, topology-based routing derives from well-known protocols that knew large success in the wired Internet, mostly from distance vector protocols, although some link state protocols can also be found. The challenge posed to topology-based routing for wireless ad hoc networks is to efficiently find good routes between any two nodes in an environment where participants have only a partial (and possibly outdaded) view of the network topology.

In position-based routing, nodes use positional information to forward packets according to their limited network view. These algorithms require fewer topological information, often close to none, thus wasting fewer energy and memory. On the other hand, position-based routing needs a complex preliminary phase before any routing algorithm can even start. In this phase, known as "pre-processing", nodes collect and treat beacons from neighbors. Doing efficient pre-processing is a difficult problem that deserves the same, or even more attention than routing algorithms themselves. Finally, position-based routing has the additional problem of collecting positional information, as this may be expensive, unreliable or difficult to obtain.

4.1.1 Routing Metrics

In formal terms, a routing metric is a value that serves to compare routes in a network. Some routing algorithms can use almost any metric, as long as there is some notion of cost to transmit a message across a link. In a model where nodes transmit with a constant power, the minimal number of hops between two nodes is likely to represent the shortest path and require the lowest accumulated energy consumption. This rationale has been followed by a number of routing algorithms, that use the number of hops that a packet takes to reach destination as the preferred metric.

Among the reasons not to use hop count, energy is probably the most important. For instance, one may want to minimize the consumption of energy in the transmission of a packet or to avoid nodes with almost discharged batteries. There are also other factors that can assume more importance than simple hop count. For instance, shortest paths in number of hops may use links with poor quality [237] or cross congested regions, thus increasing latency, collisions and energy consumption [655].

Examples of metrics concerning the energy available at the devices that can be applied to routing algorithms are [760]:

Minimize energy consumed/packet   Forwarding a packet will consume some energy at each hop. The goal of this metric is to find the route where the sum of the energy spent by all nodes in the route is minimal [83]. Each node may spend a different amount of energy for several reasons. For congested or noisy regions, the energy consumed may be significantly higher due to the need of retransmissions. Also, some network cards can vary the transmission power (and therefore, their range) or spend a variable amount of energy depending on the card specifications. An energy consumption model for network operations is presented in [276].

Maximize time to network partition   Considering a wireless ad hoc network as a graph where devices are the vertices and the available communication links the edges, protocols using this metric extend as far as possible the lifetime of all nodes that are in the critical path: those whose disconnection will partition the network, making impossible the communication between all nodes. The work of Blough and Santi [96] considers this and related approaches.

Minimize variance in node power levels   This metric assumes that all nodes are equally relevant for the network and therefore will focus on keeping the energy level of all nodes as close as possible.

Minimize cost/packet   This metric considers that nodes may associate costs to packet forwarding using some reluctance function $f(A)$. Possible examples are the individual energy level of each node or some user defined rank of the nodes. The goal of protocols implementing this metric is to select the route with the minimal sum of the cost of all intervening nodes [760]. Chang and Tassiulas [163, 164] also consider a similar case, where they try to maximize the time until the first node spends all its energy (i.e., the lifetime of the network is the time it takes to the first node failure).

Minimize maximum node cost   Like in the previous metric, this one also considers a cost value advertised by each node which may or may not be correlated with the energy available. However, this metric will rank routes by the maximum cost advertised by participating nodes and select the one whose value is lower.

An interesting example of the application of these metrics to routing protocols is to define a metric based on a function $f(A)$ which expresses the reluctance of node A to forward packets. In [760], function f is given by $f(A) = 1/g(A)$, where $g(A) \in [0, 1]$ is the available energy of A's battery. The routing metric selects the path that minimizes the sum of $f(A)$ of all nodes in the path. As one can easily see, a node that is running out of battery has the ability to significantly increase the cost of paths crossing it and will likely be avoided. These and other metrics have been experimented in a number of protocols (e.g. [164, 514, 760]), collectively known as "power-aware" or "cost-aware" routing protocols, terms coined by Stojmenovic and Lin [775].

In most topology-based routing schemes other metrics besides hop count could fit in easily. However, this would not be so easy in position-based routing because the routing metric tends to be embedded in the algorithm. This means that nodes are usually restricted to a specific neighbor when forwarding a packet. Since hop count is the most common among all these metrics, in the text that follows we consider the routing metric to be hop count, except when we explicitly state otherwise.

4.2 Topology-Based Routing

In this section, we present algorithms that route using topological information. Topology-based routing algorithms collect knowledge of the network to determine the best route. The chapter reviews some of the most cited and popular topology-based routing protocols for MANETs. Not surprisingly, all fit into the uniform category of the taxonomy proposed by Liu and Kaiser [523]. In their survey, the authors divide routing protocols under the main categories of uniform and non-uniform, depending on the similarity of the roles that nodes play in the network. By definition, MANETs are decentralized and infrastructureless what makes non-uniform algorithms inadequate.

Uniform routing protocols are further divided into reactive or proactive, respectively if they compute routes *before* or *after* routing requests. Of those surveyed, one of the earliest routing protocols for MANETs, the Destination Sequence Distance Vector (DSDV) [662] and the Optimized Link State Routing Protocol for Ad Hoc Networks (OLSR) [183, 182] are proactive protocols. The remaining, namely the Dynamic Source Routing (DSR) [429], Ad Hoc On-demand Distance Vector (AODV) [664] and Dynamic MANET On-demand Routing Protocol (DYMO) [155] are all reactive protocols.

4.2.1 Proactive Routing Protocols

Like in most routing protocols for wired networks (e.g. RIP [548] or OSPF [591]), nodes in proactive protocols keep a routing table with the next hop of every possible packet destination. The challenge is to maintain an accurate routing table given the frequent changes in the topology that result from node movement. The problem is amplified by the limited battery of the devices, which claims for a conservative exchange of messages.

Destination-Sequenced Distance-Vector. The Destination-Sequenced Distance-Vector [662] (DSDV) is a distance vector routing protocol based on RIP [548]. Authors argue that RIP works well in small networks and is thus pretty adequate for the proposed environment. Hence, DSDV takes RIP as the starting point, but includes some fixes to prevent distance vector looping problems.

Nodes advertise their routing tables to neighbors, including the following information for each of the routing table entries:

- the destination's address;
- the number of hops needed to reach destination;
- a sequence number.

The sequence number prevents looping problems in DSDV. Sequence numbers increase monotonically and are either generated by the destination of a path or by some neighbor node detecting that such destination is gone. The former generates new even sequence numbers, while the latter "breaks" a route to a destination with the following odd sequence number. Nodes flood throughout the network both kinds of advertisement. Flooding is controlled by means of two rules: i) nodes do not accept updates with lower sequence numbers; and ii) nodes only accept updates with equal sequence numbers if these updates indicate a better path. Updates with higher sequence numbers are always accepted. These rules prevent the count-to-infinity problem of distance vector protocols, caused by nodes that become unreachable.

If a node is suspected to have failed, its neighbors immediately generate new odd sequence numbers with infinity cost. This odd number immediately follows the last even number they propagated from the failing node. According to the rules above, this new sequence number will propagate until it reaches the entire network. If a node A fails, neighbors that are upstream of A must interrupt all the paths not destined to, but passing by A, by putting a special indication in the routing table. Along the path, each node will check the odd sequence number, verify that this number is higher than the previously existing sequence number and update the routing table to reflect the infinity cost.

As soon as some previously crashed node D resumes (or if it has just moved), it restarts generating sequence numbers larger than any of the sequence numbers used by its neighbors to destroy paths pointing to D. Consequently, neighbors will once again rebuild their routes to D. It is also possible that a route update with an odd sequence number finds a larger even sequence number. This would mean that the supposedly failed node would have become reachable again (because it moved back to its original position). In this case, the node receiving the old odd sequence number should trigger a route update broadcast to inform about the existing route to the (supposedly) unreachable node.

New routes are immediately broadcast by the nodes. In contrast, and to save resources, route updates advertising a shorter length are postponed. To help in this task, nodes keep data about routing settling times for each destination. In this way, when a node receives a path update for a destination, it is able to compute an adequate scheduling time to broadcast the update. This time should be shorter for very stable routes and longer for very unstable routes (e.g. suppose that there are several alternative paths to a destination and that the order in which the node receives these paths is always varying from one sequence number to the next). Additionally, nodes will also send their routing tables on a periodical basis. This can take two forms: incremental updates where they send only part of the routing table to inform of shorter paths or newer routing information (in the form of highest sequence numbers); and full dumps with the entire routing table.

Optimized Link State Routing. The Optimized Link State Routing Protocol for Ad Hoc Networks (OLSR) [183, 182] is a link state protocol. In these protocols, nodes collect topological information of the entire network, thus enabling local computation of optimal paths. While this approach automatically precludes the appearance of routing cycles, it brings the cost of extensive flooding. To mitigate this problem, OLSR uses a sort of clustered approach, where each node elects some proxies to forward its messages.

OLSR uses two main kinds of messages: HELLO and Topology Control (TC). In a HELLO message nodes beacon their own presence and list the neighbors they can hear. HELLO messages are not propagated by neighbors. However, one should notice that HELLO messages enable nodes to collect complete topological information of their two-hop neighborhood (to simplify, we omit some details necessary to ensure bi-directional links). Nodes use this two-hop view of the network to define a subgraph for broadcasting, where only a subset of its one-hop neighbors, designated "Multipoint relays" (MPRs), propagate a packet to reach all the two-hop neighbors. To have some insight on the effectiveness of this scheme, one should compare Fig. 4.1(a), where all nodes participate in flooding, with Fig. 4.1(b) [811], where only the MPRs (with a lighter filling) do the propagation. Each node has a different list of MPRs, but the combined effect is that flooding can still reach every single node of the network, while only a small subset of the network actually does the propagation.

Topology Control (TC) messages serve to spread topological information all over the network. TC messages carry the list of "MPR Selectors" of a given node. The MPR Selectors of some node, say A, are the nodes that picked A as their MPR. Unlike HELLO messages, MPRs propagate TC messages, which therefore can reach the entire network. In this way, each node in the network will receive the TC from all MPRs, which includes the list of their MPR Selectors. This enables any single node in the network to learn the entire topology, thus being able to create optimal paths to any destination.

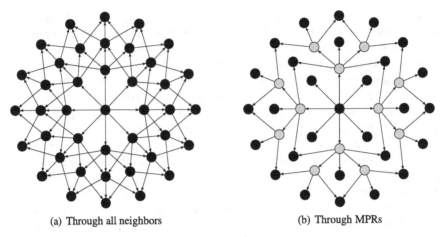

(a) Through all neighbors   (b) Through MPRs

Fig. 4.1 Alternatives for flooding

4.2.2 Reactive Protocols

Reactive, or on-demand, routing protocols learn routes exclusively for the destinations requested by the application. To save resources, the majority of these protocols keep a route cache that avoids repetitive discovery operations, which as it is shown below, can be significantly expensive.

Two major lines of research have been proposed for reactive protocols. In one, the list of nodes to be traversed is included in every data packet. This is the case of the Dynamic Source Routing (DSR) protocol. In the other, each node receiving a packet decides, using local information, the node to where the packet will be forwarded. Ad Hoc on Demand Distance Vector (AODV) protocol implements this option.

Dynamic Source Routing. The Dynamic Source Routing Protocol (DSR) [429] is an on-demand, hop count-based, source routing protocol. Successive Internet Drafts (e.g. [430]) have described and enhanced DSR. DSR relies on three control messages: ROUTE REQUEST, ROUTE REPLY and ROUTE ERROR, used on two operations: route discovery and route maintenance.

In DSR, nodes keep a route cache where they store all active paths they have discovered. A route is learned by inspecting the headers of all packets heard by the node, including snooped packets, i.e. packets that are transmitted by nodes in range to third parties. Note that since DSR is a source routing protocol, the source node is responsible for tagging all data packets with the entire route. The DSR packet header includes an integer pointing to the current address of the packet. This pointer is incremented at each hop.

In order to send packets, a DSR node first inspects its route cache. If the destination is not there, the protocol triggers a route discovery by broadcasting a ROUTE REQUEST message with its own address, the intended destination address and a

Fig. 4.2 A route discovery
propagation simplification
using discrete events

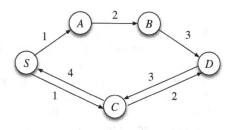

unique (in the scope of the sender) identifier. Upon reception of a ROUTE REQUEST for the first time, each node will look for the destination address in its route cache. If no route to the destination is known, the node will append its own address to the request and re-broadcast it. Therefore, a ROUTE REQUEST message accumulates the path it has traversed. If a route to the destination exists in the node's route cache or if the recipient is the destination, the node will send a ROUTE REPLY to the source node containing a route from the source to the destination.

If the ROUTE REPLY is produced by some intermediary node I, the route included is the concatenation of two components: i) the route from the source of the ROUTE REQUEST to I (available in the ROUTE REQUEST message) and ii) the route from I to the destination (extracted from I's route cache). In a ROUTE REPLY produced by the destination, the route reproduces the path accumulated on the ROUTE REQUEST message.

Route replies are sent point-to-point. If the medium supports bi-directional links, the ROUTE REPLY follows an inverted sequence of the nodes accumulated in the ROUTE REQUEST to reach the source node. Otherwise, the destination will start a new ROUTE REQUEST towards the destination and piggyback the discovered route on it.

Figure 4.2 shows a ROUTE REQUEST and ROUTE REPLY propagation from source node S to destination node D. The figure represents messages as arrows. Numbers show the sequence in which nodes transmit the messages (node S transmits to A and C, then A and C transmit to B and to destination D, respectively, etc.).

In DSR, each intermediary node is responsible of forwarding packets to the next hop in the source route. Acknowledgments can be provided implicitly by lower layer's MAC functions (such as the acknowledgment of IEEE802.11 DCF [396]), by overhearing the retransmission of the packet by the next hop, or explicitly by the protocol. DSR nodes could also request for an explicit acknowledgment from the next hop, by setting a specific bit in the packet header. When a node detects a link failure to the next hop, it is required to notify the source node of the packet by issuing a ROUTE ERROR message. The ROUTE ERROR identifies the packet failed to be forwarded and the link that caused the error. If an alternative route exists in the route cache of the intermediate node, the node may replace the original route with the one in its route cache and retry to forward the packet. Notifications of "salvaged" packets are included in the ROUTE ERROR message.

Like any other packet, ROUTE ERROR packets are equally snooped by nodes. Nodes hearing the ROUTE ERROR message (including the ROUTE ERROR destination) invalidate all routes in their caches that use this link. The source of the data packet either selects an alternative route, if one is available in its cache, or issues a new ROUTE REQUEST. In the latter, the ROUTE REQUEST message will indicate the failed link. This will be used by nodes hearing the message to update their route caches and prevents replies with the newly invalidated route.

The movement of nodes may result in the establishment of a link between two nodes that previously needed to be connected by an intermediary hop. This condition will be noticeable when a node snoops a packet whose source route includes its address as one of the forthcoming hops. In this case, a shorter route will be achieved by removing from the source route all hops between the hop that sent the message and the node that snooped the message. Nodes in this condition are requested to send an unsolicited ROUTE REPLY to the source of the packet containing the shorter route.

Ad Hoc On-Demand Distance-Vector Routing. DSDV is not adequate for large wireless ad hoc networks [664]. One disadvantage of DSDV is that it is proactive, thus continuously reacting to network changes to build new routes. In DSDV, control message overhead grows as $O(n^2)$ to an n-node network [664]. Additionally, most routing information collected by nodes is probably useless, because it will never be solicited. To overcome the problems of DSDV, Perkins et al. proposed the reactive Ad Hoc On-Demand Distance-Vector Routing (AODV) [664] protocol.

AODV is a pure distance vector protocol. Each node keeps a routing table with the distance and next hops from itself to any given destination that it hears about. To send a message to some node D, source node S must first discover an appropriate route if it does not have one. Similarly to DSR, AODV uses ROUTE REQUEST and ROUTE REPLY messages for that purpose. Each node that receives a ROUTE REQUEST from S and does not know about destination D, re-broadcasts the ROUTE REQUEST. ROUTE REQUEST and ROUTE REPLY messages do not store collected paths. To create a reverse-routing path for ROUTE REPLY messages, nodes receiving ROUTE REQUEST messages keep information of the previous hop for a limited period of time. When a node that knows the next hop to D receives a ROUTE REQUEST message, it sends a ROUTE REPLY message back to the source using the reverse-path. As the ROUTE REPLY propagates back to the source, nodes receiving it set up forward pointers to destination.

Source nodes issuing a ROUTE REQUEST packet always add a sequence number to the request, called "source sequence number". Nodes use this sequence number exclusively to maintain fresh reverse-route paths. The source node also adds a "destination sequence number" to ensure that it receives up to date paths. A node having a path to D with a sequence number lower than the destination sequence number, must not use its recorded route and must propagate the ROUTE REQUEST. Usually, as there are several paths from S to D, several ROUTE REPLY packets will travel (backwards) along paths set up by ROUTE REQUEST packets. They are propagated only if the sequence number is the same *and* if they contain shorter paths (either with fewer hops or with some lower cost regarding a different metric); re-

Fig. 4.3 DYMO collects information of intermediate nodes

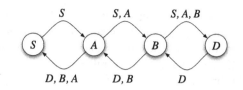

maining ROUTE REPLY packets are discarded. However, as soon as S receives the first ROUTE REPLY, it may start sending packets to D, using some available path that may be improved afterwards.

To remove reverse paths set up in the ROUTE REQUEST phase, nodes use timers, called "route request expiration timers". To maintain accurate routing tables, nodes also keep timeouts for their routing table entries. When these timers expire, nodes simply drop the routing entry. This soft-state approach is particularly important in wireless environments, because routes are expected to change frequently. Whenever a node sends a new packet to some destination, it resets the timeout for that entry.

Whenever a link becomes broken, either because some node moved or left the network, neighbors must invalidate all entries in their routing table passing through the broken link. Then, they must prepare a ROUTE ERROR message to the neighbors that may be affected by the failure. The ROUTE ERROR message can be either broadcast, if the failure affects many neighbors or unicast if it affects only one. Nodes receiving a ROUTE ERROR from node F with a list of unreachable destinations will purge from their routing tables those destinations in the list that use F as next hop. It is up to the sources to trigger new route discovery procedures to find alternative routes toward the destination.

In an optional mechanism designated as local repair, the intermediate node upstream the link break can issue a local ROUTE REQUEST to the destination. In this case, the sequence number for the destination should be increased by one to prevent other intermediary nodes from creating a route loop. While waiting for the ROUTE REPLY, the intermediary node must buffer all incoming paths for the destination. If after a certain interval, no ROUTE REPLY is received, the intermediary node should send the ROUTE ERROR message described above. Otherwise, buffered packets should be sent using the newly repaired route.

Dynamic MANET On-demand Routing Protocol. Dynamic MANET On-demand (DYMO) Routing Protocol [155] is a reactive protocol developed by some of the authors of AODV. DYMO improves AODV by adding more topological information to ROUTE REQUEST and ROUTE REPLY packets. Consider the case of Fig. 4.3 [764]. In AODV, intermediate nodes become aware of the route to source node S and to destination D. By keeping track of the complete route information, in DYMO, all nodes in the path learn about every other node in the path. In Fig. 4.3, information of D, B and A is increasingly added to the ROUTE REPLY. The same happens in the ROUTE REQUEST. Hence, unlike AODV, node S learns about B, while node D learns about node A. Authors call "path accumulation" to this technique of adding intermediate nodes to the packets.

ROUTE ERROR packets are used to notify sources they no longer have a (valid) path entry to some destination. On reception of this error, the source must delete its routing entry and request a new one. Moreover, nodes also issue immediate ROUTE ERROR packets whenever they become aware that some link in a recently used route is broken.

4.2.3 Discussion on Topology Based Routing Protocols

There are many studies available in the literature that compare the performance of different routing protocols. Comparisons between the AODV reactive protocol and the DSDV proactive protocol have shown that the latter requires more memory at the participants and, in some system models, more control messages [664]. On the other hand, reactive protocols impose a significant latency to the delivery of the first packet to an unknown destination.

AODV and DSR are two of the most frequently cited routing protocols for MANETs. Interestingly, both are reactive and use a hop count metric by default. The most relevant difference between AODV and DSR is that DSR includes the path to follow in each packet (source route), while AODV delegates the route determination to each intermediary hop (distance vector). It is not hard to find scenarios, with different movement patterns and nodes speeds, where any of them outperforms the other in terms of packet delivery ratio or bandwidth [120, 211, 212, 276, 427, 500].

DSR presents variable header length at both control and data packets in opposition to AODV. This can be a drawback for DSR due to: i) a routing overhead (in terms of packet size) increasing with the network diameter; and ii) disturbances of some upper layer protocols (like TCP) which rely on the MTU of the link layer protocol to define the size of their own segments.

The source routing nature of DSR, together with nodes eagerness in populating their route cache with new routes, may provide hints of the network topology to applications and protocols at different levels of the communication stack [36, 134, 536]. Under certain circumstances, it may also reduce the number of route discovery procedures. However, DSR specification does not include any mechanisms to purge staled routes from the cache until they are detected to be failed by some node. Experimental results [212] have shown that it is frequent for nodes running DSR to begin by trying a number of staled routes before issuing a new ROUTE REQUEST. These attempts consume bandwidth and energy at different nodes and may inclusively pollute route caches of other nodes. Interestingly, path accumulation in DYMO improves AODV in a way that bears some resemblances to a source routing protocol like DSR. This enables nodes in the middle of the path to learn about other intermediate nodes, thus paying a low cost to populate routing tables of nodes. The downside of doing this is that ROUTE REQUEST and ROUTE REPLY packets become larger. Seada *et al.* [738] show that path accumulation enables nodes to dis-

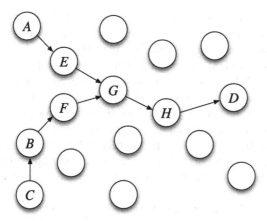

Fig. 4.4 The unfairness problem

cover network topology much faster, which, on turn, enables them to use shorter paths.

Source routing and distance vector protocols can also be compared by the complexity associated to the management of their routing tables. AODV routing table can be trivially sorted by destination address. Additional columns need to include a time-stamp, a sequence number and the next hop address. DSR in turn requires a much more complex internal representation of the routing table. Stored information are sequences of hops, which represent subsets of a graph defined by all nodes in the wireless ad hoc network. To retrieve the most adequate route, complex computations may be required.

Load Balancing and Fairness in Routing Protocols. Scenarios where nodes exhibit a negligible relative movement to each other (for instance, when users are in a library or a train) are good examples to illustrate how route caching of reactive protocols may impact load distribution. For example, consider the topology depicted in Fig. 4.4 and assume that node A initially broadcasts a ROUTE REQUEST for node D. The ROUTE REPLY containing $A \rightarrow E \rightarrow G \rightarrow H \rightarrow D$ will be snooped by node F. If later node B also broadcasts a route request for node D, F will use the cache to reply with the $F \rightarrow G \rightarrow H \rightarrow D$ route. Finally, by snooping the packets sent by B, C can also learn the route $B \rightarrow F \rightarrow G \rightarrow H \rightarrow D$ to reach D. Although several alternative routes with the same hop count exist from node C to node D, C will use the same route for any of the nodes B, F, G, H or D, which will result in an unfair traffic concentration at these nodes.

Unfairness may seriously impact the performance of wireless ad hoc networks. If participants do not use the network to achieve a common goal, they may begin to present selfish behaviour to preserve the batteries of their devices. In addition, an unfairly overused node is more likely to be congested. Congestion results in increased latency on message propagation and, depending of the media, may increase collisions which waste both bandwidth and battery power. In load-aware routing protocols like ABR [809] and DLAR [498], intermediate hops append load information to the route discovery messages. In the Hotspot Mitigation Protocol (HMP) [497] and in the extension to DSR proposed by Hu and Johnson [383] congested nodes temporarily suspend their normal route discovery behavior by ignoring incoming route requests destined to other nodes. This all-or-nothing approach may disrupt the communication between two endpoints if no alternative route exists. To circumvent this problem, both include a special flag in their messages. In [383], the flag should be activated in the ROUTE REQUEST packet header if no reply to a previous ROUTE REQUEST was received. An undesirable side-effect of this approach is the duplication of the route requests, even in the cases where no valid (congested or not) route exists.

Congestion is evaluated by load metrics such as the number of bytes per unit of time. On the other hand, unfairness may not be detected by congestion indicators if a device is being more solicited than others in the long term. Metrics that capture more adequately these scenarios are those that relate the traffic at each participant. In [580], both congestion and fairness indicators are weighted to delay the propagation of the ROUTE REQUEST message proportionally to the severity of the unfairness condition at each node. Fairness is estimated at each node by the ratio of the number of packets transmitted by the node and by all nodes within transmission range.

Implementations. Currently, there are several implementations of the topology-based schemes reviewed. For instance, OLSR has implementations for Windows, Linux and for simulators like Network Simulator 2 (ns–2) [803]. Examples are OOLSR [643] by the Hipercom team at INRIA or Ad Hoc Wireless Mesh Routing Daemon (OLSRD), currently supported by the OLSR-NG project [640]. One can easily find (e.g. [641]) many other implementations of OLSR, especially for ns–2. AODV and MAODV also have many implementations [4], including for Linux or TinyOS. There are also many implementations of AODV for ns–2. We can find a similar scenario for DSR, with many actual implementations, including some for ns–2 [243].

4.3 Position-Based Routing

4.3.1 Determining Nodes Positions

In position-based routing, nodes use positional information to forward packets. Usually nodes restrict collection of topology information to their immediate neighborhood (one or two hops). Then, they pick among the small number of known neighbors the one that seems to be closest to the destination according to the particular metric of the algorithm. The upside of position-based routing is the reduced amount of state needed, which makes networks specially fit for highly dynamic environments. On the downside, nodes need to determine their own position, position of their neighbors and position of destination, which may not be trivial.

To determine their own positions, nodes can use the Global Positioning System (GPS), if they are outdoors. As an alternative, wireless nodes can use techniques based on signal strength information, available in the standard IEEE 802.11 technology [630, 354]. A completely different approach is followed in [688] and [856], where nodes create logical instead of geographical positions.

To determine location of their neighbors, nodes usually exchange a small number of beacon packets. Finally, nodes still need to determine the position of the destination. Reasonability of this assumption depends on the concrete network. Usually, the problem of determining the network address of the destination is separated from the routing problem (take the Domain Name Service, DNS, as an example). However, in practice, it may be difficult to use a different layer to provide a location service atop of a wireless network. For this reason, there are solutions that integrate routing with the location problem, e.g., the Grid Location Service [512]. In the case of position-based distributed hash tables (DHT), e.g. [691], nodes can use the hash function to determine the destination.

4.3.2 General Greedy Routing Algorithms

Greedy Algorithm. Among all geographical and even topological routing algorithms, Greedy [280] is arguably one of the simplest. If node N wants to forward a message to destination D it just compares the position of all its neighbors and picks the one closest to D. Greedy algorithm does neither need any infrastructure, nor does it need any organized graph to take a message to destination. In the path to destination, each node along the way just forwards the message to whichever node is closest to D, if progress is made. Figure 4.5 illustrates the idea. Dashed circles show the communication range of nodes. Unfortunately, and despite the advantage of being very simple, Greedy also has some shortcomings. The most important one is that it may fail to find a path to destination, even though such path may exist. Routing failures will usually happen in regions with low node density, as these regions can have voids between the forwarding node and destination. This can cause a situation where some forwarding node M finds itself to be a local minimum on the path from S to D, as in Fig. 4.6.

Compass Routing Algorithm. Another simple and intuitive approach to route a packet in a wireless ad hoc network uses directions instead of distances. A node using Compass Routing Algorithm [477] picks for next hop the neighbor in the direction closest to D. This approach is similar to what humans sometimes do when finding their way in an unknown city: if they know the approximate direction of the point they are heading, they pick the street that keeps them close to that direction. Consider the case in Fig. 4.7, where F is forwarding the packet to destination D. In this case, it picks node N for its next hop, because this node minimizes the angle

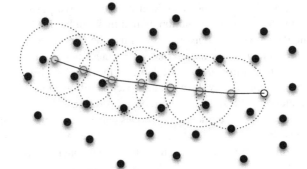

Fig. 4.5 Greedy routing algorithm

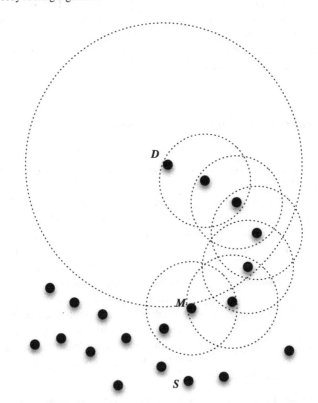

Fig. 4.6 Greedy fails due to local minimum M

$\angle NFD$ defined at F. One question that naturally emerges is which one is better: Greedy or Compass? According to some experiments included in [515], by Li et al., there is no definite answer. Unfortunately, Compass also keeps many of the problems of Greedy and brings one more with it. Like Greedy, it can follow suboptimal

Fig. 4.7 Compass routing algorithm

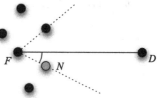

Fig. 4.8 Compass creates a loop

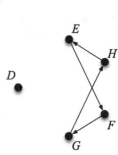

paths, because it also follows the greedy approach of picking the exact neighbor that seems to be better. A worse problem, which does not exist in Greedy, is that Compass can trap a packet in a cycle of nodes, as depicted in Fig. 4.8. This makes nodes waste their precious energy and bandwidth. To overcome this problem, either nodes or packets keep a short memory of paths or, in alternative, packets carry a time to live field.

GEDIR Algorithm. The GEographical DIstance Routing [776] (GEDIR) algorithm extends the Greedy routing algorithm with a very simple idea: a packet in a local minimum should get an extra chance to escape that minimum and travel to another node. In Greedy when a packet reaches a local minimum M it is discarded, as M cannot find any neighbor closer to D than M itself. GEDIR slightly relaxes this rule, by allowing M to forward the packet to whichever neighbor it has closest to D, disregarding its own distance to D. In fact, M allows the packet to move away from target node D. In simple terms, a packet is forwarded to whichever neighbor is closest to D (even if it moves to a farther point). We illustrate this in Fig. 4.9, where S is the source node and N is more distant to D than M. In [776], authors prove that GEDIR can have loops of length at most two, which are trivial to overcome. This means that GEDIR is better than Greedy as it works exactly the same when there are no local minima, but it can also overcome some of these local minima. In the case of Fig. 4.9, while GEDIR forwards the message to N, Greedy algorithm will simply drop the message. In a case where N can find a path to D (e.g., along the dashed curve), GEDIR will converge, while Greedy will not. Unfortunately, it is also trivial to find many situations where GEDIR fails to reach destination, even though some path exists (for instance, when it falls in a loop of two nodes).

Comparing between Greedy, GEDIR and Compass, we can say that GEDIR is usually better than Greedy, as it will try to route in situations where Greedy gives up. The cost is that it may create a loop if routing is actually impossible. However,

Fig. 4.9 Next hop of GEDIR

Fig. 4.10 Next hops of several routing algorithms

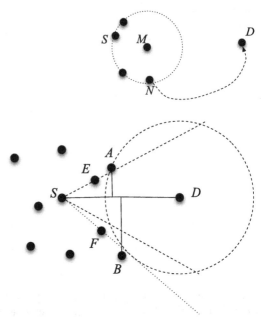

this cost is negligible, as the loop can have only two hops and is trivial to detect. Compass is different, because it uses direction instead of distance to pick paths. The problem with Compass is to know when to stop, as there is no obvious situation where the packet should be discarded, as in Greedy (although we could easily add one), and loops can have many hops, thus being difficult to prevent.

More Routing Algorithms. Besides Greedy, Compass or GEDIR, there are many more routing algorithms. In Fig. 4.10, we present some of them. For instance, the Most Forward within Radius [792] (MFR), would pick node B to send a message from S to D. This algorithm tries to maximize progress, which is interpreted as the projection along the line segment SD. Another routing algorithm is Random Compass [108]. Random Compass is similar to Compass with the following difference. While Compass picks the single node that minimizes the angle to destination $\angle NFD$ in Fig. 4.7, Random Compass considers two options one in each direction. In Fig. 4.10, S can randomly pick its next hop to be either E or F as these define the smallest angles with respect to SD in opposite directions. The idea is to introduce randomization to avoid loops.

All the algorithms that we presented here, Greedy, Compass, GEDIR or MFR try to minimize the hops used to reach destination. However, there are also many other routing algorithms that try to minimize different metrics. Some of them may seem to be very counterintuitive from the perspective of minimizing hop count. For instance, Nearest Closer (NC) Algorithm [775] picks for the next hop, the closest node that is closer to destination than the current node, i.e., it picks the closest node that ensures progress. If nodes can adjust their transmission power, this will reduce power as we discussed in Sect. 4.1.1. There are many more routing algorithms that take energy into account, for instance [712, 775, 83, 163, 164].

Another common characteristic of all the algorithms we reviewed here is that all of them search for destination using a single copy of the message. There are many algorithms that use multiple copies and consequently multiple paths to reach the destination. The simplest of these algorithms is perhaps flooding, where each node along the path sends the packet to all its neighbors (except the one where the message came from).

4.3.3 Face Routing

Most of the previous algorithms, like Greedy, Compass, MFR, etc., are greedy, in the sense that they try to maximize or minimize some metric without any consideration for the future outcome of the path. They can take some decisions that seem to be good for the next hop, but may prove totally inadequate afterward. For instance, the neighbor closest to destination might be a local minimum without any better option to reach the destination.

We need to add something to greedy selection, if we want position-based routing to converge. A well-known solution comes from a method based on "right-hand" routing, an idea derived from finding one's way in a maze by never lifting the right hand from the wall [104]. If instead of a maze, we imagine the contour of a face in a graph, like in Fig. 4.11, it is easy to see that a packet will return to the source node after going through each one of the edges of the face. With the aid of some rules, we can force the packet to visit each one of the faces in sequence until it reaches destination.

FACE-1. One of the algorithms based on the right-hand rule is Compass II [477], later renamed as FACE-1 in [109]. Algorithm 4.1 describes FACE-1, while Fig. 4.11 illustrates one particular case for source node S and destination node D. This graph has five different faces, f_i, $i = 1 \ldots 5$ from S to D. Points P_i are the intersections of the edges of the face with line segment SD. Routing procedure in FACE-1, starts by defining a line segment between source S and destination D. The idea is to force the packet to go through all the faces that intersect SD (known as r in Algorithm 4.1), until it reaches destination. To understand how FACE-1 works, assume that the packet is inside the first face, f_1. Current face is deemed as f. We set P to be the initial source node S. In general, P represents the starting node in the current face f. Starting in P, nodes forward the packet from edge to edge (either clockwise or counterclockwise) until it reaches some edge e that intersects r in point P'. Now, if P' is closest to D than P, the forwarding node will register P' as the new P in the packet. In Fig. 4.11, these points form the sequence S, P_1, P_2, P_3, P_4 and D, where the dashed line SD intersects the edges that limit the faces. When the packet returns to the source node of face f, P holds the edge of f that intersects SD closest to D. Then, the source node sends the packet to find again the edge of point P, where it

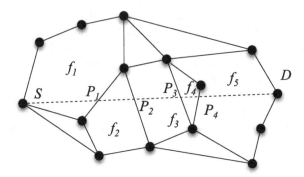

Fig. 4.11 Sequence of faces using FACE-1 routing algorithm

Algorithm 4.1: Algorithm FACE-1

---

constant $S \leftarrow$ source ;
constant $D \leftarrow$ destination ;
constant $r \leftarrow$ line segment SD ;
$P \leftarrow S$;
while $P \neq D$ **do**
    $f \leftarrow$ first face that intersects line segment PD from P to D ;
    forall *edges e of f* **do**
        if *e intersects r at point P' and P' is closer to D than P* **then**
            $P \leftarrow P'$;
        end
    end
    traverse f again until reaching edge where P was found ;
end

---

should switch to the next face. In the graph of Fig. 4.11, the packet will find points P_1, P_2, P_3, P_4 and it will go through faces f_1, f_2, ..., f_5.

FACE-2. One shortcoming of FACE-1 lies in the number of times a packet contours the edges of a face. In the particular case of Fig. 4.11, as soon as the packet reaches intersection point P_i, $i = 1, \ldots, 4$, it could immediately change to the next face, as the remaining travel around edges of the face turns out to be pointless. To improve this behavior, Bose et al. [109], proposed FACE-2, which switches face whenever the packet reaches a new intersection, provided that such intersection is closer to destination than the previous one. This process is successively repeated for each new face f. We formalize FACE-2 in Algorithm 4.2 running in graph G. Bose et al. [109] show that, in general, FACE-2 is more efficient than FACE-1. However, there are cases where FACE-1 works better than FACE-2, as it is not difficult to devise a graph with concave faces intersecting SD several times. In such cases, coming back to the source node of the face, as in FACE-1, turns out to be useful. An interesting pathological example is given in [109], where, in a snake-like graph, the packet ends up going back and forth several times along the same edges, due to the wrong decisions of FACE-2, that would not occur in FACE-1. While FACE-1

Algorithm 4.2: Algorithm FACE-2

---

constant $S \leftarrow$ source ;
constant $D \leftarrow$ destination ;
$P \leftarrow S$;
while $P \neq D$ **do**
    $f \leftarrow$ face of G with P on its boundary that intersects PD ;
    traverse f until reaching an edge UV that intersects PD at some point $P' \neq P$;
    $P \leftarrow P'$;
end

Fig. 4.12 Packet from S will never reach D

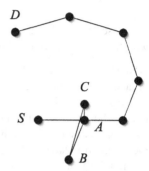

reaches destination node D in at most $3n$ hops,[1] where n is the number of edges of the graph, the upper bound for FACE-2 can raise to $\Omega(n^2)$. This is a clear sign that it is difficult if not impossible to devise an algorithm that fits well under all possible cases with such a limited amount of information as FACE-1 and FACE-2 use.

According to experimental results [109], neither FACE-1, nor FACE-2 are very efficient, as they require a large number of hops to reach destination, although there are some better paths available. As we shall discuss next, we will need to complement the use of a face algorithm with a greedy algorithm, to increase efficiency in routing.

Planarity of the Routing Graph. One distinctive feature of Fig. 4.11 is that the graph is planar, i.e., there are no edges that intersect. As a matter of fact, it is easy to show that right-hand routing algorithms require planar graphs. Refer to Fig. 4.12. In FACE-2, the packet would go through $S - A - C - B - A - S - A - \cdots$ in an infinite loop caused by the intersection of edges $SA - BC$. Other graphs with intersections could also cause the failure of FACE-1 and FACE-2. Formally, one should notice the difference between planarity of a graph and planarity of *the embedding* of a graph. According to [870], an *embedding* of a graph is a representation in a space in a way that preserves its connectivity. To understand the difference, we depict in Fig. 4.13 two embeddings of the same graph in the plane, of which only one is planar. In the case of wireless ad hoc networks, when we refer to the planarity of a graph, we mean the planarity of the actual embedding of the graph, as we cannot change the positions of nodes or edges.

---

[1] Under a slight modification of the Algorithm 4.1.

Fig. 4.13 Two different embeddings of the same graph

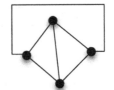

4.3.4 Hybrid Position-Based Routing Algorithms

In this section, we review some routing algorithms that mix the right-hand routing with the greedy approach. We call them "hybrid".

GEDIR—FACE 2—GEDIR (GFG). The first authors to suggest a hybrid approach for routing were Bose et al. [109] in the same paper where they presented FACE-2. GFG will successively commute between GEDIR and FACE-2 until it reaches destination. Since GEDIR performs better in denser graphs, the algorithm starts in this mode. When the packet reaches a local minimum M, GFG interrupts the standard behavior of GEDIR and resumes using FACE-2 instead (in practice, this implies that we could use GEDIR or Greedy indistinguishably as their only difference never shows up). When FACE-2 comes in, it will route the packet until the packet reaches some node N such that N is closer to destination than M is. At this point, GFG commutes back to GEDIR again.

Greedy Perimeter Stateless Routing (GPSR). Another algorithm that explores the duality between the efficient and the reliable approaches is Greedy Perimeter Stateless Routing Algorithm [444] (GPSR). This algorithm is a variant of GFG, where routing takes place using the greedy algorithm. When GPSR reaches a local minimum, it commutes to face, reverting back to greedy as soon as possible. There is however one very subtle difference between GFG and GPSR that ends up having a considerable impact (at least from a theoretical perspective, while this is arguably not that important in practice). As we showed in Algorithm 4.2, FACE-2 algorithm only changes face when it reaches a point $P' \neq P$; when this change occurs it resets P. Given the destination D, FACE-2 cares for intersections with PD. Unlike this, GPSR cares about intersections with SD, where S is the source node, where the face traversal began. In Figs. 4.14 and 4.15, which are similar to figures in [294, 460], we show a planar graph where GPSR fails to deliver a message, while GFG succeeds. In GPSR, a packet always changes face when it crosses SD. That is what happens with edge AC. Hence, since it is using right-hand routing it picks edge AB, then BC, CA, etc., going in an infinite cycle ABC inside face f_2. The packet cannot escape this face, because it does not intersect SD at any other point closer to D than P is (P' is farther away so no face change occurs there). Similarly, GFG changes the face at A. However, the algorithm states that the next face is the one that intersects PD. Consequently, it is still the same face f_1 and the next hops will go through C, E, F and finally D. Frey and Stojmenovic [294] show that GFG always converges in planar graphs. The interesting thing about this case is that for

Fig. 4.14 GPSR fails in this
graph

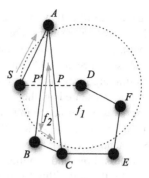

Fig. 4.15 GFG works in this
graph

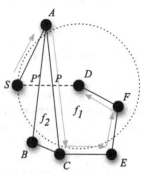

a couple of years convergence of GPSR and of many other related protocols was
accepted as a fact.

AFR and GOAFR⁺. To understand one of the problems with face routing, consider the following dilemma. Suppose that two foreigners reach the Great Wall of
China and they want to find an entrance. Should they turn left or should they turn
right? If they pick right they may end up walking a very long distance having a close
door to the left, but picking left can offer them the reciprocal scenario. Fortunately,
there is an acceptable solution to this dilemma that allows travelers to reach a door
walking at most 9 times their initial distance from that door. It may look counterintuitive, but the idea is to pick one of the directions (left or right) and walk one unit
of distance. Then take the reverse distance and walk two units to the opposite side
counting from the original position. Then, again, double the distance (from the origin) in the reverse direction, until reaching a door [42]. Walking 9 times more than
the distance needed may seem to be bad, but compare this with the lack of any limit
if the traveler keeps in the wrong direction (e.g., the door could be only 1 Km to the
left, while to the right there could be some other door only at 50 Km or 500 Km or
even more). Wireless ad hoc networks pose a similar problem, specially in the outer
face. A packet in the wrong direction may be forced to travel a long distance before
reaching its target.

To overcome this problem, Kuhn et al. [481] took the idea of using a hybrid routing algorithm a step further to present the "Adaptive Face Routing" (AFR). AFR is

guaranteed to achieve a worst case cost of $O(c^2)$, if c is the cost of the best path. Similarly to algorithms like GFG or GPSR, AFR combines greedy routing with face routing to ensure routing convergence without compromising performance. As we shall see in Sect. 4.3.5, since the worst-case cost for any localized position-based routing algorithm is $\Omega(c^2)$, AFR is asymptotically optimal. Following this work, authors presented another algorithm, GOAFR$^+$ (pronounced as "gopher-plus") that improves performance in random graphs, while maintaining the asymptotically optimal properties of AFR [482]. Unfortunately, for the reasons shown by Frey and Stojmenovic [294], there are graphs where these algorithms fail to converge. Nevertheless, Frey and Stojmenovic also introduce a small variation to GOAFR$^+$ that enables this algorithm to converge.

4.3.5 Pre-processing Algorithms

Since routing algorithms based on the right-hand rule need planar graphs to converge, we need to ensure the graph is planar before routing starts. This basically consists of pruning some edges from the original graph in a process known as "topology control". Unfortunately, this task is very difficult in practice, but there are some approaches that yield planar graphs, at least in theory. Implicitly, this assumes that despite being fallible, these approaches can nevertheless have reasonable results in practice. In general terms, we can define the notion of routing scheme as being comprised of two parts (see, for instance, [289]): i) a distributed algorithm, here known as the routing algorithm, running at every node, which is responsible for determining the output port (i.e., the next hop) of a packet; and ii) a pre-processing algorithm that, given the initial connection graph G, must create whatever information is necessary to the routing algorithm (e.g., routing tables or a subgraph of G). Up to this point we focused on routing algorithms. We shall now review some of the pre-processing algorithms. However, before we do that, we need to introduce some related concepts. First, it would make no sense to have very simple routing algorithms running on top of a very complex pre-processing scheme. Therefore, we formalize some strict constraints to the pre-processing algorithm (and to the routing algorithm as well) in the definition of "Localized Routing Scheme". Second, we need to have some idea of how does the initial routing graph looks like, before pruning any edges. Usually authors assume the "Unit Disk Graph" model (*UDG*), which we present in this section.

Localized Routing Scheme. Some types of wireless nodes have very limited memory, processing power and even tighter limitations on available energy. Moreover, existing network bandwidth must be divided by all the nodes within reach in a setting where topology might change very fast, e.g., due to mobility of nodes. This originates a strong motivation to use simple and economical routing schemes. For instance, to conserve resources, nodes should exchange the fewest possible messages and should store (either locally or in the packet) only a small amount of state. To satisfy these properties, Kranakis et al. [477] created the notion of localized routing scheme, with the following properties: *i*) the routing scheme uses information of the node where it runs, *ii*) information of neighbors that can be reached in up to *n* hops (being *n* a constant) and *iii*) information of a constant *k* number of additional nodes. A routing scheme is said to be *n*-localized if *n* is the smallest constant that satisfies condition *ii*.

The Unit Disk Graph Model. Many algorithms in position-based routing work under the assumption that nodes can see all their neighbors within some radius *r*. Formally, this means that nodes P and Q are neighbors if and only if $|PQ| \leq r$. For simplicity, authors usually set *r* to be 1. When $r = 1$, we get a model know as Unit Disk Graph (*UDG*). *UDG* is a pretty convenient model in theory, because some algorithms are guaranteed to work when it holds. For instance, many routing algorithms based on the right-hand rule [104] can ensure delivery of messages (if a path exists), but only when working on top of a planar graph. Therefore, to run right-hand routing algorithms, we usually use a pre-processing algorithm to remove some of the existing edges of *UDG*, such that the final routing graph is planar (see examples below). As long as the initial graph is *UDG* we can remove all the intersections, otherwise some intersections may subsist (unlike this, most of the routing algorithms we reviewed so far, like Greedy, GEDIR or Compass, just to state a few, can run even if the *UDG* model does not hold, because delivery is not guaranteed in general graphs[2]).

Unfortunately, *UDG* is a gross simplification of reality as we should not expect communication range of nodes to be a perfect circle. Additionally, estimation of nodes positions can also create some problems. If a node considers itself to be in a wrong position and beacons that position to its neighbors, it may end up seeing a completely unexpected set of neighbors, thus distorting *UDG*. Given the inherent limitations of *UDG* (see also [151, 303]), some authors tried to generalize position-based routing to more realistic settings. One of these is presented ahead. Other authors, like Kim et al. in [461] try to eliminate any restrictions on the connectivity of the nodes, while, for instance Lim et al. [520] try to mitigate the uncertainty on the location estimation. Nevertheless and despite the limitations of *UDG*, most related bibliography assumes this model and henceforth, we will usually assume *UDG* in our descriptions.

Gabriel Graph. Consider nodes A and B from the initial node set V. If the circle whose diameter is AB, $d(A, B)$ is empty of other nodes from V, then, edge AB belongs to the Gabriel graph. Formally, edge AB belongs to the Gabriel graph if and only if $\nexists X \in V \setminus \{A,B\} | X \in d(A,B)$. In Fig. 4.17, we depict Gabriel edge AB. The gray area must be empty of nodes. The set of all Gabriel edges defines the Gabriel graph (*GG*). Since we apply Gabriel algorithm to *UDG* we get a subgraph of *UDG*. This is often called "constrained Gabriel graph" due to the fact that all edges have length of at most 1.

We do not prove it here, but it is straightforward to show that under the *UDG* model it is always possible to create the *GG*. Barrière et al. [64] proved that the

---

[2] However, Bose and Morin [108] showed that Greedy and Compass algorithms converge in particular settings like Delaunay triangulation and others.

Fig. 4.16 This intersection is
impossible if $R/r \leq \sqrt{2}$

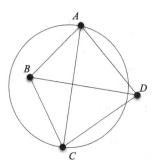

Gabriel Graph can also be created under a model that is more complex than *UDG*,
called $R/r \leq \sqrt{2}$. In this model, authors consider two circles for node transmission
capacity: a smaller circle with radius r and a larger circle with radius R, which
correspond, respectively, to the minimum and maximum admissible transmission
range. This means that messages broadcast by some node must reach all other nodes
at a distance of up to r and may or may not reach nodes that are at a distance of up
to R—i.e., transmission range does not need to be a circle, but must be within two
circles. Here, we only show why is it always possible to remove the intersections
from the *GG* in a localized fashion. Assume that two edges AC and BD intersect as
depicted in Fig. 4.16. At least one of the angles of $ABCD$ must be greater or equal
to $\pi/2$. Without loss of generality, assume that $\angle ABC \geq \pi/2$. Clearly, $B \in d(A, C)$
and hence neither A nor C are aware of B or else AC would not be a Gabriel edge.
Therefore, $\|AB\| > r \wedge \|BC\| > r$. This implies that $\|AC\| > \sqrt{2} \cdot r \geq R$. From
definition, A and C could not see each other, which is a contradiction. Therefore,
graph can be made planar if $R/r \leq \sqrt{2}$.

Relative Neighborhood Graph. In the relative neighborhood graph (*RNG*), an
edge AB exists when there is no third node $C \in V$ such that edges AC and BC
are both shorter than AB. Formally, edge AB belongs to the Relative Neighborhood
Graph if and only if $\nexists X \in V\backslash\{A, B\}|\ \|AX\| < \|AB\| \wedge \|BX\| < \|AB\|$. Again, we
are interested in the constrained *RNG* graph, which only has edges with length at
most 1, resulting from *UDG*. Both *GG* and *RNG* are planar. Both are connected as
long as the initial graph is also connected. Figure 4.18 shows the gray area that must
be free of nodes. The "exclusion zone" of the *RNG* contains the corresponding zone
of the *GG*, which means that the restricted *RNG* is a subgraph of the restricted *GG*
and of the *UDG*. *RNG* is used in [740]. Before we proceed, we need to introduce a
couple of notions that will help us to assess the quality of *GG* and *RNG*.

Spanning-ratio. The routing algorithm will resort to local information stored in
the routing table of the node. Hence, whether or not the routing scheme is localized
depends on the pre-processing algorithm. Since we want to create planar subgraphs
of *UDG*, we must attend to the trade-offs and metrics involved. On one hand we
need to use economical pre-processing algorithms, on the other hand, the resulting
routing graph must ensure good routing performance. One criterion used to evaluate
the pre-processing step is called "spanning-ratio", which assesses the *quality* of the

Fig. 4.17 In a *GG*, the gray area must be clear

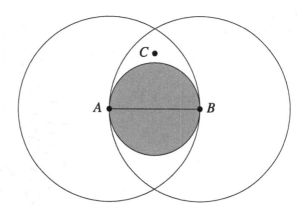

Fig. 4.18 In a *RNG*, the gray area must be clear

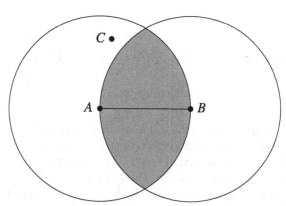

subgraph created. Consider that G is the initial graph (e.g., *UDG*) and H the subgraph created by the pre-processing algorithm. H is said to be a "t-spanner of G" if and only if:

$$\max_{\forall S, D \in V} \left\{ \frac{\|\Pi_H(S, D)\|}{\|\Pi_G(S, D)\|} \right\} \leq t$$

This means that for all nodes S and D, shortest path between S and D in H, $\Pi_H(S, D)$, is at most t times longer than in G, $\Pi_G(S, D)$. t is known as the "length stretch factor". t gives a notion of the quality of the subgraph. The smaller it is, the better. When the graph G is the complete Euclidean graph determined by V, the above expression defines an "Euclidean t-spanner".

Competitive-ratio. Spanning-ratio can help to determine the quality of a pre-processing algorithm, but the overall behavior of the routing scheme goes beyond the subgraph where it runs. The main point is still to use routing schemes that select paths almost as good as the shortest path. The "competitive-ratio" is used as an

Fig. 4.19 Variation of the
lower bound graph

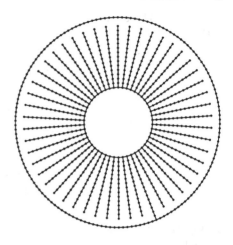

accurate measure of the quality of the routing scheme (RS) and is defined as:

$$\text{competitive-ratio(RS)} = \max_{\forall S, D \in V} \left\{ \frac{\|A_G(S, D)\|}{\|\Pi_G(S, D)\|} \right\}$$

$\|A_G(S, D)\|$ is the length of the shortest path between S and D, found by the
routing scheme A in graph G; again, $\|\Pi_G(S, D)\|$ is the length of the shortest path,
between the same pair of nodes, existing in G. A routing scheme is "t-competitive"
if its competitive-ratio is t. This is a worst-case definition, because t is determined by
the pair of nodes for which results are worst. Sometimes, the name "stretch factor"
is also used instead of "competitive-ratio".

One interesting known fact is that no localized routing scheme can be t-compe-
titive for any constant t. Kuhn et al. [481] showed that if c is the cost of the best
path for a given pair of nodes, the cost for any localized position-based routing
scheme can grow to $\Omega(c^2)$. More precisely, there are graphs where any deterministic (randomized) position-based routing scheme has a(n) (expected) cost of $\Omega(c^2)$.
Furthermore, this applies to the number of links traversed, to Euclidean distance or
to energy spent transmitting the packet. The reason for this is illustrated in Fig. 4.19,
which shows a variation of the "lower bound graph". Dots represent nodes, while
lines represent links connecting nodes. In *UDG*, the distance between nodes of the
inner circumference is precisely 1. Therefore, these nodes are connected. However,
the other nodes of the radial lines can only communicate with the immediate neighbors in their own radial, because the distance to nodes of other radials is greater
than 1. Another key aspect of this graph is that there is only one radial giving access
to the outer circumference. Having access to local information only, a node cannot
know the topology and, as a consequence, it cannot know which of the radials connects to this circumference. This means that in the average, a packet needs to try
half of the radials before hitting the right one. Hence, if the shortest path (e.g., in
number of links) to some node in the outer circumference is c, a localized routing

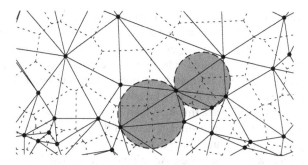

Fig. 4.20 Delaunay Triangulation, Voronoi tessellation and the empty circumcircle

scheme will end up using $\Omega(c^2)$ links to reach the only node that gives way to the outer circumference.

Delaunay Triangulations for Wireless Networks. The problem with both *GG* and *RNG* is that neither one of them is a good spanner of the initial connection graph [254], namely of *UDG*. In other words, the spanning-ratio of both *GG* and *RNG* is quite high (respectively $\sqrt{n}$ and $n - 1$, where n is the number of edges of the initial *UDG*). This is a very high price to pay for guaranteed delivery with right-hand routing algorithms. To improve this result we need to create better spanner graphs with more edges. In $\mathbb{R}^2$, the densest graphs we can get are triangulations. One possibility is to use the Delaunay triangulation, because it is easy to compute by distributed nodes and it ensures a good spanning-ratio. The *Delaunay triangulation (DT)* of a node set V, also represented as *Del(V)*, is the set of edges satisfying the "empty circle" property: edge AB belongs to the triangulation if and only if there is a circle containing A and B, but not containing any other node. An important property of *Del(V)* states that the circumcircle of a triangle does not contain any node of V. To this property we call the "empty circumcircle" property. The *DT* has an associated dual concept, called the "Voronoi tessellation" (or diagram). According to [870], the Voronoi tessellation with n generating points is a partition of the space into convex polygons, such that each polygon contains exactly one of the generating points and any point in the polygon is closer to its generator than to any other generator. This means that the polytope of node N is comprised of the points that are closer to N than to any other node of the node set V. We call "cell" to the Voronoi polygon of node N. Figure 4.20 shows the Voronoi tessellation of a random node set (in dashed lines). The figure also shows the relation between the Voronoi tessellation and the Delaunay triangulation (solid lines): two nodes share a Delaunay edge if and only if their Voronoi cells have a common border. We can also see the empty circumcircle property for two triangles, as no fourth node is inside the depicted circumcircles.

Although Delaunay triangulations have many other applications (e.g. Computer Graphics), here, we are interested in their advantages for routing purposes [108, 515, 304, 850, 489].

Localized Delaunay Triangulations.   Unfortunately, under the *UDG* model, a complete Delaunay triangulation may not exist, because some edges may be longer than 1. Even if all the Delaunay edges were shorter than 1, creating a Delaunay triangulation would make a routing scheme non-localized, because the empty circumcircle property requires information of nodes that may be close in terms of Euclidean distance, but that may be very far away in hops.

Despite these inconveniences, this triangulation still owns some attractive properties. For instance, if nodes have similar views of their neighborhood, they can deterministically compute the same triangulation. This may save many steps to reach a form of agreement among the nodes. Additionally, it is possible to create variants of the Delaunay triangulation that, unlike *GG* or *RNG* are good spanners of *UDG*. Probably the simplest variation of the Delaunay triangulation is the Unit Delaunay triangulation (*UDel*). *UDel* results from the intersection of the Delaunay triangulation with the *UDG* graph. $UDel(V) = Del(V) \cap UDG(V)$, meaning that *UDel* is the Delaunay triangulation with edges that have length at most 1.[3]

Unfortunately, one is not able to know which edges are actually Delaunay edges without gathering information from arbitrarily distant nodes and consequently *UDel* is also impossible to build in a localized fashion. To overcome this problem Li et al. in [515] proposed to build a super-graph of *UDel*, called "*k*-localized Delaunay graph over a node set *V*", $LDel^{(k)}(V)$. $LDel^{(k)}(V)$ is comprised of two types of edges (not longer than 1):

- all edges from the *GG*;
- edges of all triangles ABC for which there are no nodes inside the circumcircle ABC, reachable by A, B or C in k or fewer hops.

Li et al. proved that $LDel^{(k)}(V)$ is planar for $k \geq 2$, but edges may intersect for $k = 1$. Building $LDel^{(2)}(V)$ is still not very efficient and therefore authors sometimes resort to *PLDel*(V) [515, 489], which stands for "Planar Localized Delaunay Triangulation". *PLDel*(V) is a planar graph comprised of all triangles of $LDel^{(1)}(V)$, *except* intersecting triangles that do not belong to $UDel^{(2)}(V)$. Li et al. [515] proved that $UDel(V)$ is a $(4\sqrt{3}\pi)/9$-spanner of $UDG(V)$ and that $LDel^{(k)}(V) \supseteq UDel(V)$. Hence *PLDel*(V) and $LDel^{(k)}(V)$, for all k, are also $(4\sqrt{3}\pi)/9$-spanners of $UDG(V)$. In other words, these triangulations are good for routing.

Restricted Delaunay Graph (*RDG*).   Just for the sake of showing an algorithm that creates triangulations we present here the algorithm of Gao et al. [304]. This algorithm creates a graph called "Restricted Delaunay Graph" (*RDG*). *RDG* is simply any planar super-graph of *UDel*. Hence, $RDG(V) \supseteq UDel(V)$, which means that *RDG* is also a $(4\sqrt{3}\pi)/9$-spanner of $UDG(V)$. Although relatively simple to build, the *RDG* graph of Gao et al. is possibly expensive in terms of communication. After exchanging information of its own position with its neighbors, each node uses an additional communication step to broadcast its own view of the Delaunay triangulation. This serves to make the triangulation between nodes converge and to eliminate possible intersections.

---

[3] Another possible name for this would be the "constrained Delaunay triangulation".

Algorithm 4.3: Algorithm that creates RDG

$E(A) \leftarrow \{AB | AB \in UDel(A)\}$;
forall *edge* $AB \in E(A)$ **do**
    forall $C \in N(A)$ **do**
        if $A, B \in N(C)$ *and* $AB \notin UDel(C)$ **then**
            Delete edge AB from $E(A)$;
        end
    end
end

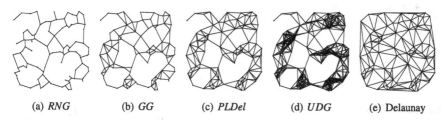

  (a) *RNG*   (b) *GG*   (c) *PLDel*   (d) *UDG*   (e) Delaunay

Fig. 4.21 Example of graphs

Let $UDel(A)$ be the Unit Delaunay triangulation computed by node A. Algorithm 4.3 shows the pseudo-code that node A must execute after this final communication step. Basically, A deletes edge AB if there is some node C that simultaneously knows A and B and that does not include AB in its triangulation. This means that AB is not a Delaunay edge.

We illustrate in Fig. 4.21, the results achieved by different pre-processing algorithms we described. From left to right, one can see that graphs become denser. The exception is the Delaunay triangulation, which is not a subgraph of UDG, because it includes long edges that cannot exist in a wireless ad hoc network. By comparing these graphs, one can easily see that, by being denser, UDG provides more routing options than $PLDel$,[4] which on turn provides more options than GG, while RNG is the worst.

4.3.6 Pitfalls

It is important to include here some remarks about position-based routing. To start, as we said before, the UDG assumption is often too optimistic. In practice nodes may be unable to detect all their neighbors even though they can be quite close. For instance, in Fig. 4.22, we show a simple case where some terrain obstacles prevent two pairs of nodes from detecting the intersection. This is only one of the problems. As shown in the interesting work of Kim et al. [460], there are many other problems caused by a planarization algorithm like the Gabriel graph. They consider the fol-

---

[4] However, since UDG is not planar it cannot guarantee message delivery.

Fig. 4.22 *A*, *B*, *X* and *Y* fail
to detect intersection

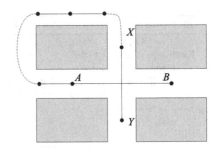

Fig. 4.23 Obstruction in
GG planarization creates
unidirectional edge

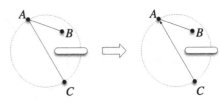

Fig. 4.24 Obstruction in
GG planarization creates
disconnection

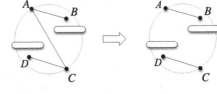

Fig. 4.25 Location error in
GG planarization creates
unidirectional edge

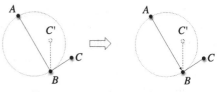

lowing cases: unidirectional links caused by obstacles after planarization (Fig. 4.23);
links disconnected by obstacles after planarization (Fig. 4.24); unidirectional links
caused by location errors (Fig. 4.25) and disconnected links caused by location errors (Figs. 4.26). All these cases can result in routing problems. In some of these
cases, the pre-processing algorithm ends up creating a partition in a graph that was
initially connected. As a consequence, for many source-destination pairs, routing
will no longer be possible after the pre-processing algorithm runs.

4.4 Summary and Outlook

Topology-based routing evolved in the direction of reducing node and bandwidth requirements: from proactive routing schemes like DSDV, we reached reactive routing
in AODV or DSR, where nodes only try to discover routing paths when they actually want to send a packet to some unknown destination. Still, all of them are similar
in the sense that they need to collect topological information from nodes that may

Fig. 4.26 Location error in
GG planarization creates
disconnection

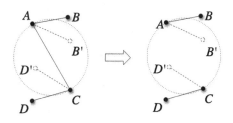

be arbitrarily distant. This inevitably imposes a high cost to maintain paths to distant nodes: control cost clearly grows for larger distances and for larger networks, but resources can simply not grow in a shared communication media. Topological routing protocols were one of the first topics to be addressed in wireless ad hoc networks and this is still an active field of research. Many directions are currently being followed. Some of them, like power-aware or load-balancing have been briefly addressed in this chapter. There are however others. Examples are hybrid routing protocols, that combine both proactive and reactive approaches (e.g. the Zone Routing Protocol [352]) and multicast protocols. For the latter subject, the interested reader is referred to works like the Multicast Ad hoc On-Demand Distance Vector Protocol (MAODV) [719], On-demand Multicast Routing Protocol (ODMRP) [501] or Mobicast [385] for an introduction on the subject. It should be noted that the first two are topology-based routing algorithms, while the latter makes use of positional information.

In position-based routing, we reviewed algorithms like Greedy that looks for the neighbor that is closest to destination (and closer than itself) or Compass that looks for the neighbor with smallest angle with destination. These and other algorithms require that the node looks up the best candidate among its hopefully small list of neighbors. Unfortunately neither Greedy nor Compass ensure routing convergence in arbitrary graphs and the same goes for many other related routing algorithms. To overcome this problem, one can resort to right-hand routing, but this solution comes with two costs: planarization of the graph makes it sparser, thus worse for routing; and, additionally, right-hand routing performs worse than Greedy, Compass and other related algorithms. We reviewed several algorithms to planarize the initial (arbitrary) graph: the Relative Neighborhood Graph, the Gabriel Graph, Delaunay triangulations and some variations. Finally, to avoid the bad performance of right-hand routing one can use hybrid approaches that mix greedy with right-hand routing.

The idea that position-based protocols scale better than their topology-based counterparts was experimentally confirmed in [409, 512]. Moreover, position-based protocols are much simpler than topology-based ones. For these reasons, they seem to be a very promising approach for routing. On the negative side, they require more technology to determine location (e.g., GPS) and they can be seriously affected by location errors or distortions in the range of communication. Additionally, it is worthwhile observing that topology-based approaches typically pose no restrictions about the communication range, while position-based algorithms often depend on rather unrealistic assumptions of circular communication range, like the *UDG* model.

Part II
Communication Models

Depending on the way mobile devices are deployed and used, different interaction patterns are established among them. A communication model captures what is common to a representative set of interaction patterns. Thus, these models make the bridge between networking issues and middleware-specific issues. The second part of the book gathers chapters presenting key communication models used in ad hoc networks.

- In Chap. 5, J. Aspnes and E. Ruppert introduce a theoretical model for a collection (or population) of tiny mobile agents that interact with one another to carry out a computation. Their chapter surveys results that describe what can be computed in various versions of the population protocol model.
- Opportunistic networking relies on the idea of exploiting the mobility of nodes to provide connectivity in scenarios where the source and destination nodes might never be connected to the same network at the same time. Routing issues specific to opportunistic networking are discussed in Chap. 6 by M. Conti, J. Crowcroft, S. Giordano, J. Crowcroft, P. Hui, H. Nguyen, and A. Passarella.
- Chapter 7 describes an emerging network architecture, known as mesh network, that is characterised by the partial coverage of the network with infrastructure support. That is, a mesh network can be considered as a mix between a network-centric architecture and a network-eccentric architecture. This type of networks is discussed in Chap. 7 by J. Ishmael and N. Race.

Chapter 5
An Introduction to Population Protocols

James Aspnes, *Yale University, USA*
Eric Ruppert, *York University, Canada*

5.1 Introduction

Population protocols are used as a theoretical model for a collection (or population) of tiny mobile agents that interact with one another to carry out a computation. The agents are identically programmed finite state machines. Input values are initially distributed to the agents, and pairs of agents can exchange state information with other agents when they are close together. The movement pattern of the agents is unpredictable, but subject to some fairness constraints, and computations must eventually converge to the correct output value in any schedule that results from that movement. This framework can be used to model mobile ad hoc networks of tiny devices or collections of molecules undergoing chemical reactions. This chapter surveys results that describe what can be computed in various versions of the population protocol model.

First, consider the basic population protocol model as a starting point. A formal definition is given in Sect. 5.2. Later sections describe how this model has been extended or modified to other situations. Some other directions in which the model could be extended as future work are also identified.

The defining features of the basic model are:

- Finite-state agents. Each agent can store only a constant number of bits in its local state (independent of the size of the population).
- Uniformity. Agents in the same state are indistinguishable and a single algorithm is designed to work for populations of any size.
- Computation by direct interaction. Agents do not send messages or share memory; instead, an *interaction* between two agents updates both of their states according to a joint transition table. The actual mechanism of such interactions is abstracted away.
- Unpredictable interaction patterns. The choice of which agents interact is made by an adversary. Agents have little control over which other agents they interact with. (In some variants of the model, the adversary may be limited to pairing only agents that are adjacent in an *interaction graph*, typically representing distance

constraints.) A global *fairness condition* is imposed on the adversary to ensure the protocol makes progress.

- Distributed inputs and outputs. The input to a population protocol is distributed across the initial states of the entire population. The output is distributed to all agents.
- Convergence rather than termination. Agents cannot, in general, detect when they have completed their computation; instead, the agents' outputs are required to converge after some finite time to a common, correct value.

The population protocol model [24] was designed to represent sensor networks consisting of very limited mobile agents with no control over their own movement. It also bears a strong resemblance to models of interacting molecules in theoretical chemistry [314, 315].

The population protocol model was inspired in part by the work of Diamadi and Fischer [226] on trust propagation in a social network. The *urn automata* of [20] can be seen as a first draft of the model that retained in vestigial form several features of classical automata: instead of interacting with one another, agents could interact only with a finite-state controller, complete with input tape. The motivation given for the current model in [24] was the study of sensor networks in which passive agents were carried along by other entities; the canonical example was a flock of birds with a sensor attached to each bird. The name of the model was chosen by analogy to *population processes* [66] in probability theory.

A population protocol often looks like an amorphous soup of lost, nearly mindless, anonymous agents blown here and there at the whim of the adversary. Although individual agents lack much intelligence or control over their own destinies, the population as a whole is nonetheless capable of performing significant computations. For example, even the simplest model is capable of solving some practical, classical distributed problems like leader election, majority voting, and organizing agents into groups. Some extensions of the model are much more powerful—under some conditions, they provide the same power as a traditional computer with the same total storage capacity. Some examples of simple population protocols are given in Sect. 5.2.1.

Much of the work so far on population protocols has concentrated on characterizing what predicates (i.e., boolean-valued functions) on the input values can be computed. This question has been resolved for the basic model, and studied for several different variants of the model and under various assumptions, such as a bounded-degree interaction graph or random scheduling.

The worst-case interaction graph for computation turns out to be a complete graph, since any other interaction graph can simulate a complete interaction graph by shuffling states between the nodes [24]. In a complete interaction graph, all agents with the same state are indistinguishable, and only the counts of agents in each state affect the outcome of the protocol. The set of computable predicates in most variants of the basic model for such a graph is now known to be either exactly equal to or closely related to the set of *semilinear* predicates, those definable in *first-order Presburger arithmetic* [316, 683]. These results, which originally appeared in [21, 24, 25, 27, 22, 23, 221], are summarized in Sects. 5.3, 5.4, 5.5, 5.7

and 5.9. Sometimes the structure of incomplete interaction graphs can be exploited to simulate a Turing machine, which implies that a restricted interaction graph can make the system stronger than a complete interaction graph.

Several extensions of the basic model have been considered that are intended to reflect the limitations and capabilities of practical systems more accurately. The basic model requires coordinated two-way communication between interacting agents; this assumption is relaxed in Sect. 5.4. Work on incorporating agent failures into the model are discussed in Sects. 5.7 and 5.9. Versions of the model that give agents slightly increased memory capacity are discussed in Sect. 5.8.

More recent work has concentrated on performance. Because the weak scheduling assumptions in the basic model allow the adversary to draw out a computation indefinitely, the worst-case adversary scheduler is replaced by a random scheduling assumption, where the pair of agents that interacts at each step is drawn uniformly from the population as a whole. This gives a natural notion of *time* equal to the total number of steps to convergence and *parallel time* equal to the average number of steps initiated by any one agent (essentially the total number of steps divided by the number of agents).

As with adversarial scheduling, for random scheduling the best-understood case is that of a complete interaction graph. In this case, it is possible to simulate a register machine, where subpopulations of the agents hold tokens representing the various register values in unary. It is not hard to implement register operations like addition, subtraction, and comparison by local operations between pairs of agents; with the election of a leader, one can further construct a finite-state control. The main obstacle to implementing a complete register machine is to ensure that every agent completes any needed tasks for each instruction cycle before the next cycle starts. In [24], this was handled by having the leader wait a polynomial number of steps on average before starting the next cycle, a process which gives an easy proof of polynomially-bounded error but which also gives an impractically large slowdown. Subsequent work has reduced the slowdown to polylogarithmic by using epidemics both to propagate information quickly through the population and to provide timing [29, 28]. These results are described in more detail in Sect. 5.6.

5.2 The Basic Model

In the basic population protocol model, a collection of agents are each given an input value, and agents have pairwise interactions in an order determined by a scheduler, subject to some fairness guarantee. Each agent is a kind of finite state machine and the program for the system describes how the states of two agents can be updated by an interaction. The agents are reliable: no failures occur. The agents' output values change over time and must eventually converge to the correct output value for the inputs that were initially distributed to the agents.

A protocol is formally specified by

- Q, a finite set of possible states for an agent,
- Σ, a finite input alphabet,

- ι, an input map from Σ to Q, where $\iota(\sigma)$ represents the initial state of an agent whose input is σ,
- ω, an output map from Q to the output range Y, where $\omega(q)$ represents the output value of an agent in state q, and
- $\delta \subseteq Q^4$, a transition relation that describes how pairs of agents can interact.

A computation proceeds according to such a protocol as follows. The computation takes place among n *agents*, where $n \geq 2$. Each agent initially has an input value from Σ. Each agent's initial state is determined by applying ι to its input value. This determines an initial configuration for an execution. A *configuration* of the system can be described by a vector of all the agents' states. Because agents with the same state are indistinguishable in the basic model, each configuration could also be viewed as an unordered multiset of states.

An execution of a protocol proceeds from the initial configuration by interactions between pairs of agents. Suppose two agents in states q_1 and q_2 meet and have an interaction. They can change into states q_1' and q_2' as a result of the interaction if (q_1, q_2, q_1', q_2') is in the transition relation δ. Note that interactions are in general asymmetric, with one agent (q_1) acting as the *initiator* of the interaction and the other (q_2) acting as the *responder*. Another way to describe δ is to list all possible interactions using the notation $(q_1, q_2) \rightarrow (q_1', q_2')$. (By convention, there is a null transition $(q_1, q_2) \rightarrow (q_1, q_2)$ if no others are specified with (q_1, q_2) on the left hand side.) If there is only one possible transition $(q_1, q_2) \rightarrow (q_1', q_2')$ for each pair (q_1, q_2), then the protocol is *deterministic*. If C and C' are configurations, $C \rightarrow C'$ means C' can be obtained from C by a single interaction of two agents. In other words, C contains two states q_1 and q_2 and C' is obtained from C by replacing q_1 and q_2 by q_1' and q_2', where (q_1, q_2, q_1', q_2') is in δ. An *execution* of the protocol is an infinite sequence of configurations $C_0, C_1, C_2, \ldots$, where C_0 is an initial configuration and $C_i \rightarrow C_{i+1}$ for all $i \geq 0$. Thus, an execution is a sequence of snapshots of the system after each interaction occurs. In a real distributed execution, interactions between several disjoint pairs of agents could take place simultaneously, but when writing down an execution those simultaneous interactions can be ordered arbitrarily. The notation $\overset{*}{\rightarrow}$ represents the transitive closure of $\rightarrow$, so $C \overset{*}{\rightarrow} C'$ means that there is a fragment of an execution that goes from configuration C to configuration C'.

The order in which pairs of agents interact is unpredictable: think of the schedule of interactions as being chosen by an adversary, so that protocols must work correctly under any schedule the adversary may choose. In order for meaningful computations to take place, the adversarial scheduler must satisfy some restrictions; otherwise it could, for example, divide the agents into isolated groups and schedule interactions only between agents that belong to the same group.

The *fairness* condition imposed on the scheduler is quite simple to state, but is somewhat subtle. Essentially, the scheduler cannot avoid a possible step forever. More formally, if C is a configuration that appears infinitely often in an execution, and $C \rightarrow C'$, then C' must also appear infinitely often in the execution. Another way to think of this is that anything that always has the potential to occur eventually does occur: it is equivalent to require that any configuration that is always reachable is eventually reached.

At any point during an execution of a population protocol, each agent's state determines its output at that time. If the agent is in state q, its output value is $\omega(q)$. Thus, an agent's output may change over the course of an execution. The fairness constraint allows the scheduler to behave arbitrarily for an arbitrarily long period of time, but does require that it behave nicely eventually. It is therefore natural to phrase correctness as a property to be satisfied eventually too. For example, the scheduler could schedule only interactions between agents 1 and 2, leaving the other $n - 2$ agents isolated, for millions of years, and it would be unreasonable to expect any sensible output during the period when only two agents have undergone state changes. Thus, for correctness, all agents must produce the correct output (for the input values that were initially distributed to the agents) at some time in the execution and continue to do so forever after that time.

In general, the transition relation can be non-deterministic: when two agents meet there may be several possible transitions they can make. This non-determinism sometimes comes in handy when describing protocols. However, it is not a crucial assumption: using a bit of additional machinery, agents can simulate a nondeterministic transition function by exploiting the nondeterminism of the interaction schedule. (See [21] for details.)

To summarize, a protocol computes a function f that maps multisets of elements of Σ to Y if, for every such multiset I and every fair execution that starts from the initial configuration corresponding to I, the output value of every agent eventually stabilizes to $f(I)$.

5.2.1 Examples of Population Protocols

Example 5.1. Suppose each agent is given an input bit, and all agents are supposed to output the 'or' of those bits. There is a very simple protocol to accomplish this: each agent with input 0 simply outputs 1 as soon as it discovers that another agent had input 1. Formally, $\Sigma = Y = Q = \{0, 1\}$ and the input and output maps are the identity functions. The only interaction in δ is $(0, 1) \rightarrow (1, 1)$. If all agents have input 0, no agent will ever be in state 1. If some agent has input 1 the number of agents with state 1 cannot decrease and fairness ensures that it will eventually increase to n. In both cases, all agents stabilize to the correct output value.

Example 5.2. Suppose the agents represent dancers. Each dancer is (exclusively) a leader or a follower. Consider the problem of determining whether there are more leaders than followers. Let $Y = \{0, 1\}$, with 1 indicating that there are more leaders than followers. A centralized solution would count the leaders and the followers and compare the totals. A more distributed solution is to ask everyone to start dancing with a partner (who must dance the opposite role) and then see if any dancers are left without a partner. This cancellation procedure is formalized as a population protocol with $\Sigma = \{L, F\}$ and $Q = \{L, F, 0, 1\}$. The input map ι is the identity, and the output map ω maps L and 1 to 1 and maps F and 0 to 0. The transitions of δ are

$$(L, F) \rightarrow (0, 0),$$
$$(L, 0) \rightarrow (L, 1),$$
$$(F, 1) \rightarrow (F, 0) \quad \text{and}$$
$$(0, 1) \rightarrow (0, 0).$$

The first rule ensures that, eventually, either no L's or no F's will remain. At that point, if there are L's remaining, the second rule ensures that all agents will eventually produce output 1. Similarly, the third rule takes care of the case where F's remain. In the case of a tie, the last rule ensures that the output stabilizes to 0.

It may not be obvious why the protocol in Example 5.2 must converge. Consider, for example, the following transitions between configurations, where in each configuration, the agents that are about to interact are underlined.

$$\{\underline{L}, L, \underline{F}\} \rightarrow \{\underline{0}, \underline{L}, 0\} \rightarrow \{\underline{1}, L, \underline{0}\} \rightarrow \{0, \underline{L}, 0\} \rightarrow \{\underline{0}, L, \underline{1}\} \rightarrow \{0, L, 0\}.$$

Repeating the last four transitions over and over yields a non-converging execution in which every pair of agents interacts infinitely often. However, this execution is not fair: the configuration $\{0, L, 1\}$ appears infinitely often and $\{0, L, 1\} \rightarrow \{1, L, 1\}$, but $\{1, L, 1\}$ never appears. This is because the first two agents only interact at "inconvenient" times, i.e., when the third agent is in state 0. The definition of fairness rules this out. Thus, in some ways, the definition of fairness is stronger than saying that each pair of agents must interact infinitely often. (In fact, the two conditions are incomparable, since there can be fair executions in which two agents never meet. For example, an execution where every configuration is $\{L, L, L\}$ and all interactions take place between the first two agents is fair.)

Exercise 5.1. Show the protocol of Example 5.2 converges in every fair execution.

The definition of fairness was chosen to be quite weak (although it is still strong enough to allow useful computations). Many models of mobile systems assume that the mobility patterns of the agents follow some particular probability distribution. The goal of the population protocol model is to be more general. If there is an (unknown) underlying probability distribution on the interactions, which might even vary with time, and that distribution satisfies certain independence properties and ensures that every interaction's probability is bounded away from 0, then an execution will be fair with probability 1. Thus, any protocol will converge to the correct output with probability 1. So the model captures computations that are correct with probability 1 for a wide range of probability distributions, even though the model definition does not explicitly incorporate probabilities.

Other predicates can be computed using an approach similar to Example 5.2.

Exercise 5.2. Design a population protocol to determine whether more than 60% of the dancers are leaders.

Exercise 5.3. Design a population protocol to determine whether more than 60% of the dancers dance the same role.

Some predicates, however, require a different approach.

Example 5.3. Suppose each agent is given an input from $\Sigma = \{0, 1, 2, 3\}$. Consider the problem of computing the sum of the inputs, modulo 4. The protocol can gather the sum (modulo 4) into a single agent. Once an agent has given its value to another agent, its value becomes null, and it obtains its output value from the eventually unique agent with a non-null value. Formally, let $Q = \{0, 1, 2, 3, \perp_0, \perp_1, \perp_2, \perp_3\}$, where $\perp_v$ represents a null value with output v. Let $\iota(v) = v$ and $\omega(v) = \omega(\perp_v) = v$ for $v = 0, 1, 2, 3$. The transition rules of δ are $(v_1, v_2) \rightarrow (v_1 + v_2, \perp_{v_1+v_2})$ and $(v_1, \perp_{v_2}) \rightarrow (v_1, \perp_{v_1})$, where v_1 and v_2 are 0, 1, 2 or 3. (The addition is modulo 4.) Rules of the first type ensure that, eventually, at most one agent will have a non-null value. Since the rules maintain, as an invariant, the sum of all non-null states (modulo 4), the unique remaining non-null value will be the sum modulo 4. The second type of rule then ensures that all agents with null states eventually converge to the correct output.

In some cases, agents may know when they have converged to the correct output, but in general they cannot. While computing the 'or' of input bits (Example 5.1), any agent in state 1 knows that its state will never change again: it has converged to its final output value. However, no agent in the protocol of Example 5.3 can ever be certain it has converged, since it may be that one agent with input 1 has not yet taken part in any interactions, and when it does start taking part the output value will have to change.

Two noteworthy properties of the population protocol model are its uniformity and anonymity. A protocol is *uniform* because its specification has no dependence on the number of agents that take part. In other words, no knowledge about the number of agents is required by the protocol. The system is *anonymous* because the agents are not equipped with unique identifiers and all agents are treated in the same way by the transition relation. Indeed, because the state set is finite and does not depend on the number of agents in the system, there is not even room in the state of an agent to store a unique identifier.

5.3 Computability

Just as traditional computability theory often restricts attention to decision problems, one can restrict attention to computing predicates, i.e., functions with range $Y = \{0, 1\}$, when studying what functions are computable by population protocols. There is no real loss of generality in this restriction. For any function f with range Y, let $P_{f,y}$ be a predicate defined by $P_{f,y}(x) = 1$ if and only if $f(x) = y$. Then, f is computable if and only if $P_{f,y}$ is computable for each $y \in Y$. The "only if" part of this statement is trivial. For the converse, a protocol can compute all the predicates $P_{f,y}$ in parallel, using a separate component of each agent's state for each y.

(Note that it only makes sense to talk about computing a function using a population protocol if the function has a finite range Y.) This will eventually give each agent enough information to output the value of the function f.

For the basic population protocol model, there is an exact characterization of the computable predicates. To describe this characterization, some definitions and notation are required. A multiset over the input alphabet Σ can also be thought of as a vector with $d = |\Sigma|$ components, where each component is a natural number representing the multiplicity of one input character. For example, the input multiset $\{a, a, a, b, b\}$ over the input alphabet $\Sigma = \{a, b, c\}$ can be represented by the vector $(3, 2, 0) \in \mathbb{N}^3$. Let $(x_1, x_2, \ldots, x_d) \in \mathbb{N}^d$ be a vector that represents the input to a population protocol. Here, d is the size of the input alphabet, Σ. A *threshold predicate* is a predicate of the form $\sum_{i=1}^{d} c_i x_i < a$, where $c_1, \ldots, c_d$ and a are integer constants. A *remainder predicate* is a predicate of the form $\sum_{i=1}^{d} c_i x_i \equiv a \pmod{b}$, where $c_1, \ldots, c_d$, a and $b > 0$ are integer constants. Angluin et al. [24] gave protocols to compute any threshold predicate or remainder predicate; the protocols are generalizations of those in Examples 5.2 and 5.3. They use the observation that addition is trivially obtained by renaming states: to compute $A + B$ from A and B, just pretend that any A or B token is really an $A + B$ token. Finally, one can compute the and or the or of two of these predicates by running the protocols for each of the basic predicates in parallel, using separate components of the agents' states, and negation simply involves relabeling the output values. Thus, population protocols can compute any predicate that is a boolean combination of remainder and threshold predicates. Surprisingly, the converse also holds: these are the *only* predicates that a population protocol can compute. This was shown for the basic model by Angluin, Aspnes, and Eisenstat [25].

Before discussing the proof of this result, there are two alternative characterizations of the computable predicates that are useful in understanding the result. These characterizations are also used in the details of the proof.

The first is that the computable predicates are precisely the *semilinear predicates*, defined as follows. A *semilinear set* is a subset of $\mathbb{N}^d$ that is a finite union of *linear sets* of the form $\{\mathbf{b} + k_1\mathbf{a}_1 + k_2\mathbf{a}_2 + \cdots + k_m\mathbf{a}_m \mid k_1, \ldots, k_m \in \mathbb{N}\}$, where $\mathbf{b}$ is a d-dimensional base vector, and $\mathbf{a}_1$ through $\mathbf{a}_m$ are basis vectors. See Figs. 5.1a and 5.1b for examples when $d = 2$. A *semilinear predicate* on inputs is one that is true precisely on a semilinear set. See Fig. 5.1c for an example.

To illustrate how semilinear predicates characterize computable predicates, consider the examples of the previous paragraph. Membership in the linear set S of Fig. 5.1a can be described by a boolean combination of threshold and remainder predicates: $(x_2 < 6) \wedge \neg(x_2 < 5) \wedge (x_1 \equiv 0 \pmod{2})$. Similarly, the linear set T of Fig. 5.1b can be described by $\neg(2x_1 - x_2 < 6) \wedge \neg(x_2 < 2) \wedge (2x_1 - x_2 \equiv 0 \pmod{6})$. The semilinear set $S \cup T$ of Fig. 5.1c is described by the disjunction of these two formulas.

A second alternative characterization of semilinear predicates is that they can be described by first-order logical formulas in Presburger arithmetic, which is arithmetic on the natural numbers with addition but not multiplication [683]. Thus, for example, the set T of Fig. 5.1b can be described by $\neg(x_1 + x_1 - x_2 < 6) \wedge$

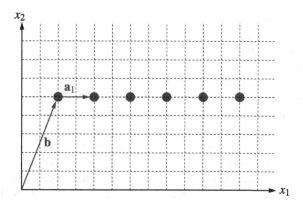

Fig. 5.1a A linear set $S = \{\mathbf{b} + k_1\mathbf{a}_1 \mid k_1 \in \mathbb{N}\}$

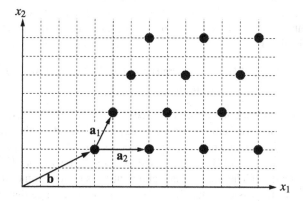

Fig. 5.1b A linear set $T = \{\mathbf{b} + k_1\mathbf{a}_1 + k_2\mathbf{a}_2 \mid k_1, k_2 \in \mathbb{N}\}$

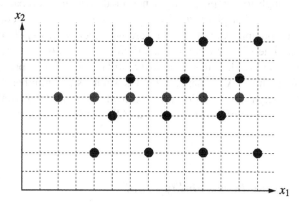

Fig. 5.1c A semilinear set $S \cup T$

$\neg(x_2 < 2) \wedge \exists j (x_1 + x_1 - x_2 = j + j + j + j + j + j)$. Presburger arithmetic allows for *quantifier elimination*, replacing universal and existential quantifiers with formulas involving addition, $<$, the equivalence mod b predicates for each constant b, and the usual logical connectives $\wedge$, $\vee$, and $\neg$. For example, eliminating quantifiers from the formula for T yields $\neg(x_1 + x_1 - x_2 < 6) \wedge \neg(x_2 < 2) \wedge (x_1 + x_1 - x_2 \equiv 0 \pmod 6))$, which can be computed by a population protocol, as mentioned above.

The proof that only semilinear predicates are computable is obtained by applying results from partial order theory. The proof is quite involved, but the essential idea is that, like finite-state automata, population protocols can be "pumped" by adding extra input tokens that turn out not to affect the final output. By carefully considering exactly when this is possible, it can be shown that the positive inputs to a population protocol (considered as sets of vectors of natural numbers) can be separated into a collection of cones over some finite set of minimal positive inputs, and that each of these cones can be further expressed using only a finite set of basis vectors. This is sufficient to show that the predicate corresponds to a semilinear set as described above [25, 27]. A sketch of this argument is given in Sect. 5.3.1. The full characterization is:

Theorem 5.1 ([24, 25, 27]). *A predicate is computable in the basic population protocol model if and only if it is semilinear.*

Similar results with weaker classes of predicates hold for restricted models with various forms of one-way communication [22]; Sect. 5.4 describes these results in more detail. Indeed, these results were a precursor to the semilinearity theorem of [25]. The journal paper [27] combines and extends the results of [25] and [22].

A useful property of Theorem 5.1 is that it continues to hold unmodified in many simple variants of the basic model. The reason is that any change that weakens the agents can only decrease the set of computable predicates, while any model that is still strong enough to compute congruence modulo k and comparison can still compute all the semilinear predicates. So the semilinear predicates continue to be those that are computable when the inputs are not given immediately but stabilize after some finite time [21] or when one agent in an interaction can see the other's state but not vice versa [27], as in each case it is still possible to compute congruence and threshold in the limit. A similar result holds when a small number of agents can fail [221]; here a slight modification must be made to allow for partial predicates that can tolerate the loss of part of the input. All of these results are described in later sections.

5.3.1 Sketch of the Impossibility Proof

The proof that all predicates computable in the basic population protocol model are semilinear is quite technical. To give a flavor of the results, here is a simplified version, a *pumping lemma* that says that any predicate stably computed by a population protocol is a finite union of *monoids*: sets of the form

$$\{\mathbf{b} + k_1\mathbf{a}_1 + k_2\mathbf{a}_2 + \cdots \mid k_i \in \mathbb{N} \text{ for all } i\},$$

where the number of terms may be infinite (this is the first step in proving the full lower bound in [25], where the number of generators for each monoid is also shown to be finite). The main tool is Higman's Lemma [369], which states that any infinite sequence $\mathbf{x}_1, \mathbf{x}_2, \ldots$ in $\mathbb{N}^d$ has elements $\mathbf{x}_i, \mathbf{x}_j$ with $\mathbf{x}_i \leq \mathbf{x}_j$ and $i < j$, where comparisons between vectors are done componentwise. It follows from Higman's Lemma that (a) any subset of $\mathbb{N}^d$ has finitely many minimal elements (Dickson's Lemma), and (b) any infinite subset of $\mathbb{N}^d$ contains an infinite ascending sequence $\mathbf{a}_1 < \mathbf{a}_2 < \mathbf{a}_3 \ldots$.

For the proof, configurations are represented as vectors of counts of agents in each state. Thus, a configuration is a vector in $\mathbb{N}^{|Q|}$, just as an input is a vector in $\mathbb{N}^{|\Sigma|}$. Using Dickson's Lemma, it can be shown that the set of output-stable configurations of a population protocol, where all agents agree on the output and continue to agree in all successor configurations, is semilinear. The proof is that if some configuration C is *not* output-stable, then there is some submultiset of agents $\mathbf{x}$ that can together produce an agent with a different output value. But since any $\mathbf{y} \geq \mathbf{x}$ can also produce this different output value, the property of being non-output-stable is closed upwards, implying that there is a finite collection of minimal non-output-stable configurations. Thus, the set of non-output-stable configurations is a finite union of cones, so both it and its complement—the output-stable configurations—are semilinear. (The complement of a semilinear set is also semilinear.)

Unfortunately this is not enough by itself to show that the input configurations that eventually reach a given output state are also semilinear. The second step in the argument is to show that when detecting if a configuration $\mathbf{x}$ is output-stable, it suffices to consider its truncated version

$$\tau_k(x_1, x_2, \ldots, x_m) = (\min(x_1, k), \min(x_2, k), \ldots, \min(x_m, k)),$$

provided k is large enough to encompass all of the minimal non-output-stable configurations as defined previously. The advantage of this step is that it reduces the set of configurations that must be considered from an infinite set to a finite set.

For each configuration $\mathbf{c}$ in this finite set, define the set of *extensions* $X(\mathbf{c})$ of $\mathbf{c}$ by

$$X(\mathbf{c}) = \{\mathbf{x} \mid \exists \mathbf{d} \text{ such that } \mathbf{c} + \mathbf{x} \overset{*}{\to} \mathbf{d} \text{ and } \tau_k(\mathbf{d}) = \tau_k(\mathbf{c})\}.$$

Intuitively, this means that $\mathbf{x}$ is in $X(\mathbf{c})$ if $\mathbf{c}$ can be "pumped" by $\mathbf{x}$, with the extra agents added in $\mathbf{x}$ disposed of in the coordinates of $\mathbf{c}$ that already have k or more agents. It is not hard to show that extensions are composable: if $\mathbf{x}, \mathbf{y}$ are in $X(\mathbf{c})$, then so is $\mathbf{x} + \mathbf{y}$. This shows that $\mathbf{b} + X(\mathbf{c})$ is a monoid for any configurations $\mathbf{b}$ and $\mathbf{c}$.

Finally, given any predicate that can be computed by a population protocol, these extensions are used to hunt for a finite collection of monoids whose union is the set Y of all inputs that produce output 1. The method is to build up a family of sets of the form $\mathbf{x} + X(\mathbf{c})$ where $\mathbf{x}$ is an input and $\mathbf{c}$ is an output-stable configuration reachable from that input. In more detail, order Y so that $\mathbf{y}_i \leq \mathbf{y}_j$ implies $i \leq j$; let $B_0 = \emptyset$; and compute B_i as follows:

- If $\mathbf{y}_i \in \mathbf{x} + X(\mathbf{c})$ for some $(\mathbf{x}, \mathbf{c}) \in B_{i-1}$, let $B_i = B_{i-1}$.
- Otherwise, construct B_i by adding to B_{i-1} the pairs

  - $(\mathbf{y}_i, s(\mathbf{y}_i))$, and
  - $(\mathbf{y}_i, s(\mathbf{c} + \mathbf{y}_i - \mathbf{x}))$ for all $(\mathbf{x}, \mathbf{c}) \in B_{i-1}$ with $\mathbf{x} \leq \mathbf{y}_i$,

  where $s(\mathbf{z})$ is any stable configuration reachable from $\mathbf{z}$.

Finally, let $B = \bigcup B_i$.

Then $\{\mathbf{b} + X(\mathbf{c}) \mid (\mathbf{b}, \mathbf{c}) \in B\}$ covers Y because a set containing $\mathbf{y}_i$ was added to B_i if $\mathbf{y}_i$ was not already included in one of the sets in B_{i-1}. Furthermore, none of these sets contain anything outside Y. The proof of this last fact is that because $\mathbf{b} \xrightarrow{*} \mathbf{c}, \mathbf{z} \in \mathbf{b} + X(\mathbf{c})$ implies $\mathbf{z} \xrightarrow{*} \mathbf{z}'$ for some $\mathbf{z}' \in \mathbf{c} + X(\mathbf{c})$ (just run the $\mathbf{b} \xrightarrow{*} \mathbf{c}$ computation, ignoring any agents in $\mathbf{z} - \mathbf{b}$). But then $\mathbf{z}$ converges to same output as $\mathbf{c}$, by the definition of $X(\mathbf{c})$, the construction of B_i only includes vectors $\mathbf{c}$ that are successors of inputs in Y, so this output value is positive. It follows that B gives a representation of Y as a union of monoids, one for each element of B. It remains to show that B is finite.

To do so, suppose B is infinite. Use Higman's Lemma to get an increasing sequence $\mathbf{b}_1 < \mathbf{b}_2 < \cdots$ such that $(\mathbf{b}_i, \mathbf{c}_i) \in B$ for some $\mathbf{c}_i$. Use Higman's Lemma *again* to get an infinite subsequence $(\mathbf{b}_{i_j}, \mathbf{c}_{i_j})$ where both the $\mathbf{b}$ and $\mathbf{c}$ components are increasing. Because these components are increasing, they eventually reach the bound imposed by truncation: for some i_j, $\tau_k(\mathbf{c}_{i_{(j+1)}}) = \tau_k(\mathbf{c}_{i_j})$. But then $\mathbf{b}_{i_{(j+1)}} - \mathbf{b}_{i_j}$ is in $X(\mathbf{c}_{i_j})$, so $\mathbf{b}_{i_{(j+1)}}$ cannot be in B, a contradiction.

This argument showed that any stably computable set has a finite cover by monoids of the form

$$\{\mathbf{b} + k_1 \mathbf{a}_1 + k_2 \mathbf{a}_2 + \cdots \mid k_i \in \mathbb{N} \text{ for all } i\}.$$

An immediate corollary is that any infinite stably computable set Y can be pumped: there is some $\mathbf{b}$ and $\mathbf{a}$ such that $\mathbf{b} + k\mathbf{a}$ is in Y for all $k \in \mathbb{N}$. Sadly, this is not enough to exclude some non-semilinear sets like $\{(x, y) \mid x < y\sqrt{2}\}$. However, with substantial additional work these bad cases can be excluded as well; the reader is referred to [25, 27] for details.

5.4 One-Way Communication

In the basic population protocol model, it is assumed that two interacting agents can simultaneously learn each other's state before updating their own states as a result of the interaction. This requires two-way communication between the two agents. Angluin et al. [27] studied several weaker interaction models where, in an interaction, information flows in one direction only. A *receiver* agent learns the state of a *sender* agent, but the sender learns nothing about the state of the receiver. The power of a system with such one-way communication depends on the precise nature of the communication mechanism.

The model is called a *transmission* model if the sender is aware that an interaction has happened (and can update its own state, although the update cannot depend on the state of the receiver). In an *observation* model, on the other hand, the sender's state is observed by the receiver, and the sender is not aware that its state has been observed. Another independent attribute is whether an interaction happens instantaneously (*immediate transmission* and *immediate observation* models) or requires some interval of time (*delayed transmission* and *delayed observation* models). The *queued transmission* model is similar to the delayed transmission model, except that receivers can temporarily refuse incoming messages so that they are not overwhelmed with more incoming information than they can handle. The queued transmission model is the closest to traditional message-passing models of distributed computing.

The weakest of these one-way models is the delayed observation model: Agents can observe other agents' input symbols to determine whether each input symbol is present in the system or not. If an agent ever sees another agent with the same input symbol as itself, it learns that there are at least two copies of that symbol, and can tell every other agent this fact. Thus, delayed observation protocols can detect whether the multiplicity of any particular input symbol is 0, 1 or at least 2, so a protocol can compute any predicate that depends only on this kind of information. Nothing else can be computed. For example there is no way for the system to determine whether some input symbol occurs with multiplicity at least 3. Intuitively, this is because there is no way to distinguish between a sequence of observations of several agents with the same input and a sequence of observations of a single agent.

The immediate observation model is slightly stronger: protocols in this model can count the number of agents with a particular input symbol, up to any constant threshold. For example, a protocol can determine whether the number of copies of input symbol a is 0, 1, 2, 3 or more than 3. Consequently, any predicate that depends only on this kind of information can be computed. A kind of pumping lemma can be used to show that no other predicates are computable.

Angluin et al. also showed that the immediate and delayed transmission models are equivalent in power. They gave a characterization of the computable predicates that shows the power of these models is intermediate between the immediate observation model and the standard two-way model.

Finally, the queued transmission model is equivalent in power to the standard two-way model: any protocol designed for the two-way model can be simulated using queued transmission and vice versa. This holds even though the set of configurations reachable from a particular initial configuration of a protocol in the queued transmission model is in principle unbounded; the ability to generate large numbers of buffered messages does not help the protocol, largely because there is no guarantee of where or when they will be delivered.

5.5 Restricted Interaction Graphs

In some cases the mobility of agents will have physical limitations, and this will limit the possible interactions that can occur. An *interaction graph* represents this information: nodes represent agents and edges represent possible interactions. The basic model corresponds to the case where the graph is complete. In this model, a configuration is always represented as a vector of n states. (The agents are no longer indistinguishable, so one cannot use a multiset.) If C and C' are configurations, $C \rightarrow C'$ means that C' can be obtained from C through a single interaction of *adjacent* agents, and the definitions of executions and fairness are as before, using this modified notion of a step.

Having a non-complete (but connected) interaction graph does not make the model any weaker, since adjacent agents can swap states to simulate free movement [24]. For some interaction graphs, the model becomes strictly more powerful. For example, consider a straight-line graph. It is not difficult to simulate a linear-space Turing machine by using each agent to represent one square of the Turing machine tape. This allows computation of any function or predicate that can be computed by a Turing machine using linear space. Many such functions are not semilinear and thus not computable in the complete interaction graph of the basic model. For example, a population protocol can use standard Turing machine methods to compute a multiplication predicate over the input alphabet $\{a, b, c\}$ that is true if and only if the number of a's multiplied by the number of b's is equal to the number of c's.

In addition to computing predicates on the inputs to agents, it also makes sense in this model to ask whether properties of the interaction graph itself can be computed by the agents in the system. Such problems, which were studied by Angluin et al. [21], could have useful applications in determining the network topology induced by an ad hoc deployment of mobile agents. This section describes some of their results.

As a simple example, one might want to determine whether the interaction graph has maximum degree k or more, for some fixed k. This can be done by electing a single moving leader token. Initially, all agents hold a leader token. When two leader tokens interact, the tokens coalesce, and when a leader agent interacts with a non-leader agent the leader token may change places. To test the maximum degree, the leader may instead choose to mark up to k distinct neighbors of its current node. By counting how many nodes it successfully marks, the leader can get a lower bound on the degree of the node.

A complication is that the leader has no way to detect when it has interacted with all neighbors of the current node. The best it can do is nondeterministically wait for some arbitrary but finite time before gathering in its marks and trying again. In doing so it relies on the fairness condition to eventually drive it to a state where it has correctly computed the maximum degree (or determined that it is greater than k). To accomplish the unmarking, the leader keeps track of how many marks it has placed, so that it can simply wait until it has encountered each marked neighbor again. During the initial leader election phase, two leaders deploying marks could interfere with each other. To handle this, the survivor of any interaction between two leaders collects all outstanding marks from both and resets its degree estimate.

A similar mechanism can be used to assign unique colors to all neighbors of each node in a bounded-degree graph: a wandering *colorizer* token deploys pairs of marks to its neighbors and recolors any it finds with the same color. Once this process converges, the resulting *distance-2 coloring* (so called because all nodes at distance 2 have distinct colors) effectively provides local identifiers for the neighbors of each node. These can be used to carry out arbitrary distributed computations using standard techniques (subject to the $O(1)$ space limit at each node). An example given in [21] is the construction of a rooted spanning tree, which can be used to simulate a Turing machine tape (as in the case of a line graph) by threading the Turing machine tape along a traversal of the tree (a technique described earlier for self-stabilizing systems by Itkis and Levin [407]). It follows that arbitrary properties of bounded-degree graphs that can be computed by a Turing machine using linear space can also be computed by population protocols.

5.6 Random Interactions

An alternative assumption that also greatly increases the power of the model is to replace the adversarial (but fair) scheduler of the basic model with a more constrained interaction pattern. The simplest such variant assumes *uniform random interactions*: each pair of agents is equally likely to interact at each step.

Protocols for random scheduling were given in the initial population protocol paper of Angluin et al. [24], based in part on similar protocols for the related model of urn automata [20]. The central observation was that the main limitation observed in trying to build more powerful protocols in the basic model was the inability to detect the absence of agents with a particular state. However, if a single leader agent were willing to wait long enough, it could be assured (with reasonably high probability) that it would meet every other agent in the population, and thus be able to verify the presence or absence of particular values stored in the other agents by direct inspection. The method used was to have the leader issue a single special marked token to some agent; when the leader encountered this special agent k times in a row it could be reasonably confident that the number of intervening interactions was close to $\Theta(n^{k+1})$. This is sufficient to build unary counters supporting the usual increment, decrement, and zero test operations (the last probabilistic). With counters, a register machine with an $O(\log n)$ bit random-access memory can be simulated using a classic technique of Minsky [578].

The cost of this simulation is a polynomial blowup for the zero test and a further polynomial blowup in the simulation of the register machine. A faster simulation was given by Angluin, Aspnes, and Eisenstat [28], based on epidemics to propagate information quickly through the population. This simulation assumes a single designated leader agent in the initial configuration, which acts as the finite-state controller for the register machine. Register values are again stored in unary as tokens scattered across the remaining agents.

To execute an operation, the leader initiates an epidemic containing an operation code. This opcode is copied through the rest of the population in $\Theta(n \log n)$ interactions on average and with high probability; the latter result is shown to follow by a reduction to a concentration bound for the coupon collector problem due to Kamath et al. [440]. Arithmetic operations such as addition, comparison, subtraction, and multiplication and division by constants can be carried out by the non-leader agents in $O(n \log^c n)$ interactions (or $O(\log^c n)$ parallel time units) each, where c is a constant. Some of these algorithms are quite simple (adding A to B requires only adding a new B token to each agent that already holds an A token, possibly with an additional step of unloading extra B tokens onto empty agents to maintain $O(1)$ space per agent), while others are more involved (comparing two values in [28] involves up to $O(\log n)$ alternating rounds of doubling and cancellation, because simply having A and B tokens cancel each other as in Example 5.2 might require as many as $\Theta(n^2)$ expected interactions for the last few survivors to meet). The most expensive operation is division, at $O(n \log^5 n)$ interactions (or $O(\log^5 n)$ parallel time units).[1]

Being able to carry out individual arithmetic operations is of little use if one cannot carry out more than one. This requires that the leader be able to detect when an operation has finished, which ultimately reduces down to being able to detect when $\Theta(n \log n)$ interactions have occurred. Here the trick of issuing a single special mark is not enough, as the wait needed to ensure a low probability of premature termination is too long.

Instead, a *phase clock* based on successive waves of epidemics is used. The leader starts by initiating a phase 0 epidemic which propagates through the population in parallel to any other activity. When the leader meets an agent that is already infected with phase 0, it initiates a phase 1 epidemic that overwrites the phase 0 epidemic, and similarly with phase 2, 3, and so on, up to some fixed maximum phase $m - 1$ that is in turn overwritten by phase 0 again. Angluin et al. show that, while the leader might get lucky and encounter one of a small number of newly-infected agents in a single phase, the more typical case is that a phase takes $\Theta(n \log n)$ interactions before the next is triggered, and over m phases the probability that all are too short is polynomially small. It follows that for a suitable choice of m, the phase clock gives a high-probability $\Theta(n \log n)$-interaction clock, which is enough to time the other parts of the register machine simulation.

A curious result in [28] is that even though the register machine simulation has a small probability of error, the same techniques can compute semilinear predicates in polylogarithmic expected parallel time with no error in the limit. The trick is to run a fast error-prone computation to get the answer quickly most of the time, and then switch to the result of a slower, error-free computation using the mechanisms of [24] after some polynomially long interval. The high time to converge for the second algorithm is apparent only when the first fails to produce the correct answer;

---

[1] While the conference version of [28] claimed $O(n \log^4 n)$ interactions, this was the result of a calculation error that has been corrected by the authors in the full version of the paper.

but as this occurs only with polynomially small probability, it disappears in the expectation.

This simulation leaves room for further improvement. An immediate task is to reduce the overhead of the arithmetic operations. In [29], the same authors show how to drop the cost of the worst-case arithmetic operation to $O(n \log^2 n)$ interactions by combining a more clever register encoding with a fast *approximate majority* primitive based on dueling epidemics. This protocol has only three states: the decision values x and y, and b (for "blank"). When an x token meets a y token or vice versa, the second token turns blank. When an x or y token meets a blank agent, it converts the blank token to its own value. Much of the technical content of [29] involves showing that this process indeed converges to the majority value in $O(n \log n)$ interactions with high probability, which is done using a probabilistic potential function argument separated into several interleaved cases. The authors suggest that simplifying this argument would be a very useful target for future research. It is also possible that further improvements could reduce the overhead for arithmetic operations down to the $O(n \log n)$ interactions needed simply for all tokens to participate.

A second question is whether the distinguished leader in the initial configuration could be replaced. The coalescing leader election algorithm of [24] takes $\Theta(n^2)$ interactions to converge, which may dwarf the time for simple computations. A heuristic leader-election method is proposed in [29] that appears to converge much faster, but more analysis is needed. The authors also describe a more robust version of the phase clock of [28] that, by incorporating elements of the three-state majority protocol, appears to self-stabilize in $O(n \log n)$ interactions once the number of leaders converges to a polynomial fraction, but to date no proof of correctness for this protocol is known.

5.7 Self-stabilization and Related Problems

A series of papers [23, 26, 282] have examined the question of when population protocols can be made self-stabilizing [228], or at least can be made to tolerate input values that fluctuate over some initial part of the computation. Either condition is a stronger property than the mere convergence of the basic model, as both require that the population eventually converge to a good configuration despite an unpredictable initial configuration. Many of the algorithms designed to start in a known initial configuration (even if it is an inconvenient one, with, say, all agents in the same state) will not work if started in a particularly bad one. An example is leader election by coalescence: this algorithm can reduce a population of many would-be leaders down to a single unique leader, but it cannot create a new leader if the initial population contains none.

Angluin et al. [23] gave the first self-stabilizing protocols for the population protocol model, showing how to carry out various tasks from previous papers without assuming a known initial configuration. These include a distance-2 coloring protocol for bounded-degree graphs based on local handshaking instead of a wandering colorizer token (which is vulnerable to being lost). Their solution has each node track whether it has interacted with a neighbor of each particular color an odd or even number of times; if a node has two neighbors of the same color, eventually its count will go out of sync with that of one or the other, causing both the node and its neighbor to choose new colors. This protocol is applied in a framework that allows self-stabilizing protocols to be composed, to give additional protocols such as rooted spanning tree construction for networks with a single special node. This last protocol is noteworthy in part because it requires $O(\log D)$ bits of storage per node, where D is the diameter of the network; it is thus one of the earliest examples of pressure to escape the restrictive $O(1)$-space assumption of the original population protocol model. Other results in this paper include a partial characterization of which network topologies do or do not support self-stabilizing leader election.

This work was continued by Angluin, Fischer, and Jiang [26], who considered the issue of solving the classic *consensus problem* [656] in an environment characterized by unpredictable communication, with the goal of converging to a common consensus value at all nodes eventually (as in a population protocol) rather than terminating with one. The paper gives protocols for solving consensus in this stabilizing sense with both crash and Byzantine failures. The model used deviates from the basic population protocol model in several strong respects: agents have identities (and the $O(\log n)$-bit memories needed to store them), and though the destinations to which messages are delivered are unpredictable, communication itself is synchronous.

Fischer and Jiang [282] return to the anonymous, asynchronous, and finite-state world of standard population protocols to consider the specific problem of leader election. As observed above, a difficulty with the simple coalescence algorithm for leader election is that it fails if there is no leader to begin with. Fischer and Jiang propose adding to the model a new *eventual leader detector*, called Ω?, which acts as an oracle that eventually correctly informs the agents if there is no leader. (The name of the oracle is by analogy to the classic eventual leader *election* oracle Ω of Chandra and Toueg [160].) Self-stabilizing leader election algorithms based on Ω? are given for complete interaction graphs and rings. Curiously, the two cases distinguish between the standard global fairness condition assumed in most population protocol work and a local fairness condition that requires only that each action occurs infinitely often (but not necessarily in every configuration in which it is enabled). The latter condition is sufficient to allow self-stabilizing leader election in a complete graph but is provably insufficient in a ring. Many of these results are further elaborated in Hong Jiang's Ph.D. dissertation [422].

5.8 Larger States

The assumption that each agent can only store $O(1)$ bits of information is rather restrictive. One direction of research is to slowly relax this constraint to obtain other models that are closer to real mobile systems while still keeping the model simple enough to allow for a complete analysis.

5.8.1 Unique Identifiers

As noted in Sect. 5.2, the requirements that population protocols be independent of n and use $O(1)$ space per agent imply that agents cannot have unique identifiers. This contrasts with the vast majority of models of distributed computing, in which processes do have unique identifiers that are often a crucial component of algorithms. Guerraoui and Ruppert investigated a model, called *community protocols*, that preserve the tiny nature of agents in population protocols, but allow agents to be initially assigned unique identifiers drawn from a large set [338]. Each agent is equipped with $O(1)$ memory locations that can each store an identifier. It is assumed that transition rules cannot be dependent on the values of the identifiers: the identifiers are atomic objects that can only be tested for equality with one another. (For example, bitwise operations on identifiers are not permitted.) This preserves the property that protocols are independent of n. They gave the following precise characterization of what can be computed in this model.

Theorem 5.2 ([338]). *A predicate is computable in the community protocol model if and only if it can be computed by a nondeterministic Turing machine that uses $O(n \log n)$ space and permuting the input characters does not affect the output value.*

The necessity of the second condition (symmetry) follows immediately from the fact that the identifiers cannot be used to order the input symbols. The proof that any computable predicate can be computed using $O(n \log n)$ space on a nondeterministic Turing machine uses a nondeterministic search of the graph whose nodes are configurations of the community protocol and whose edges represent transitions between configurations.

Conversely, consider any symmetric predicate that can be computed by a nondeterministic Turing machine using $O(n \log n)$ space. The proof that it can also be computed by a community protocol uses Schönhage's pointer machines [735] as a bridge. A pointer machine is a sequential machine model that runs a program using only a directed graph structure as its memory. A community protocol can emulate a pointer machine by having each agent represent a node in the graph data structure. Some care must be taken to organize the agents to work together to simulate the sequential machine. It was known that a pointer machine that uses $O(n)$ nodes can simulate a Turing machine that uses $O(n \log n)$ space [836].

It follows that the restriction that agents can use their additional memory space only for storing $O(1)$ identifiers can essentially be overcome: the agents can do just as much as they could if they each had $O(\log n)$ bits of storage that could be used arbitrarily.

5.8.2 Heterogeneous Systems

One interesting direction for future research is allowing some heterogeneity in the model, so that some agents have more computational power than others. As an extreme example, consider a network of weak sensors that interact with one another, but also with a base station that has unlimited capacity.

Beauquier et al. [70] studied a scenario like this, focusing on the problem of having the base station compute n, the number of mobile agents. They replaced the fairness condition of the population protocol model by a requirement that all pairs of agents interact infinitely often. They considered a self-stabilizing version of the model, where the mobile agents are initialized arbitrarily. (Otherwise the problem can be trivially solved by having the base station mark each mobile agent as it is counted.) The problem cannot be solved if each agent's memory is constant size: they proved a tight lower bound of n on the number of possible states the mobile agents must be able to store.

5.9 Failures

The work described so far assumes that the system experiences no failures. This assumption is somewhat unrealistic in the context of mobile systems of tiny agents, and was made to obtain a clean model as a starting point. Some work has studied fault-tolerant population protocols, although this topic is still largely unexplored.

5.9.1 Crash Failures

Crash failures are a relatively benign type of failure: faulty agents simply cease having any interactions at some time during the execution. Delporte-Gallet et al. [221] examined how crash failures affect the computational power of population protocols. They showed how to transform any protocol that computes a function in the failure-free model into a protocol that can tolerate $O(1)$ crash failures. However, this requires some inevitable weakening of the problem specification.

To understand how the problem specification must change when crash failures are introduced, consider the majority problem described in Example 5.2. This problem was solved under the assumption that there are no failures. Now consider a version of the majority problem where up to 5 agents may crash. Consider an execution with m followers and $m + 5$ leaders. According to the original problem specification, the output of any such execution must be 1. Suppose, however, that the agents associated with 5 of the $m+5$ leaders crash before having any interactions. There is no way that the non-faulty agents can distinguish such an execution from a failure-free execution involving m followers and m leaders. In the latter execution, the output must be 0. So, the majority problem, in its original form, cannot be solved when crash failures occur. Nevertheless, it is possible to solve a closely related problem. Suppose there are preconditions on the problem, requiring that the margin of the majority is at least 5. More precisely, it is required that either the number of leaders exceeds the number of followers by more than 5 or the number of followers exceeds the number of leaders by at least 5. Under this precondition, it can be shown that the majority problem becomes solvable even when up to 5 agents may crash.

The above example can be generalized in a natural way: to solve a problem in a way that tolerates up to f crash failures, where f is a constant, there must be a precondition that says the removal of f of the input values cannot change the output value. It is not difficult to see that such a precondition is necessary. To prove that this is sufficient to make the predicate computable in a fault-tolerant way (assuming that the original predicate is computable in the failure-free model), Delporte-Gallet et al. [221] designed an automatic transformation that converts a protocol P for the failure-free model into a protocol P' that will tolerate up to f failures.

The transformation uses replication. In P', agents are divided (in a fault-tolerant way) into $\Theta(f)$ groups, each of size $\Theta(n/f)$. Each group simulates an execution of P on the entire set of inputs. Each agent of P' can store, in its own memory, the simulated states of $O(f)$ agents of P, since f is a constant, so each group of $\Theta(n/f)$ agents has sufficient memory space to collectively simulate all agents of P. To get a group's simulation started, agents within the group gather the initial states (in P) of *all* agents. Up to f agents may crash before giving their initial states to anyone within that group, but the precondition ensures that this will not affect the output of the simulated run. Thus, any group whose members do not experience any crashes will eventually produce the correct output. It follows that at least $f + 1$ of the $2f + 1$ groups will converge on the correct output, and any non-faulty agent can compute this value by remembering the output value of the last agent it saw from each group and taking the majority value. (If the range of the function to be computed is larger than $\{0, 1\}$, a larger number of groups must be used.)

A variant of the simulation handles a combination of a constant number of transient failures (where an agent spontaneously changes state) and crash failures [221]. It can also be used in the community protocol model described in Sect. 5.8.1 [338].

5.9.2 Byzantine Failures

An agent that has a Byzantine failure may behave arbitrarily: it can interact with all other agents, pretending to be in any state for each interaction. This behavior can cause havoc in a population protocol since none of the usual techniques used in distributed computing to identify and contain the effects of Byzantine agents can be used. Indeed, it is known that no non-trivial predicate can be computed by a population protocol in a way that tolerates even one Byzantine agent [338]. Two ways of circumventing this fact have been studied.

In the community protocol model of Sect. 5.8.1, some failure detection is possible, provided that the agent identifiers cannot be tampered with. Guerraoui and

Ruppert give a protocol that solves the majority problem, tolerating a constant number of Byzantine failures, if the margin of the majority is sufficiently wide [338]. (In defining this model, the fairness condition has to be altered to exclude Byzantine agents.)

Byzantine agents also appear in the random-scheduling work of [29], where it is shown that the approximate majority protocol quickly converges to a configuration in which nearly all non-faulty agents possess the correct decision value despite the actions of a small minority of $o(\sqrt{n})$ Byzantine agents. Here there is no extension of the basic population protocol model to include identifiers, but the convergence condition is weak, and the Byzantine agents can eventually—after exponential time—drive the protocol to any configuration, including stable configurations in which no agent holds a decision value. Determining the full power of random scheduling in the presence of Byzantine agents remains open.

5.10 Relations to Other Models

There are other mathematical models that bear some similarities to population protocols. Techniques or results from those models might prove useful in studying population protocols.

The *cellular automata* model of von Neumann [844] also models computation by a collection of communicating finite automata. However, the agents lack any mobility. In the classical version of this model, identical agents are arranged in a highly symmetric, regular, constant-degree graph (such as a grid) and each agent updates its state based on a snapshot of all of its neighbours' states. This model assumed all agents run synchronously, but some researchers have studied an asynchronous version of the model, defined by Ingerson and Buvel [404], in which a single agent updates its state in each step. The way in which this agent is chosen varies. Their work (and most that followed) is experimental. In each interaction, only one agent's state is updated (as in the immediate observation model discussed in Sect. 5.4). However, their model still assumes that an agent can learn the states of all its neighbours simultaneously, in contrast to the pairwise interactions that form the basis of population protocols.

Inspired by biological processes, Păun defined *P systems* [653] to model a collection of mobile finite-state agents that interact with one another and also with *membranes*. The membranes divide space into regions and groups of agents within a single region have interactions. The system specifies a set of interaction rules for each region. These interactions can create new agents, destroy existing agents, change agents' states, cause agents to cross a membrane into an adjacent region, or even dissolve a membrane to merge two adjacent regions. The basic model is synchronous, and the choice of which interactions happen in each round of computation is highly constrained by priorities assigned to each rule. Unlike typical distributed models of computation, where algorithms must compute correctly in all possible executions, the emphasis here is on nondeterministic computation: there should exist a correct execution for each input. These factors make P systems a very powerful computational model. However, a simplified version of this model may be appropriate for modelling mobile systems where the algorithm has some coarse-grained control over the mobility pattern of the agents (controlling which region the agent is in, without controlling its position within that region).

For the probabilistic variants of the population protocol model, where interactions are scheduled according to a probability distribution, each configuration of the system creates a probability distribution on the set of possible successor configurations. Thus, a population protocol can be modelled as a *Markov chain* in a straightforward way. Researchers have studied some classes of Markov chains that are similar to population protocols. For example, *Markov population processes* [66, 463] (which inspired the name of population protocols), model a collection of agents finite state agents but instead of pairwise interactions, a step in the process can be either the birth or death of an agent (in some particular state), or the spontaneous change of an agent from one state to another. The probabilities of these events can depend on the relative numbers of agents in each state, but agents in the same state are treated as indistinguishable, as in population protocols. Population processes have been used to model biological populations, rumour spreading and problems in queueing theory.

5.11 Summary and Outlook

Population protocol models are a fairly recent development. Some of the most basic questions about them have been answered, but there also remain a great number of open questions. There are many ways in which the population protocol model could be further extended to open new avenues of research.

So far, work on random interactions has focussed on the uniform model, where all pairs of agents are equally likely to interact at each step. This may not be a very realistic probability distribution for many systems. One way to make it more realistic (without making it impossibly difficult to analyze) might be to look at a uniform distribution within a system with a very regular interaction graph (instead of a complete graph). For example, could the simulation of a linear-space Turing machine for bounded-degree interaction graphs (described in Sect. 5.5) be made efficient in this case? Modelling the probabilistic movement of agents explicitly would also be of great interest, but would probably require substantial technical machinery.

Existing characterizations of what can be done focus on problems that require agreement (i.e., all agents stabilizing to the same output). Many problems that are of interest in distributed computing do not have this property. Agents may need to produce different outputs (as in leader election) or the output may have to be distributed across the entire system (just as the input is distributed). For example, Angluin et al. [24] describe a simple algorithm for dividing an integer input by a constant. In this case the result cannot be represented in a single agent; instead, the number of agents that output 1 stabilizes to the answer. Other examples of problems that cannot be captured by function computation are the spanning tree construction and the node colouring algorithm described in Sect. 5.5. The problem of characterizing exactly which problems of this more general type can be solved is open. Coping with failures in those kinds of computations is also not well-understood.

The model of population protocols was intentionally designed to abstract away many genuine issues in mobile systems, to obtain a model that could be theoretically analysed. Once this model is well understood, it would be desirable to begin augmenting the model to handle some of those issues. The restriction to a constant amount of memory space per agent may be overly strict: even though the model is intended to describe extremely weak agents where one should be as parsimonious as possible with memory requirements, agents with slightly larger memory capacities could also be considered. The design of algorithms that would respect other real-world constraints on such agents is also an interesting topic: for example, how can the algorithms minimize the number of interactions required in order to preserve power to extend the lifetime of the batteries used by the agents. See also Sect. 5.8.2 for a discussion of the effects of assuming heterogeneity, a common feature of practical mobile systems.

Many known population protocols require strong assumptions about the initial configuration; for example, the register machine simulation of [28] requires an initial designated leader agent, and will generally not recover from erroneous configurations (even those reachable with low probability) in which the phase clock is corrupted. It is an interesting question of whether such protocols can be made more robust, or whether the price of high performance is vulnerability to breakdown.

Acknowledgments

James Aspnes was supported in part by NSF grant CNS-0435201. Eric Ruppert was supported in part by the Natural Sciences and Engineering Research Council of Canada. A preliminary version of this survey appeared in [32].

Chapter 6
Routing Issues in Opportunistic Networks

Marco Conti, *IIT-CNR, Italy*
Jon Crowcroft, *Univ. of Cambridge, Computer Lab, UK*
Silvia Giordano, *SUPSI-DTI, Switzerland*
Pan Hui, *Univ. of Cambridge, Computer Lab, UK*
Hoang Anh Nguyen, *SUPSI-DTI, Switzerland*
Andrea Passarella, *IIT-CNR, Italy*

6.1 Introduction

The opportunistic networking idea stems from the critical review of the research field on Mobile Ad hoc Networks (MANET). After more than ten years of research in the MANET field, this promising technology still has not massively entered the mass market. One of the main reasons of this is nowadays seen in the lack of a *practical* approach to the design of infrastructure-less multi-hop ad hoc networks [186, 185]. One of the main approaches of conventional MANET research is to design protocols that mask the features of mobile networks via the routing (and transport) layer, so as to expose to higher layers an Internet-like network abstraction. Wireless networks' peculiarities, such as mobility of users, disconnection of nodes, network partitions, links' instability, are seen—as in the legacy Internet—as exceptions. This often results in the design of MANET network stacks that are significantly complex and unstable [107].

Opportunistic networking constitutes a medium-term application of general-purpose MANET for providing connectivity opportunities to pervasive devices when no direct access to the Internet is available. Pervasive devices, equipped with different wireless networking technologies, are frequently out of range from a network but are in the range of other networked devices, and sometime cross areas where some type of connectivity is available (e.g. Wi-Fi hotspots). Thus, they can opportunistically exploit their mobility and contacts for data delivery [185]. Opportunistic networks [657] thus aim at building networks out of mobile devices carried by people, possibly without relying on any pre-existing infrastructure. Moreover, opportunistic networks look at mobility, disconnections, partitions, etc. as *features* of the networks rather than exceptions. Actually, mobility is exploited as a way to bridge disconnected "clouds" of nodes and enable communication, rather than a drawback to be dealt with. More specifically, in opportunistic networking no assumption is made on the existence of a complete path between two nodes wishing to communicate. Source and destination nodes might never be connected to the same network, at the same time. Nevertheless, opportunistic networking techniques allow such nodes to exchange messages. By exploiting the *store-carry-and-forward* paradigm [273], intermediate nodes (between source and destination) store messages when no forwarding opportunity towards the final destination exists, and exploit any future contact opportunity with other mobile devices to bring the messages closer and closer to the destination. This approach to build self-organising infrastructure-less wireless networks turns out to be much more practical than the conventional MANET paradigm. Indeed, despite the fact that opportunistic network research is still in its early stages, the opportunistic networking concept is nowadays exploited in a number of concrete applications (see [657]).

Until now, most of the research efforts in the opportunistic networking area focus on routing & forwarding issues,[1] due to the inherent complexity of the problem [902, 657]. Therefore, in Sect. 6.2 we provide a brief survey of the main routing approaches available in the literature. Specifically, we categorise protocols based on the amount of *context information* they exploit, by identifying three main classes, i.e., context-oblivious, partially context-aware and fully context-aware protocols. The main idea behind using context information is to enable routing protocols to learn the network state, autonomically adapt to its dynamic evolution, and thus optimise their operations.

In the next part of the chapter (Sect. 6.3) we present the main routing approaches falling in the partially and fully context-aware classes, investigated within the European Haggle Project (http://www.haggleproject.org). Haggle is a 4-year project funded under the FET-SAC Proactive Initiative by the European Commission, which studies innovative networking paradigms and architectures for opportunistic networks. We firstly provide a general presentation of the routing approaches investigated in the project, highlighting the differences and complementarity among the protocols. Then, we describe three protocols in details, namely Bubble Rap, Hi-BOp and Propicman. These protocols are evaluated via simulation in comparison with reference alternative solutions available in the literature (mainly, Epidemic and PROPHET, which are described in Sect. 6.2). Finally, conclusions are drawn in Sect. 6.4.

6.2 Routing in Opportunistic Networks

Routing in opportunistic networks is surely one of the most compelling challenge. The design of efficient routing strategies for opportunistic networks is generally a complicated task due to the absence of knowledge about the topological evolution of the network. Routing performance improves when more knowledge about the expected topology of the network can be exploited [411]. Unfortunately, this kind of knowledge is not easily available, and a trade-off must be met between performance and knowledge requirement. A key piece of knowledge to design efficient

---

[1] As will be clear in the following, in opportunistic networks the routing and forwarding tasks are strictly intertwined and usually performed at the same time. Therefore, hereafter we use the terms routing and forwarding interchangeably.

routing protocols is information about the *context* in which the users communicate. Context information, such as the users' working address and institution, the probability of meeting with other users or visiting particular places, can be exploited to identify suitable forwarders based on context information about the destination. In the following of this section we classify the main routing approaches proposed in the literature based on the amount of knowledge about the context of users they exploit. We specifically identify three classes, corresponding to *context-oblivious, partially context-aware,* and *fully context-aware* protocols. Approaches in the context-oblivious class neglect the context information, and thus work in the very same way independently of the context. Oblivious approaches do not assume any knowledge about the environment or the user behaviour, and look at optimisations of dissemination schemes such as Epidemic routing. The other approaches assume that some knowledge about the environment and/or the users' behaviour can be learnt by nodes themselves through autonomic features, and exploit this knowledge to drive the forwarding process. The ones in the partially context-aware class use only part of the context information, thus for some aspects can benefit from the knowledge of the context information, but for several other aspects they behave like the approaches of the context-oblivious class. The approaches in the fully context-aware class exploit the context information as much as possible.

6.2.1 Context-Oblivious Routing

Routing techniques in this class basically exploit some form of flooding. The heuristic behind this policy is that, when there is no knowledge of a possible path towards the destination nor of an appropriate next-hop node, a message should be disseminated as widely as possible. Protocols in this class might be the only solution when no context information is available. Clearly, they generate a high overhead (as we also highlight in the performance evaluation section), may suffer high contention and potentially lead to network congestion [425]. To limit this overhead, the common technique is to control flooding by either limiting the number of copies allowed to exist in the network, or by limiting the maximum number of hops a message can travel. In the latter case, when no relaying is further allowed, a node can only send directly to the destination when (in case) it is met.

The most representative protocol of this type is Epidemic Routing (Epidemic for short) [833]. Whenever two nodes come into communication range they exchange summary vectors that contain a compact unambiguous representation of the messages currently stored in the local buffers. Then, each node requests from the other the messages it is currently missing. The dissemination process is somehow bounded because each message is assigned a hop count limit giving the maximum number of hops it is allowed to traverse till the destination. When the hop count limit is set to one, the message can only be sent directly to the destination node.

Dissemination-based algorithms also include network-coding-based routing [865], which takes an original approach to limit message flooding. Just to give a classical example, let A, B, and C, be the only three nodes of a string network, such as any message travelling between A and C has to be relayed by B. Let node A generate message a addressed to node C, and node C generate the message c addressed to node A. In a conventional forwarding scheme node B has to relay message a to C and message c to A. In network coding, node B broadcasts a single packet containing $a \oplus c$. Once received $a \oplus c$, both nodes A and C can decode the messages. In general, network coding-based routing outperforms flooding, as it is able to deliver the same amount of information with fewer messages injected into the network. A more extended survey about network coding techniques can be found in [658].

An alternative, drastic way of reducing the overhead of Epidemic without relying on network coding is implemented by Spray&Wait [769]. Message delivery is subdivided in two phases: the spray phase and the wait phase. During the spray phase multiple copies of the same message are spread over the network both by the source node and those nodes that have first received the message from the source node itself. This phase ends when a given number of copies, say L, have been disseminated in the network. Then, in the wait phase each node holding a copy of the message (i.e., each relay node) stores its copy and eventually delivers it to the destination when (in case) it comes within reach. The analytical model derived in [769] shows that L can be chosen based on a target average delay. The spray phase may be performed in many ways. Under the assumption that nodes movements are i.i.d., the *Binary* Spray and Wait policy is the best one in terms of delay. Any node (including the sender) holding n copies ($n > 1$) of the message hands over $\lfloor \frac{n}{2} \rfloor$ copies to the first encountered node, and keeps the remaining copies for itself. When a node is left with only one copy of the message, it switches to direct transmission and only transmits the message to the final destination node when (if) it is met.

6.2.2 Partially Context-Aware Routing

Partially context-aware protocols exploit some particular piece of context information to optimise the forwarding task. The main difference with fully context-aware protocols is the fact that the latter usually provide a full-fledged set of algorithms to gather and manage *any* type of context information, while the former are customised for a specific type of context information.

Probabilistic ROuting Protocol using History of Encounters and Transitivity (PROPHET [521]) is one of the most popular examples of protocols falling in this class. PROPHET is an evolution of Epidemic that introduces the concept of delivery predictability. The delivery predictability is the probability for a node to encounter a certain destination. The delivery predictability for a destination increases when the node meets the destination, and decreases (according to an ageing function) between meetings. A transitivity law is also included in the algorithm, such that if node A frequently meets node B, and node B frequently meets node C, then nodes A and C have high delivery predictability to each other. The PROPHET forwarding algorithm is similar to Epidemic except that, during a contact, nodes also exchange their delivery predictability to destinations of messages they store in their buffers, and messages are requested only if the delivery predictability of the requesting node is higher than that of the node currently storing the message.

The context information used by PROPHET is the frequency of meetings between nodes. The same type of context information is also used by MV [126] and MaxProp [125], which, in addition, also exploit information about the frequency of visits to specific physical places. Other protocols use the time lag from the last meeting with a destination to estimate the probability of delivering the messages. The bottomline idea (thoroughly investigated in [334]) is that the decreasing gradient of the time lag identifies a suitable path towards the destination. Examples of protocols exploiting this piece of context information are Last Encounter Routing [334] and Spray&Focus [769].

In MobySpace Routing [504] the mobility pattern of nodes is the context information used for routing. The protocol builds up a high dimensional Euclidean space, named MobySpace, where each axis represents a possible contact between a couple of nodes and the distance along an axis measures the probability of that contact to occur. Two nodes that have similar sets of contacts, and that experience those contacts with similar frequencies, are close in the MobySpace. The best forwarding node for a message is the node that is as close as possible to the destination node in this space. Obviously, in the virtual contact space just described, the knowledge of all the axes of the space also requires the knowledge of all the nodes that are circulating in the space. This full knowledge, however, might not be required for successful routing.

The final example we mention is Bubble Rap [389, 392], in which the context information is the social community users belong to. Bubble Rap is described in details in Sect. 6.3.2.

6.2.3 Fully Context-Aware Routing

Fully context-aware protocols not only exploit context information to optimise routing, but also provide general mechanisms to handle and use context information. The advantage of this approach is to be much more general than the approaches mentioned in Sect. 6.2.2. Indeed, these routing protocols can be used with *any* set of context information, thus allowing the system to be customised to the particular environment it has to operate in. To the best of our knowledge, three protocols only fall in this category, i.e., Context-Aware Routing (CAR [605]), History Based Opportunistic Routing (HiBOp [101, 103]) and Probabilistic Routing Protocol for Intermittently Connected Mobile Ad hoc Network (Propicman [626]).

CAR assumes an underlying MANET routing protocol that connects together nodes in the same MANET cloud. To reach nodes outside the cloud, a sender looks for the node in its current cloud with the highest probability of delivering the message successfully to the destination. This node temporarily stores the message, waiting either to get in touch with the destination itself, of to enter a cloud with other nodes with higher probability of meeting the destination. Therefore, nodes in CAR compute delivery probabilities proactively, and disseminate them in their ad hoc cloud. Note that context information is exploited to evaluate probabilities just for those destinations each node is aware of (i.e., that happen to have been co-located in the same cloud at some time). The main focus of CAR is on defining algorithms to combine context information (which is assumed available in some way) to compute delivery probabilities. Specifically, a multi-attribute utility-based framework is defined to this end. The framework is general enough to accommodate for different types of context information. As an example, in [605] authors use residual battery life, the rate of connectivity change and the probability of meeting between nodes as context information.

With respect to CAR, HiBOp and Propicman are more general, as they does not necessarily require an underlying MANET routing protocol, and are able to exploit context information also for those nodes that have never been within the same cloud. Furthermore, the definition and management of context information is not addressed in CAR, while it is a core part of HiBOp and Propicman. Finally, and most importantly, CAR does not capture, in the context definition, any information about the users social behavior, which results in [101, 103] and [626] demonstrate being a particularly valuable piece of information to design an efficient routing scheme.

HiBOp and Propicman are described in details in Sects. 6.3.3 and 6.3.4, respectively.

6.3 Haggle and its "Informed" Approach to Routing

The FET-SAC European Haggle project is currently investigating a number of routing schemes for opportunistic networks. They fall into two of the classes presented above: the *oblivious* and *non-oblivious* approaches. In the following of this section we present the main non-oblivious protocols investigated in Haggle, as oblivious protocols are more suitable for data dissemination than for messaging applications,[2] which is the target application of our evaluation part.

6.3.1 General Presentation of the Non-oblivious Approaches

The motivation behind the three approaches (Bubble Rap, HiBOp, Propicman) is similar. Naive protocols (such as Epidemic Routing) can easily saturate network and nodes' resources pretty fast. This not only reduces the network efficiency, but also has detrimental effects on the delivery performance (e.g., it has been shown that taking into consideration realistic CSMA MAC protocols, approaches such as Epi-

---

[2] We do not discuss here opportunistic network applications, as this chapter is mainly centred on the networking and routing aspects of opportunistic networks. The interested readers can find more information about the application aspects on the Haggle project web site: www.haggleproject.org.

demic achieve extremely high delay and message loss, simply because they generate too much traffic in the network [767]).

A mobile network has a dual nature: it is both a physical network and at the same time it is also a social network. A node in the network is a mobile device, and is also associated with a mobile human. Many MANET and some Delay Tolerant Network (DTN) routing algorithms [432, 521] provide forwarding by building and updating routing tables whenever mobility occurs. We believe this approach is not cost effective for an opportunistic network, since mobility is often unpredictable, and topology changes can be rapid. Rather than exchange much control traffic to create unreliable routing structures, we prefer to search for some characteristics of the network which are less volatile than mobility. An opportunistic network is formed by people. Those people's social relationships may vary much more slowly than the topology, and therefore can be used for better forwarding decisions. The main idea of context-aware forwarding is looking for nodes that show increasing *match* with known context attributes of the destination. High match means high similarity between node's and destination's contexts and, therefore, high probability for the node to bring the message in the destination's community (possibly, to the destination). We also consider that people are not likely to move around randomly. Rather, they move in a predictable fashion based on repeating behavioral patterns at different timescales (day, week, and month). If a node has visited a place several times before, it is likely that it will visit this location again in the future. Therefore, Bubble Rap, HiBOp, and Propicman exploit additional information about the context users are embedded in to make the forwarding function more efficient.

Bubble Rap assumes that relationships among users follow a precise model of social structure. Specifically, they assume that nodes are clustered in cliques (representing social groups), and that nodes social connectivity degrees (within each clique) are highly non homogeneous. In other words, the number of social links each node has towards other nodes in its clique is highly variable, and is distributed according to power laws (which have been observed in real social networks). Cliques can communicate thanks to shared members (i.e., users being part of different social groups).

The main idea behind Bubble Rap is automatically inferring the parameters of the underlying social structure, and exploiting the structure properties to select paths. To this end, Bubble Rap dynamically identifies users' communities, and ranks the nodes "sociability" (measured as the number of links) inside each community. These building blocks allow Bubble Rap to derive a model of the users' social structure, which is then exploited to forward data. Messages are pushed up in the starting community towards higher-rank nodes (i.e., more sociable users), until a contact with the destination's community is found. Pushing messages up in the senders' community ranking stores messages in more popular nodes, that have more chances to get in touch with the destination community.

Among the three protocols, Bubble Rap is the only one that assumes and explicitly derives a model of the users' social network structure. HiBOp and Propicman infer, as a side effect, social relationships between nodes from context information dynamically gathered at each node. However, they do not include an explicit representation of the underlying social network.

Propicman stores at each node a profile of the node's user. Profiles are exploited upon contact opportunities to forward messages. The main idea is to look for increasing matches between the profile of the destination user, and the profile of encountered users. When a user showing a higher match is encountered, the message(s) addressed to the destination are handed over to that node. A distinctive features of the Propicman approach with respect to Bubble Rap and HiBOp is exploiting decision trees to select next hops. Propicman idea is thus looking for users with increasing commonality (based on users' profiles) with the destination. The HiBOp approach is partly similar. However, with respect to Propicman, HiBOp manages context information quite differently. HiBOp distinguishes between different contexts (the context of the node, the context of its neighbourhood, the historical context). To select next hops, HiBOp looks (as Propicman does) at context information related to the user of the encountered node, but also at historical context information related to that user. In other words, HiBOp is able to understand if a node is a good forwarded based also on the similarity between the destination, and the context (the set of nodes) the encountered user is typically in touch with. This allows HiBOp not only to forward through users similar to the destination, but also through users often in touch with users similar to the destination. Thanks to this feature, HiBOp infers social relationships between users and users' communities, and implicitly learns the network social structure defined by the users' habits. Finally, HiBOp includes mechanisms to control the messages' replication rate, by dynamically selecting the number of copies in the network.

With respect to Bubble Rap, HiBOp and Propicman do not assume a precisely defined model of the social relationships among users. The underlying structure is automatically learnt based on the historical context information gathered by nodes (e.g., cliques—if present—are implicitly derived in HiBOp and Propicman because clique members show in their historical context the clique's characteristic attributes). While the Bubble Rap approach is more efficient when the model of the social relationships match the real social structure, HiBOp and Propicman are more flexible, as they are able to learn generic structures of social relationships between users.

Finally, as far as security and privacy issues, Propicman includes self-defined security/privacy mechanisms, while both Bubble Rap and HiBOp rely on the solutions developed within other parts of the Haggle Project's activities.

6.3.2 Bubble Rap

In Bubble Rap forwarding, we focus on two specific aspects of society: community [619] and centrality [292]. *Community* is an important attribute of opportunistic networks. Cooperation binds, but also divides human society into communities. Human society is structured. Within a community, some people are more popular, and interact with more people than others (i.e. have high *centrality*); we call them hubs. Popularity ranking is one aspect of the population. For an ecological community, the idea of correlated interaction means that an organism of a given type is more likely to interact with another organism of the same type than with a randomly chosen member of the population [638]. This correlated interaction concept also applies to human, so we can exploit this kind of community information to select forwarding paths.

Introduction to Algorithm

Bubble Rap combines the knowledge of community structure with the knowledge of node centrality to make forwarding decisions. There are two intuitions behind this algorithm. Firstly, people have varying roles and popularity in society, and these should be true also in the network—the first part of the forwarding strategy is to forward messages to nodes which are more popular than the current node. Secondly, people form communities in their social lives, and this should also be observed in the network layer—hence the second part of the forwarding strategy is to identify the members of destination communities, and to use them as relays. Together, we call this Bubble Rap forwarding.

For this algorithm, we make two assumptions:

- Each node belongs to at least one community. Here we allow single node communities to exist.
- Each node has a global ranking (i.e. global centrality) across the whole system, and also a local ranking within its local community. It may also belong to multiple communities and hence may have multiple local rankings.

Forwarding is carried out as follows. If a node has a message destined for another node, this node first *bubbles* the message up the hierarchical ranking tree using the global ranking, until it reaches a node which is in the same community as the destination node. Then the local ranking system is used instead of the global ranking, and the message continues to bubble up through the local ranking tree until the destination is reached or the message expires. This method does not require every node to know the ranking of all other nodes in the system, but just to be able to compare ranking with the node encountered, and to push the message using a greedy approach. In order to reduce cost, we also require that whenever a message is delivered to the community, the original carrier can delete this message from its buffer to prevent further dissemination. This assumes that the community member can deliver this message. We call this algorithm Bubble Rap, using the metaphor of *bubble* for a community.

The forwarding process fits our intuition and is taken from real life experiences. First you try to forward the message via surrounding people more popular than you, and then you bubble it up to well-known popular people in the wider-community, such as a postman. When the postman meets a member of the destination community, the message will be passed to that community. The first community member

---

Algorithm 6.1: Bubble Rap

---

begin
    foreach *EncounteredNode_i* **do**
        if *(LabelOf(currentNode) == LabelOf(destination))* **then**
            if *(LabelOf(EncounteredNode_i) == LabelOf(destination))* **and**
            (LocalRankOf(EncounteredNode_i) > LocalRankOf(currentNode)) **then**
                EncounteredNode_i.addMessageToBuffer(*message*)
            end
        end
        else
            if *(LabelOf(EncounteredNode_i) == LabelOf(destination))* **or**
            (GlobalRankOf(EncounteredNode_i) > GlobalRankOf(currentNode)) **then**
                EncounteredNode_i.addMessageToBuffer(*message*)
            end
        end
    end
end

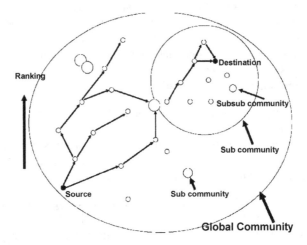

Fig. 6.1 Illustration of the Bubble algorithm

who receives the message will try to identify more popular members within the community, and bubble the message up again within the local hierarchy, until the message reaches a very popular member, or the destination itself, or the message expires. Figure 6.1 illustrates the Bubble Rap algorithm and Algorithm 6.1 summarise the operations in a flat community (not hierarchical) space.

Evaluation of Algorithm

To evaluate the Bubble Rap algorithm, we use 4 experimental data sets gathered by us for a period of 2 years in Hong Kong, Cambridge UK, Infocom05 conference and Infocom06 Conference referred to as *Hong Kong, Cambridge, Info-*

com05, *Infocom06* respectively, and one other dataset from the MIT Reality Mining Project [244], referred to as *Reality*. Here we show *Reality* experiment as an example, and the results about other experiments can be found in [389, 392].

To evaluate the forwarding algorithm, we extract a 3 week session during term time from the whole 9 month dataset. Emulations were run over this dataset with uniformly generated traffic. There is a total 8 groups within the whole dataset. We observed that within each individual group, the node centrality demonstrate diversity similar to the *Cambridge* case.

In order to evaluate different forwarding algorithms, we use *HaggleSim* emulator [390]. For each emulation, 1000 messages are created, uniformly sourced between all node pairs. Each emulation is repeated 20 times with different random seeds for statistical confidence. For all the emulations we have conducted for this work, we have measured the following two metrics and compute the 95th percentile using *t*-distribution.

Delivery ratio: The proportion of messages that have been delivered out of the total unique messages created.

Delivery cost: The total number of messages (include duplicates) transmitted across the air. To normalize this, we divide it by the total number of unique messages created.

In *Reality*, 100 smart phones were deployed to students and staff at MIT over a period of 9 months. These phones were running software that logged contacts with other Bluetooth enabled devices by doing Bluetooth device discovery every five minutes. It was found out that people did not switch on their Bluetooth very often, so the network is relatively very sparse.

From Fig. 6.2(a) and Fig. 6.2(b), we can see that of course flooding (FLOOD) achieves the best for delivery ratio, but the cost is 2.5 times that of multiple-hop-multiple-copy (MCP) controlled flooding scheme, and 5 times that of Bubble Rap. On the other hand, wait-for-destination (WAIT) has very low cost but it only has at most 10% delivery. Bubble Rap is very close in performance to MCP in the multiple-group case as well, and even outperforms it when the time TTL of the messages is allowed to be larger than 2 weeks. However, the cost is only 50% that of MCP.[3]

In order to further justify the significance of social based forwarding, we also compare Bubble with a benchmark 'non-oblivious' forwarding algorithm, Prophet [521]. Prophet uses the history of encounters and transitivity to calculate the probability that a node can deliver a message to a particular destination. Since it has been evaluated against other algorithms before and has the same contact-based nature as Bubble Rap (i.e. does not use location information), it is a good target to compare with Bubble Rap.

---

[3] Two weeks seems to be very long, but as we have mentioned before, the *Reality* network is very sparse. we choose it mainly because it has long experimental period and hence more reliable community structures can be inferred. The evaluations here can serve as a proof of concept of the Bubble algorithm, although the delays are large in this dataset. Delay in other datasets are much lower; for example in *Infocom05*, the median delay is 3 hours.

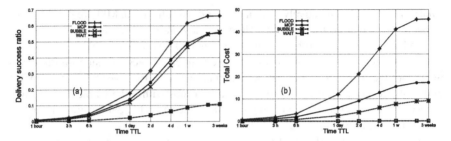

Fig. 6.2 Comparisons of several algorithms on *Reality* dataset

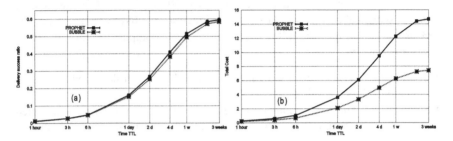

Fig. 6.3 Comparisons of Bubble Rap and PROPHET on *Reality* dataset

Prophet has four parameters. We use the default Prophet parameters as recommended in [521]. However, one parameter that should be noted is the time elapsed unit used to age the contact probabilities. The appropriate time unit used differs depending on the application and the expected delays in the network. Here, we age the contact probabilities at every new contact. In a real application, this would be a more practical approach since we do not want to continuously run a thread to monitor each node entry in the table and age them separately at different time.

Figure 6.3 (a) and (b) shows the comparison of the delivery ratio and delivery cost of Bubble Rap and Prophet. Here, for the delivery cost, we only count the number of copies created in the system for each message as we have done before for the comparison with the oblivious algorithms. We did not count the control traffic created by Prophet for exchanging routing table during each encounter, which can be huge if the system is large (PROPHET uses flat addressing for each node and its routing table contains an entry for each known node). We can see that most of the time Bubble Rap achieves a similar delivery ratio to Prophet with only half of the cost.

Considering that Bubble Rap does not need to keep and update routing table for each node pair, the improvement is significant.[4] Similar significant improvements by using Bubble Rap are observed in other datasets, demonstrating the generality of the Bubble Rap algorithm. Further details of the algorithm can also refer to [392].

---

[4] The update of the ranking has a much lower cost. As we presented in [392], it can be approximated by a exponential smoothing function of its past contact history.

Table 6.1 Identity table

Personal Information		Residence	
Name	John Doe	City	Pisa
Email	j.doe@iit.cnr.it	Street	Via Garibaldi, 2
...		...	

6.3.3 HiBOp

History Based Routing Protocol for Opportunistic Networks (HiBOp) is a fully context-aware routing protocol completely described in [101, 103]. HiBOp includes mechanisms to handle any type of context information. As a particular instance, in [101, 103] the context is assumed to be a collection of information that describes the community in which the user lives, and the history of social relationships among users. At each node, basic data used to build the context can be personal information about the user (e.g. name), about her residence (e.g. address), about her work (e.g. institution), etc. In HiBOp nodes share their own data during contacts, and thus learn the context they are immersed in. Messages are forwarded through nodes that share more and more context data with the message destination. Note that HiBOp message can contain any payload, and context information is only used to identify good *carriers* (pretty much like the destination address is used in legacy IP networks). Since users of HiBOp have possibly to share personal information, privacy issues should be considered. Privacy management in opportunistic networks is—in general—a topic still largely not addressed, and it is not the target of this section to provide complete privacy solutions for HiBOp. Some privacy solutions are given in Sect. 6.3.4. It should be noted that the set of information that is considered in [101, 103] (and that we also consider hereafter) is equivalent to personal information people advertise on their public web pages (e.g., the working institution and address) which are, therefore, not perceived as sensitive information from a privacy standpoint. Designing complete privacy solutions for HiBOp is one of the main subjects of future work.

HiBOp can be used for a number of applications of opportunistic networks in particular, and pervasive networking more in general. Beside simple messaging applications, it can be used for targeted advertisements (identifying groups of interested users through common context information, as shown in [103]), file sharing and data dissemination in general, and even emergency scenarios [123].

More in detail, HiBOp assumes that each node locally stores an Identity Table (IT), that contains personal information on the user that owns the device (an example is reported in Table 6.1). Nodes exchange ITs when getting in touch. At each node, its own IT, and the set of current neighbours' ITs, represent the *Current Context*, which provides a snapshot of the context the node is currently in.

The current context is useful in order to evaluate the *instantaneous* fitness of a node to be a forwarder. But even if a node is not a good forwarder because of its current location/neighbors, it could be a valid carrier because of its habits and past experiences. Under the assumption that humans are most of the time "predictable", it is important to collect information about the context data seen by each node in the past, and the recurrence of these data in the node's Current Context. To this end, each context attribute seen in the Current Context (i.e., each row in neighbours' ITs) is recorded in a History Table (HT), together with a Continuity Probability index, that represents the probability of encountering that attribute in the future (actually more indices are used, as described in [101, 103]).

The main idea of HiBOp forwarding is looking for nodes that show high match with known context attributes of the destination, as described in Sect. 6.3.1. Therefore, a node wishing to send a message through HiBOp specifies (any subset of) the destination's Identity Table in the message header. Any node in the path between the sender and the destination asks encountered nodes for their match with the destination attributes, and hands over the message if an encountered node shows a greater match than its own. The detailed algorithms to evaluate matches are described in [101, 103]. It is worth recalling here that matches are evaluated as delivery probabilities, and distinct probabilities are computed based on the Current Context (P_{CC}) only, and on the History (P_H) only. The final probability is evaluated via standard smoothed average, as $P = \alpha \cdot P_H + (1 - \alpha) \cdot P_{CC}, 0 \leq \alpha \leq 1$. The α parameter allows HiBOp to tune the relative importance of the Current Context and History.

In HiBOp just the source node is allowed to replicate the message, in order to tightly control the trade-off between reliability and message spread. Specifically, the source node replicates the message until the joint loss probability of nodes used for replication is below a system-defined threshold (p_l^{max}). Specifically, if $p_{(i)}$ is the delivery probability of the i-th node used for replicating the message, and k is the number of nodes used for replication, the following equation holds:

$$ k = \min \left\{ j \mid \prod_{i=0}^{j} (1 - p_{(i)}) \leq p_l^{max} \right\}. $$

Evaluation Strategy

We evaluate HiBOp in comparison with Epidemic by exploiting group-based mobility models, and specifically by using the Home-cell Community-based Mobility Model proposed in [102]. Community-based (or group) mobility models are attracting interest of researches in the opportunistic networking area, because they are suitable to realistically model the influence of social relationships between people on the user mobility patterns.

In HCMM every node belongs to a social community (group). Nodes that are in the same social community are called *friends*, while nodes in different communities are called *non-friends*. Relationships between nodes are modelled through social links (each link has an associated weight). At the system start-up all friends have a link to each other. Also two nodes that are not friends can have a link, according to the *rewiring probability* (p_r) parameter. Specifically, for each node, each link towards a friend is rewired to a non-friend with p_r probability. Initially, all nodes of the same group are placed in the same cell of a grid, which is called the "home" cell of the group.

Social links are then used to drive node movements. Nodes move in a grid, and each community is initially randomly placed in a square of the grid. Nodes' movement is made up of two component: first, a node has to select the cell towards which to move. Node selects the target cell according to the social attraction exerted by each cell on the node. Attraction between a node and a given cell is measured as the sum of the links' weights between the node and those nodes whose home is in that cell. The target cell is finally selected based on the probabilities defined by cells' attraction (i.e., if a_j is the attraction of cell j, then the probability of selecting that cell is $a_j / \sum_j a_j$). After selecting the target cell, node selects the "goal" within a cell (the precise point towards which node will be heading) according to a uniform distribution. Finally, speed is also selected accordingly to a uniform distribution within a user-specified range. HCMM also allows for collective group movements. Specifically, once every *reconfiguration period* nodes of each group select a (different) cell and move to that cell. Reconfigurations are synchronous across groups, i.e., all groups start moving to the new cell at the same time. Therefore, during reconfigurations nodes of different groups may get in touch.

In a nutshell, HCMM models the fact that humans are social (belongs to groups), move towards other people they have relationships with (most likely within their group, but also outside their group), and occasionally move collectively with their group. Furthermore, results presented in [102] show that the duration of contact and inter-contact times under HCMM are similar to those measured in real experiments, which shows that HCMM provides realistic movement traces.

Routing Sensitiveness to Social Mobility Patterns

Our aim is to understand how different humans mobility patterns impact on routing performance in opportunistic networks. We focus on Epidemic and HiBOp, to show how representative protocols belonging to opposite classes of routing schemes are sensitive to human movements' parameters.

We identify three main scenarios for our study. In the first one, we use the reconfiguration parameter to analyse the reactivity of routing protocols to sudden contacts among groups. Specifically, we focus on closed groups (i.e., $p_r = 0$), and then we force groups to collectively move with varying frequency. Messages addressed to nodes outside the group can be delivered only during contacts between different group members during collective movements. In the second scenario, we analyse the effect of social relationships between users. We want to understand how routing protocols react to different levels of users' sociability, measured as the probability of users having relationships outside their reference group. We clearly achieve this by varying the rewiring parameter (p_r). The higher p_r, the more nodes are "social", the lesser groups are closed communities. In the third scenario, we look at how protocols work in completely closed groups. In this case no rewiring nor re-

Table 6.2 Users QoS (reconf)

	reconf (s)	HiBOp	Epidemic
	2250	0 ± 0	0 ± 0
ploss (%)	9000	7.61 ± 1.49	5.56 ± 1.37
	36000	26.7 ± 1.07	25.49 ± 1.07
	2250	1158.32 ± 74.52	894.84 ± 61.02
delay (s)	9000	3525.40 ± 255.09	3172.04 ± 230.72
	36000	5732.36 ± 185.65	5562.04 ± 190.41

configurations are allowed, and we place a different group in each cell of the grid. Therefore, the only chance of delivering messages between groups is by exploiting contacts between nodes at the borders of the cells. We study the routing protocols performance as function of the nodes' transmission range. Basically, this scenario allows us to understand how protocols can exploit contacts that are not related to social relationships, but just happen because of physical co-location (e.g., contacts between people working for different companies in the same floor of a building).

We tested routing performance in terms of QoS perceived by users (average message delay and loss), and resource consumption (average buffer occupation at nodes and overall traffic overhead, also including the traffic required to exchange context information). To isolate the effect of context use, we consider an idealized configuration in which buffers and bandwidth are infinite (this setup favours Epidemic, which is more resource hungry). Finally, unless otherwise stated, our setup consists of 30 nodes evenly divided in three groups. We assume a square of size 1250 m × 1250 m, divided in a 5 × 5 grid. The default transmission range is 125 m. Unless otherwise stated 2 nodes for group generate messages, with an inter-spacing time exponentially distributed (with average 300 s). Each message is destined to a friend or to a non-friend with 50% probability. Messages are timed-out after 18000 s. Each simulation run at least for 90000 s (of simulated time). Statistics are collected by eliminating the initial transitory regime. Each setup was replicated 50 times: statistics presented hereafter are averaged over the 50 replicas, with confidence interval at 95% confidence level.

Impact of Groups' Movements. It is worth recalling that in this scenario the rewiring probability is 0, and thus, except for reconfigurations, nodes do not have chances to meet. The reconfiguration interval varies between 2250 s, 9000 s, and 36000 s. Table 6.2 shows the QoS performance as a function of the reconfiguration interval. As expected, both packet loss and delay increase with this parameter, because messages addressed outside the group of the sender are forced to wait for a reconfiguration. Note that, even though HiBOp provides higher packet loss and delay, the difference with Epidemic is quite thin. Note that, as buffers and bandwidth are not limited, Epidemic gives a reference upper bound on the performance achievable by any routing protocol. These results clearly shows that HiBOp is able to identify very good paths even during sporadic, sudden contacts during reconfigurations among nodes belonging to different groups.

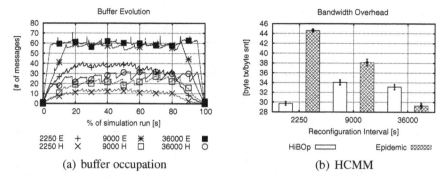

Fig. 6.4 HiBOp under reconfigurations

The good performance in terms of users QoS shown by HiBOp comes along with a drastic reduction in resource usage. Figure 6.4(a) shows the buffer occupation over time shown as a percentage of duration of a simulation run (points are average values over the replicas). HiBOp is much less greedy in spreading messages, and therefore the buffer occupation is drastically reduced. This is a general difference between Epidemic and HiBOp, which is confirmed in all scenarios we have tested. The extent of this reduction depends on the scenario, and can be as high as an order of magnitude. Finally, Fig. 6.4(b) shows the bandwidth overhead of the two protocols. It allows us to highlight a main difference between HiBOp and Epidemic, related to how they react to movement patterns. Reducing the reconfiguration interval (from 36000 s down to 2250 s) means increasing the forwarding opportunities, because nodes get in touch with more peers more frequently. Epidemic does not use these additional "connectivity resources" wisely, as it is based on flooding. Therefore, the bandwidth overhead steadily increases. HiBOp behaves radically different. When groups do not mix (reconfiguration interval equal to 36000 s) paths for messages going outside the sender's group are seldom available. In this case Epidemic uses less resources because it just floods the group. HiBOp instead periodically looks for new forwarding opportunities that are clearly unavailable. We are improving HiBOp to reduce overhead in this case. When groups mix a lot, (reconfiguration interval at 2250), nodes meet frequently, and context information is thus able to spread between groups. HiBOp immediately finds very good paths towards destinations, and does not need to replicate the messages broadly, thus resulting again in low overhead. At a reconfiguration interval of 9000 s, there is an intermediate regime in which context information about nodes outside groups is available but is not very precise, and HiBOp needs to spread messages slightly more aggressively to reach the destinations.

Impact of Users' Sociability. To understand the impact of users sociability on routing performance we vary the rewiring parameter (p_r). When a node goes to a cell different from its home it shows to nodes in the "foreign" cell context information related to its home cell, thus becoming a good next hop for messages destined to its friends. On the other hand, it roams in the foreign cell for a number of rounds and

Table 6.3 Average delay (HCMM)

	p_r	HiBOp	Epidemic
	0.03	206.66 ± 51.81	135.86 ± 18.14
delay (s)	0.1	134.58 ± 11.72	83.66 ± 7.54
	0.5	107.99 ± 7.99	75.45 ± 6.87

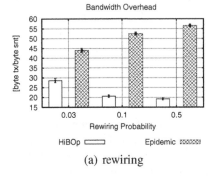

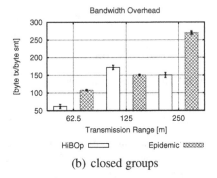

(a) rewiring   (b) closed groups

Fig. 6.5 Bandwidth overhead

collects context data about nodes in that cell. When it then comes back to the home cell, this knowledge can effectively be used for sending messages to that particular foreign cell. Indeed, that node is likely to go back to the *same* foreign cell after a while, because the social links towards nodes in that cell are still active. Clearly, when HCMM is used, the routing performance are sensitive to the users sociability, because users having social relationships with other groups are the only possible way of getting messages out of the originating group. This sensitiveness impacts differently on the resource usage of HiBOp and Epidemic, as shown by Fig. 6.5(a). Similar remarks drawn with respect to reconfiguration intervals apply also here. The higher the users sociability (high p_r), the higher the mix between nodes and the forwarding opportunities. While Epidemic naively uses all these resources spreading messages, HiBOp leverages nodes' mixing (and the resulting spread of context information) to identify good paths more and more accurately.

As far as the QoS performance figures (Table 6.3), again the packet loss is negligible, while—as expected—the average delay decreases as users become more social. However, the performance of HiBOp are still not far from the bound represented by Epidemic. It is also interesting to note that the delay of messages towards friends node tends to slightly *increase* as users become more social, because they spend (on average) more time outside their home group. However, as shown by Table 6.3, the advantage of connecting more efficiently users between groups as users become more social overwhelms the slight performance reduction experienced by friends.

Breaking Closed Groups. In this set of simulations we use a 3×3 grid with 9 groups of 5 nodes each. Just one node, located in the upper left cell sends messages,

Table 6.4 Users QoS (closed groups)

	range (m)	HiBOp	Epidemic
	62.5	65.79 ± 9.29	0 ± 0
ploss (%)	125	0 ± 0	0 ± 0
	250	0 ± 0	0 ± 0
	62.5	15579.56 ± 734.45	531.79 ± 19.14
delay (s)	125	568.08 ± 157.71	103.00 ± 2.59
	250	1.51 ± 0.64	23.35 ± 0.52

destined to a node in the lower right cell. Recall that the only way a message can reach its final destination is through edge contacts with nodes between which no social relation exists. By varying nodes' transmission range we can analyse how this edge effect impacts on forwarding. We use three values for the transmission range, i.e. 62.5 m, 125 m and 250 m. Therefore, nodes cover—on average—less than half a cell, slightly less than a cell, and one and a half cell.

The bottomline of the results is that HiBOp is not suitable for networks with no sociability. At very small transmission ranges (62.5 m) HiBOp is not able to deliver acceptable QoS (Table 6.4). HiBOp needs a minimum number of contacts between users to spread context information around. Indeed, at 125 m HiBOp restores acceptable QoS at least in terms of packet loss, and is fully effective at 250 m. Also in this case Epidemic and HiBOp behave differently with respect to the bandwidth overhead (Fig. 6.5(b)). At 62.5 m HiBOp seldom forwards messages. As context data is not circulating, nodes in the sender's group are almost all equally fit to carry the messages closer to the destination. At high transmission range the context data is circulating effectively, and therefore good paths can be identified soon. Again, note that Epidemic is not able to exploit rich connectivity scenarios without flooding the network.

6.3.4 Propicman

Introduction

Probabilistic Routing Protocol for Intermittently Connected Mobile Ad hoc Networks (Propicman [626]) also belongs to the fully context-aware routing protocols class. In Propicman, the carrier's information, called *node profile*, plays an important role in predicting the mobility of the nodes, and describe the social environment of the users and their relationships. We consider that people have repeating behavioral patterns at different timescales (day, week, and month). If a node has visited a place several times before, it is likely that it will visit this location again in the future. For example, because this person likes pasta and the place is an Italian restaurant, as described in Sect. 6.3.1.

Thus, relevant information can be deducted by the node profile. Based on the node profiles of its two-hop neighbor nodes, a sender can select, as forwarder(s), the node(s) with the highest probability of delivering the message toward the destination (*delivery probability*). The delivery probability can be considered quasi-static information, hence the mobility of a node does not really affect to the validity of Propicman selection.

Our results show that Propicman exploits effectively the mobility, as well as reduces significantly the number of nodes involved in the forwarding process. Thus the network overhead is significantly low in comparison with the other recent dissemination-oriented algorithms, such as Epidemic or Prophet as we will see the full comparison in Sect. 6.3.4.

Propicman also takes into account information privacy from design. Most of the solutions proposed to MANET routing force users to share and exchange their information during the routing process, but it is unlikely that people accept that. This is one of the main barriers to MANET acceptance. In our solution nodes can share their information in a "hidden" format, but this information can still be used for routing. Furthermore, only the destination can read and understand the message content, while this content is hidden from the intermediate nodes.

System Architecture

Goals and Design Issues of Propicman.   The goals of Propicman are to: i) efficiently distribute messages through opportunistic mobile ad hoc networks in a probabilistic fashion, ii) minimize the amount of resources consumed in delivering any single message, and iii) maximize the percentage of messages that are eventually delivered to their destination. To achieve these goals, we use the probability of nodes to meet the destination, and infer from that the delivery probability. The most innovative aspects of our approach are:

- *Routing with Zero Knowledge*: A sender does not need to have any knowledge of other nodes in the network. It bases its routing decisions only on the information it knows about the destination(s).
- *User profile*: Unlike some other protocols working with profiles, Propicman puts the different weights on the attributes of profile (herein called *evidences*). Thus user profile is a very flexible element.
- *Security within Community*: Our goal is not to develop a security mechanism that no one can bypass, but to avoid that anyone can easily get some information in the network. We aim to protect not only the message content, but also destination information inside secure communities as in Haggle project.

Propicman.   When the node S wants to send a message M to the destination D, it sends to its neighbors M's header. This header contains the information the sender knows about the destination. Based on this information, the neighbor nodes compute their delivery probability, which we assume are independents from each other. M's

Table 6.5 Common set of evidences T

Evidence Names (E)	Weights (W)
E_1	W_1
...	...
E_n	W_n

Table 6.6 A node profile (V_i can be empty)

Evidence name (E)	Value (V)	Hashed values
E_1	V_1	$H(E_1,V_1)$
...	...	...
E_n	V_n	$H(E_n,V_n)$

header is forwarded to the two-hop neighbors of S. S will send the message M only on the two-hop route(s) with the highest delivery probability, if this is higher than its own. After sending the message content, S keeps the message for the next eventual encounter. As also highlighted in research of Grossglauser on "Mobility Increases the Capacity of Ad hoc Wireless Networks" [333], two-hop routes are the most suitable choice to exploit more routing information with minimal additional costs and risks. In order to compute the delivery probability at each node, we first define a common set of evidences that is the list of evidences used in the network and the related weight (importance of this evidence in the network). Each node, will build its node profile, which is an instantiation of the common set of evidences.

Building a Node Profile. We reasonably assume that, when the sender wants to send a message, it must have some information (attributes) about the destination node. Starting from the common set of evidences **T**, each node can create its own node profile (see Table 6.5), where n is the number of evidences.

The node profile is represented by the set of evidence/value couples. For security reasons we will discuss in the next section, each evidence/value couple in the node profile is hashed by the hash function $H(E_i,V_i)$, and the results (*hashed values*) are added to the node profile as in Table 6.6.

Strategy to Select the Best Forwarder(s). Suppose that the source node S— the node holding the message and trying to choose the best route to forward the message M to the destination D—knows some information about D's profile. From this information (some evidence/value couples), S builds the message header h_M which is the concatenation of all the hashed pairs of evidences/values as follows:

$$h_M = \underset{i=1}{\overset{n}{Concatenation}}(\mathbf{H}(E_{Di},V_{Di}), MAC_S, SN) \tag{6.1}$$

It also includes the MAC address of S (MAC_S), and the sequence number of M (SN), used to avoid duplicated messages. Note that in (6.1), except for the values that S knows about D, the other values are left blank.

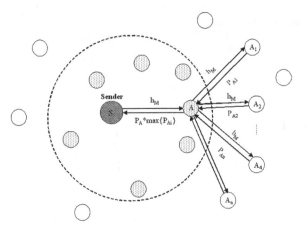

Fig. 6.6 Two-hop route probability selection

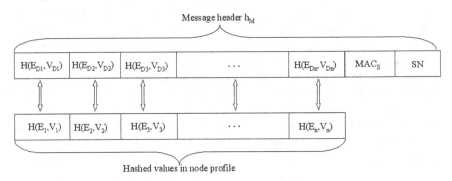

Fig. 6.7 Local matching between message header and receiver

At first, S sends the message header h_M to every neighbor node, without including the message content. Each neighbor node (for instant, A), when receiving h_M, it compares the hashed pairs of evidences/values $H(E_{Di}, V_{Di})$ in h_M with its own hashed values. The comparison is performed in the very same order of evidences in T. For each matching element, A gets from T the corresponding weight. From all the elements that matched, A computes its delivery probability to the destination as follows:

$$0 \le P_A = \frac{\sum Matched(W_i)}{\sum W_{Mi}} \le 1 \tag{6.2}$$

where $\sum Matched(W_i)$ is the sum of the weights of the evidences in A, which hashed values matched the ones in h_M, and $\sum (W_{Mi})$ is the sum of the weights of all the evidences that S knows about D (not empty in h_M).

For the next hop level (second hop—Fig. 6.6), at its turn, each neighbor of S, say A, sends h_M to every node in its neighborhood (A_j). Every neighbor node of A computes its own delivery probability as in (6.2). With all the delivery probabilities of its neighbors, A is able to compute the one with highest delivery probability, send the result back to S. Then S is able to compute the two-hop route(s) with the highest delivery probability.

$$P_{two-hop} = P_{S \to first\ hop \to second\ hop} = Max\{P_{first\ hop} * P_{second\ hop}\} \tag{6.3}$$

For every neighbor node, S calculates the maximum delivery probability of the two-hop routes as in (6.3), and chooses the two-hop route that has the highest delivery probability. Then S starts to send the message content M on the route it chosen. In this approach, all the intermediate nodes still keep the message for the next encounters. S selects only routes with delivery probability higher than or equal to the delivery probability of S. Thus, on his way to D, the message is always forwarded to node(s) with higher delivery probability. If S wants to ensure a delivery rate or when there are no constraints about the network overhead, S can choose more than one (highest probability) route, replicate and send M over those routes. This is a trade-off between data latency and network overhead. If the number of selected routes equals to 1, we have the lowest overhead, but quite long delay. This is further analyzed in Sect. 6.3.4.

Security Considerations

One of the most important issues in ad hoc networks is security. Members in ad hoc networks can be grouped in different communities. Here we assume that security measures, such as authentication, signature, etc., are implemented at community level. In our proposal, we aim to ensure message privacy and member's information within a community. When a node S wants to send a message M to another node D, it can happen that S uses some intermediate nodes. In Propicman, before sending the message content M, S sends h_M that carries some information about D (evidences), and this could be captured. However, we hide the plain text of h_M (and the message content M) to the intermediate nodes. Only the destination node D can recover and read M. The intermediate nodes are not allowed to know to whom this message is sent, and the content of the message. As mentioned in the Subsect. 6.3.4, in this work we are not aiming to build a complete security mechanism that no one can bypass, this solution is considered as a suggestion to use among members inside a community.

Hiding Destination's Information. We adopt a one-way hash function H that uses the message digest algorithm to hash the evidences of the destination node in h_M. As described in the Sect. 6.3.4, the profiles of nodes also contain the hashed values $H(E_i, V_i)$. Every time a node receives h_M, it compares all its hashed values with the hashed values in h_M. Thus, it can only "spy" some information of D in h_M if it matches its hashed values. But, it cannot know about the other non-matched pairs in h_M. And D is the only one who can get the complete information, as it is the only one who can match all the hashed values with the ones in its profile.

Encrypting/Decrypting Message Content. We also hide the message content from "unintended" nodes. Only the destination node can receive and understand (recover) the message content. As we discussed above, the information of the destination node is hidden from others by using the function H. Thus, we can make use of this fact to encrypt/decrypt the message content. The encryption scheme consists of an encryption operation and a decryption operation, where the encryption operation produces a cipher text from the message content with some keys, and the decryption operation recovers the message from the cipher text under the same keys. $F_{enc}(K, M)$ where M is the message content and K is the encryption key. We use the information (evidences/values) that the sender knows about the destination node as keys. With this technique, in a community scenario, only the destination node can decrypt the cipher message to get the plain message since it knows the keys.

$$K = \overset{n}{\underset{i=1}{Concatenation}}(E_i V_i) \qquad (6.4)$$

where n is the number of evidences in T. V_i will be empty if S does not know about E_i of D.

Simulation and Performance Evaluation

Methodology and Implementation.   We suppose that there are no battery consumption or network bandwidth constraints in the simulation. With these assumptions, we compare the routing performance of Propicman with Prophet and Dynamic Epidemic using the following two metrics:

- **Network Byte Overhead:** the ratio of the amount of control and data information (measured in bytes) that got transmitted from a node to its neighbors, over the total amount of data information (measured in bytes) that was received correctly, for the entire duration of the simulation.
- **Average End-to-End Data Delay:** the end-to-end data delay is defined as the time interval (measured in seconds) between the generation of a data packet at the source and its reception at the destination. This value is averaged over all packets correctly received at the destination.

The simulation focuses on the operation of routing protocol and does not simulate the details of the underlying layers. From investigations of daily activities in our department, the scenario is designed as a rectangular of 870×520 meters, divided into 6 areas. Each person (node) is randomly distributed across the 6 areas, with different weights for each area, depending on the person. The nodes will probably be, for most of the (daily) time, in the working room (highest probability), then in the canteen and so on. For example, node 1 will be 89% in his working room, 6% in the canteen, 3% in the rest room, 2% at the coffee machine area, and 0% in the other spaces. Note that the scenario is artificially simple. In reality, the 6 areas considered could span for a (very) large geographical area, where the number of encountered nodes is very high and epidemic routing is pretty unreliable. We did not include a specific mobility model, because, as mentioned above, the delivery probability

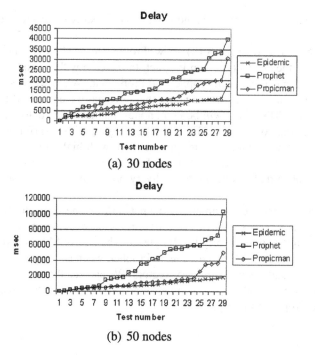

(a) 30 nodes

(b) 50 nodes

Fig. 6.8 Delay

is quasi-static information. We assume that nodes have the same cover range, and move from one area to another based on the probabilistic distribution.

Simulation Results and Analysis.   Our system relies on predictions about the future values of context evidences. However, predictions are not always reliable because of the human-based behavior exists in the prediction process. The result presented in the chapter is the one we take with the standard confidence level of 95% and the confidence interval at each test is $[X - 970, X + 970]$ (msec). We ran our probabilistic algorithm against the Prophet and Dynamic Epidemic algorithms. We evaluated the transmission delay and the network overhead of each algorithm in two different network sizes: 30 nodes and 50 nodes. We sort the results for clearly making the comparison between different approaches. As we have no bandwidth constraints, the Epidemic algorithm clearly will have the highest performance for the transmission delay. The sender in the Epidemic algorithm will "flood" all the nodes that it meets on the way of movement, and so on. This is illustrated in Fig. 6.8 (a) and Fig. 6.8 (b). Prophet uses the encounter history of nodes to predict the meeting probability between nodes. It supposes that if two nodes meet each other very often in the past, they likely meet again in the future. But in real life, this is not always correct, and needs further refinement. For this reason, we introduced the priority of information (evidences) of nodes. By putting different weights to the evidences, the selection of candidates among the neighbor nodes results more specific and accurate.

As Fig. 6.8 (a) and Fig. 6.8 (b) show, Propicman needs less time to transmit the message to the destination than Prophet. In the case with 50 nodes, we can see that there is a big gap between Prophet and Propicman. When we have more nodes (higher density), it is easier to find nodes that have more identical evidences as the destination node. Thus there will be more nodes that have high delivery probability to the destination. As clear, the Epidemic routing algorithm performs very badly for network overhead. Furthermore, in a network with high mobility, the flooding load increases with mobility. Prophet and Propicman only forward the message to the neighbor node that has the higher delivery probability, thus are no affected by mobility.

This technique reduces significantly the network overhead. Each source node in Propicman adds $32 * n + 19$ bytes of traffic in the network where n is the number of evidences in the common profile, 19 bytes are the size of MAC address of S and the sequence number of M. Meanwhile, in Prophet, when two any nodes meet, they exchange MAC address and summary vectors. We can see that the analysis results on the network overhead are also similar to the simulation results in the both cases Prophet and Propicman [626].

6.4 Summary and Outlook

This chapter discussed the advantages of opportunistic routing and presented three context-aware approaches developed within the European Haggle Project. While presenting a different way of behaving with context-awareness, these approaches (Bubble Rap, HiBOp and Propicman) show that the use of context information in opportunistic routing can be very advantageous. In fact, this permits to exploit effectively the mobility, the social connections and the community relationships, and, for this reason, context-aware approaches can reduce significantly the number of nodes involved in, as well as speed up the forwarding process. Thus, while belonging to opportunistic routing category, they perform much better when compared to context-oblivious or partially context-aware approaches such as Epidemic or Prophet. A major assumption for effectively using context information, is that people move in a predictable fashion based on repeating behavioral patterns at different timescales (day, week, and month), or that these patterns are known. This allows us to see the mobility as expression of the users social behavior. This, is in line with the most recent studies on the social contacts, also performed within the Haggle project [154].

In order to further justify the significance of social based forwarding, we have shown comparison of Bubble Rap with a benchmark 'non-oblivious' forwarding algorithm, Prophet. Most of the time Bubble Rap achieves a similar delivery ratio to Prophet with only half of the cost.

Similarly, in order to understand the impact of humans mobility on routing performance in opportunistic networks, we have shown a comparison between HiBOp and Epidemic. It is clearly illustrated a drastic reduction in resource usage.

Furthermore, with the assumption that people have repeating behavioral patterns at different timescales (day, week, and month), we can use context personal information very efficiently for selecting as forwarder(s) the node(s) with the highest probability of delivering the message toward the destination. The Propicman simulation results show that this can be very effective, with high gain in terms of packet loss and network overhead compared to both Prophet and Epidemic.

Thus, while exploiting different aspects of context information, the three approaches presented in this chapter all illustrate that the use of social information is fundamental for opportunistic networking and that this can be very advantageous in terms of performances.

As shown in [625], this has a straightforward extension in sensor networking. In [625] authors describe how sensor networks and in particular SANET, can exploit the information of mobile devices carriers, and also combine them with MULE like approaches [746, 796] to exploit the vehicular mobility. The European Haggle Project is implementing the architecture for supporting context-aware routing, with the final goal of demonstrating the effectiveness of context-aware approaches in medical scenarios and other social scenarios where the social behavior is relevant.

Acknowledgements

This work was partially funded by the IST program of the European Commission under the HAGGLE (027918) FET-SAC project.

Chapter 7
Wireless Mesh Networks

Johnathan Ishmael, *Lancaster University, UK*
Nicholas Race, *Lancaster University, UK*

7.1 Introduction

Wireless Mesh Networks have emerged as an important technology in building next-generation networks. They are seen to have a range of benefits over traditional wired and wireless networks including low deployment costs, high scalability and resiliency to faults. Moreover, Wireless Mesh Networks (WMNs) are often described as being *autonomic* with self-* (healing and configuration) properties and their popularity has grown both as a research platform and as a commercially exploitable technology.

Initially this chapter examines the challenges faced by traditional network technologies and discusses the role of WMNs in overcoming some of these problems. Following this an overview of Wireless Mesh Networks is presented, providing comparisons with similar network technologies including *sensor networks* and *Mobile Ad Hoc Networks* (MANETs). The core of the chapter then details the state-of-the-art technologies involved in the construction of WMNs, including standards based activities, academic research, commercial products and deployment testbeds.

7.2 Wireless Mesh Networks

Wireless Mesh Networks are based on the concept of ad hoc networking, in which opportunistic networks can be quickly established without fixed infrastructure. The concepts behind ad hoc networks have existed in various forms for a number of years—including the packet radio network, the survivable adaptive radio network and global mobile information system [122]. Traditionally these networks have been used in a military context due to the lack of low-cost, high-speed wireless networking devices available to general consumers. Recently the emergence of wireless technologies such as Bluetooth and IEEE 802.11 has seen increased research and development in the ad hoc networking field.

While there are numerous publications available in the field of ad hoc networking, it is seldom used outside of a laboratory environment. Previous research efforts have been for specific applications, usually military based, involving large numbers of nodes. Recent research has changed the focus of ad hoc networking to one which involves small groups of people forming opportunistic networks to share resources (e.g. a connection to the Internet). This shift in research has seen the creation of a new type of network—the Wireless Mesh Network—which is built on a mix of fixed and mobile nodes, interconnected using the principles of ad hoc networking.

7.2.1 Why Wireless Mesh Networks?

Today the Internet is considerably more accessible than a decade ago - and for many it represents an integral part of daily life. The number of people with access to the Internet has increased significantly over the past five years, also fuelled by an increase in consumer access speeds as the underlying technology has moved from baseband to broadband services. However, this technological shift to broadband has seen many people disadvantaged with a "digital-divide" emerging between those with access to high speed broadband services and those without. Rural areas have been particularly affected, as the underlying technology for broadband is heavily dependent on the distance of the subscriber to the nearest telephone exchange. Providing broadband networks to rural areas is therefore a significant challenge, with companies reluctant to invest in new infrastructure for a relatively small number of subscribers leaving communities faced with either no access, or very limited access compared to that available in urban areas.

The growth in demand for mobile services and access to data on the move is also driving developments in technology that can provide *ubiquitous* access. Wireless networks are often seen as a technology that can deliver this capability, however the typical deployment of a wireless network will consist of a series of wireless access points (APs) that rely on access to a wired infrastructure for connectivity—resulting in the coverage of AP-based wireless network that is still largely restricted by that of the wired infrastructure.

This presents three challenges for operators trying to provide coverage for rural areas, including how to:

1. Deploy a network which can deliver high bandwidth, low latency services.
2. Provide ubiquitous connectivity.
3. Remain cost effective in sparsely populated areas.

Wireless Mesh Networks (WMNs) are seen as a potential solution to this paradigm: offering a scalable, low-cost, autonomous platform capable of delivering an access network without the need to utilise any previous infrastructure. WMNs are also highly versatile and can be applied to many scenarios in which traditional networking would be difficult—the rapid deployment and zero-configuration nature of the technology makes it particularly suited for use in hostile environments, such those faced by fire crews. The robust and distributed nature of the network also provides significant benefits for security applications, such as a distributed file store for archiving CCTV footage.

7.2.2 Background

Wireless Mesh Networking is a considerably vast topic, covering a wide range of research areas. This section provides an introduction to the field, highlighting its place within modern networking.

Local Area Networks (LANs) can be connected to other networks covering a wider geographical area, such as a Metropolitan Area Network (MAN) or Wide Area Network (WAN), via the use of a router. In the case of a WAN (such as the Internet) data may traverse several routers before getting to the required destination, with each router potentially having access to several paths to that final destination. While such connectivity can result in a vast number of networks being interconnected, it is still constrained by the underlying wired infrastructure to carry the data.

To overcome the constraint of the wired infrastructure, we turn to the use of wireless devices connected to LANs, producing Wireless Local Area Networks (WLANs). These allow users access to a LAN without the constraint of a wire. A client connects wirelessly to an access point (AP) which is in turn connected to the wired infrastructure; the AP acts as a switch to the LAN. In the presence of multiple access points, a user is (in theory)[1] able to roam between access points. An example of a modern network infrastructure can be seen in Fig. 7.1. This highlights how members of a WLAN and a LAN can intercommunicate both locally and to a WAN.

7.2.3 Mesh Network Architectures

Recent research has changed the focus of ad hoc networking to one which involves small groups of people forming opportunistic networks to share resources (e.g. a connection to the Internet). This shift in research has seen the creation of a new type of network, the Wireless Mesh Network which is built on a mix of fixed and mobile nodes, interconnected using the principles of ad hoc networking.

Wireless Mesh Networks (WMNs) re-use the concepts of routing and switching from traditional networks but apply them to a Wireless LAN environment. The wireless medium provides both the core backbone of the network (with each node capable of routing and forwarding packets on behalf of other nodes) as well as offering client (i.e. end-user) access. The network infrastructure is based around nodes that can wirelessly communicate directly with at least one other node, with data

---

[1] While the IEEE 802.11 standard provides a means to roam between access points, differing implementations amongst manufacturers can sometimes prevent this from working successfully.

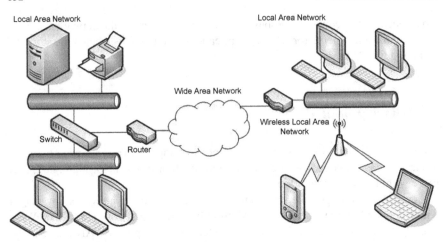

Fig. 7.1 The layout of a typical network

relayed via intermediate nodes if the destination is outside the immediate wireless coverage area. Unlike traditional (W)LAN networks, WMNs have no fixed layout or topology. The path which data takes can vary on a per-packet basis, allowing the topology of the network to change over time.

Figure 7.2 highlights a typical layout of a WMN. The mesh devices form a central infrastructure for the network, these provide backbone connectivity to clients as well as gateway and bridging solutions to other networks (including the wider Internet). Conventional clients can connect to any of these mesh devices, which will route data on behalf of the client, either directly or via a multi-hop path. The wireless mesh clients form their own ad hoc network, allowing both communication to client devices and communication amongst the wider mesh back bone.

A WMN consists of three node types: a mesh device, a mesh client and a conventional client. A mesh device functions similarly to a home wireless router, providing gateway and repeater functions. In addition to this, a mesh device usually contains multiple radio cards allowing the use of different radio channels or to access different wireless technologies. Using various radio channels is one of the methods used to increase throughput and promote redundancy. The use of different wireless technologies permits the bridging of WMN with other network types, for instance WiMAX or cellular networks. A mesh client is an end user device (e.g. a PDA or Laptop) which is capable of connecting to and utilising the network. It also has the ability to act as a router to relay data on behalf of other clients. However, it does not provide gateway functions to other networks. Finally, a conventional client represents an end user device which can connect to and utilise the network, however does not relay data for other clients.

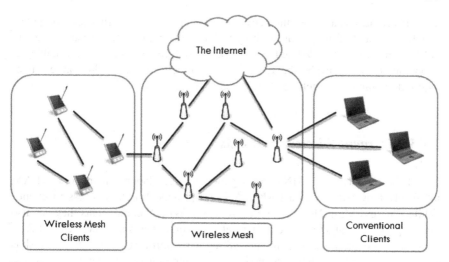

The Internet

Wireless Mesh
Clients

Wireless Mesh

Conventional
Clients

Fig. 7.2 An example of the structure of a typical Wireless Mesh Network (WMN)

7.2.4 Usage Scenarios

There are various uses for WMN technology, from small scale three node deployments to large scale, wide area networks. The establishment of flexible community networks allows sharing of local resources amongst members of a community. A community network also boasts several advantages compared to traditional broadband solutions, including using the local network to share local content (reducing the load on access networks) and distributed solutions, such as file storage and utilising multiple gateways for Internet access. An example of a community WMN can be seen in the Wray Broadband Project (Sect. 7.4.1). On a smaller scale, a home WMN can permit the interconnection between devices in the home, allowing data to stay close to the device which is communicating reducing the bottle necks involved when using a central hub. Other uses of WMNs include Metropolitan Area Networking such as that provided by Tropos Networks[2] and Pronto Networks,[3] enterprise networking and providing infrastructure for small scale ISPs.

Due to the novel nature in which WMN are deployed, they are a cost effective technology for small network operators. As the number of users on a network expands additional nodes can be added; expanding the coverage area of the network and permitting a greater number of users. They also provide the significant advantage of not requiring existing infrastructures, reducing the overhead required in creating or upgrading a network. Providing WAN access to a WMN is also a simple task, with any node able to act as a router to the wider network.

In order to reduce costs to small network operators, one of the critical functions which WMNs must support is the ability to self-adapt and self-heal. As each node on

---

[2] http://www.tropos.com/.

[3] http://www.prontonetworks.com/.

the network acts as a router, the network is susceptible to disturbances; computers which are turned off may cause routing paths on the network to vary and access to WAN routers to change. While the network is cost effective in terms of scalable deployment, it should also be able to exhibit similar behaviour in the human resources required to manage the network.

7.2.5 Divergent Wireless Networks

It could be argued that a WMN is nothing more than an elaborate name for a WLAN, ultimately they connect to a LAN/WAN and provide Wireless access to clients. While Wireless Mesh Networks do make use of WLAN components they are also made up of an amalgamation of other network components and services. WLANs are established and integrated into part of the wired infrastructure, their sole task is to deliver access to the wireless medium and interconnect it to the existing wired infrastructure. On no occasions do they exhibit self-managing behaviour; they are statically configured and operate as such. On the other hand, WMNs provide the services you would expect to see on a standard network (DHCP, DNS, etc.), the 'wireless' infrastructure as well as wireless access to the clients. They are also designed with self-managing properties in mind, able to adapt and cope with changes to the environment in which they are located.

Wireless sensor networks are typically tailored to a specific task, running on a bespoke hardware platform. The devices are usually low powered aimed and transferring specific pieces of information gathered from sensors, distributing them back to a more powerful node which stores or uploads them. MANETs on the other hand are not too dissimilar from WMNs, aimed at providing connectivity to mobile nodes in an ad hoc environment. A MANET arguably contains all the properties of a WMN, most notably the self-managing characteristics as well as the network services such as DHCP and DNS. The purpose of which is to provide connectivity amongst those on the network, potentially exploiting a shared resource (e.g. Internet connectivity). MANETs are designed to be mobile not relying on any statically deployed components, unlike WMNs. WMNs take the concepts proposed in MANETs and evolve them to form a static, semi-permanent infrastructure which takes the place of a traditional wired network. Unlike a MANET, a WMN contains both fixed and mobile nodes to construct the infrastructure and relies upon management platforms to provide aspects of user authentication, billing and configuration.

7.3 Building Wireless Mesh Networks—Research Challenges

A WMN consists of a number of devices from wireless routers to clients, in turn these devices consist of a number of components from physical hardware through to routing protocols and system management software. Due to the number of components involved in the construction of a WMN, its research is spread across a wide range of domains, from MAC layer modifications through to applications. This section details the components which constitute a WMN, highlighting notable previous work for further study and a discussion on the outstanding challenges. The section is organised based on the Open Systems Interconnection Reference Model (OSI).

7.3.1 Physical Layer

The physical layer provides the raw data transfer between wireless devices on the network, this includes modulation and encoding techniques for wireless transmission as well as directional and smart antennas. The significant factors within this layer are high throughput, localised transmission range (so as not to deafen other nodes on the network) and the compatibility of higher layer (MAC) with the properties of the physical medium.

At the physical layer, recent research has allowed the technology which drives wireless communications to provide a number of different modulation and coding rates as defined in the IEEE standards 802.11a [398], 802.11g [399] and still in discussion IEEE802.11n (draft) [397]. The standards provide the ability to modify the modes under which data is transmitted to promote link adaptation, increased data rates and error resilience especially in a multi-path environment [684, 171, 331]. In addition to the changes to the modulation schemes, research has been undertaken on multiple-antenna systems such as [597] and [95]. Multiple-antenna systems known as MIMO (multiple-in multiple-out) are another popular research topic, currently being implemented in IEEE 802.11n [397]. This permits the use of multiple antennas to both transmit and receive data using processing techniques such as Bell Labs layered Space-Time (V-BLAST) [318] to reduce multi-path fading effects and increase data rates. A further research topic in the physical layer is that of directional antennas [766], this reduces noise for other nodes on a network by minimising the transmission area to a narrow band.

7.3.2 MAC Layer

The basic function of the Medium Access Control (MAC) layer is to manage communications between wireless devices over a shared communications channel. This component and its interaction plays a vital part in WMNs due to the multi-hop nature of the network and the large number of nodes which usually participate.

In many cases ad hoc and WMNs present poor performance to the end user, especially as the number of nodes in the network increases. This scalability issue is a result of incompatibilities between the shared wireless medium, the MAC and higher level protocols. There are currently three approaches to solve this issue; Improving existing MAC protocols such as the work produced by Cali [129] and

Qiao [684], improving cross-layer designs to permit the MAC layer to interact with lower level hardware requirements and finally new MAC protocols such as that proposed in [457]. In addition to these three approaches, there exists a multi-channel MAC scheme which permits the use of multiple radio cards on varying channels to transmit and receive data which promotes higher throughput and interference avoidance [13, 5].

7.3.3 Network Layer

The role of the network layer is to provide the means of transferring data from a source to a destination, including functionality such as routing, fragmentation and reassembly, reporting delivery errors and acting upon quality of service requests from the Transport layer. The research within this field is highly active, recent years have seen published articles on over a hundred new and modified routing protocols, each with their own reason for conception. The reason for such activity within this field is down to three reasons. Firstly, there are still avenues to peruse with regards to the metric used to choose a route. Secondly, current implementations lack cross-layer communication between the MAC and network layer, permitting poor routing decisions to be made without input from other sources. Finally many of the routing protocols which are published are designed to carry out a specific task in a fixed scenario or environment. The construction of a successful routing protocol requires it to consider a wide range of scenarios, metrics and other features:

Scenario   Numerous routing protocols for mesh networks are designed for specific application scenarios, e.g. emergency services, community WMNs, long distance WMNs, and home based WMNs. In the case of a new scenario being conceived, either a new routing protocol is developed, or an ill-fitting one utilised. Any new routing protocol should have the ability to adapt to operate under several operating scenarios and environments.

Hybrid Performance metrics   The choice of performance metric used to select a route is the main focus of a routing protocol. The majority of routing protocols utilise minimum hop count to select the path. While functional it does not perform well under most conditions as it does not take into account factors such as the quality of the wireless link (e.g. loss rates, signal strength). The result is that sometimes two or three good quality links can outperform a one hop low quality link. An ideal routing protocol would consider multiple metrics including round trip time, link quality and load.

Fault tolerance   A key reason for using a WMN is that of robustness and ability to adapt to failure. This is paramount both in safety critical environments as well as Community WMNs where mesh devices are located in "exposed" environments. Should a communication path fail between two points the network should detect and adapt to this. In addition to dealing with failures, it should also consider transient and temporary failures.

Aside from the traditional ad hoc routing protocols such as DSR [429] and AODV [664] there are a number of new routing protocols designed specifically for WMNs.

Mesh devices with multiple radio cards permit an increase in bandwidth on WMNs, as well as a possible reduction in interference. The routing protocol by Draves [237] presents one such protocol, in which a new hybrid metric the weighted cumulative expected transmission time (WCETT) is proposed. This protocol uses two measurable values, link quality and minimum hop count, but measures them across multiple radios to provide a good throughput to latency trade-off. One of the aspects implemented in the WCETT metric is the expected transmission count ETX as presented in [217]. ETX aims to utilise the minimum hop count method while ensuring maximum throughput is achieved.

Multi-path routing provides the ability to reduce the impacts of faults as well as promote load balancing on the network. The concept is to select multiple paths between a source-destination pair and use all of these paths when sending packets, as discussed in [592]. Routing decisions within WMNs may make use of clustering, most commonly seen within overlay networks such as Peer to Peer (P2P) networks. Initially when a node connects to the network it is placed at the edge—as its availability (duration of time connected to the network) increases it is moved from the edge towards the centre of the network. The position of a node within the network may also depend upon its resources (e.g. processing power and available bandwidth). The result is a network in which the most resource rich devices are placed in the centre of the network, with the backbone to the network changing infrequently, while nodes connecting to central nodes are free to migrate and be mobile. An example this is presented in [75] and [873].

7.3.4 Transport Layer

Providing reliable data transfer services to higher layers of the OSI stack is the role of the transport layer. Although not strictly adhering to the OSI stack, the closest transport layer examples are Transmission Control Protocol (TCP) and User Datagram Protocol (UDP). These mechanisms are in place to counter the effects of less reliable network infrastructures. Many of the transport layer protocols which are in existence where devised without the concept of packet loss through anything other than congestion. A significant proportion of the research carried on the transport layer reflects this, specifically with the interaction between TCP and the unreliable wireless medium. The research in this area can be defined into two fields, variants on TCP, protocols for real time delivery and new protocols design specifically for mesh networks.

TCP Variants. TCP is unable to differentiate between losses caused by congestion, losses caused by failed transmissions or delays in getting access to the wireless medium. In any one of the three cases, TCP invokes congestion control mechanisms resulting in a drop in network performance. Furthermore once the transmission medium has recovered TCP is slow in resuming its previous transmission speed [876].

The most common solution to this problem is that of using a feedback mechanism [161], which provides information on the state of the channel to the sender. If a channel becomes unavailable it halts its transmission of packets as well as freezing any timers it has running. Once the link between the sender and receiver has recovered it is then able to restart transmission of packets as well as resuming any halted timers. While this solution is successful for static mesh devices, in nodes which are mobile two new problems arise. The first is that permanent link failure may occur (due to the mobile nature of the host) as discussed by in [374]. In such situations if a mobile host suffers from a failed link, TCP will assume the link is congested and reduce its re-transmission timeout value. This results in reduced performance should the link be re-established over a different path. The second problem which occurs due to mobile hosts is due to the asymmetric nature of many mesh networks. The return path for a TCP ACK may be vastly different to the sending path as discussed in [47]. As a result a TCP data or ACK packet my experience totally different latency and bandwidth issues, resulting in poor TCP performance [667]. At present this author is unaware of any attempt to fix this problem.

Providing Real-Time Data Delivery. Real-Time Transport Protocol (RTP) provides a standard packet format for delivering audio and video over the Internet and commonly utilises UDP. However, UDP cannot guarantee real time delivery due to the lack of any timing metrics. Furthermore UDP may starve TCP connections in the same network, due to a lack of congestion or rate control. While RTP has been proposed for wired networks there have been a few Real-Time data delivery protocols suggested for wireless networks and even less for ad hoc and mesh networks. Any proposed Real-Time data delivery protocol has to consider transmission failure aside from network congestion, similar to the behaviour of TCP. A selection of algorithms is discussed in [149].

7.3.5 Application Layer

The application layer is the point of interaction between a user process (program) and the network. There are already a significant number of applications which operate under Wireless Mesh Networks, some of which utilise the benefits of mesh networking. These applications can be grouped into three areas: The first is that of general Internet provision, in which mesh networks are used as an alternative to other traditional last-mile networks. Typical applications include online shopping, game playing and accessing multimedia content. At present WMNs do not provide any substantial benefits to traditional networks in urban areas. In rural locations however, they can provide an infrastructure to deliver an Internet connection, where traditional infrastructures are unable to reach. The second application area is that of distributed computing (processing and storage), this is similar to the peer to peer paradigm seen in today's Internet. A WMN provides a resource to permit data to be stored and transferred in distributed fashion, this could be for a number of applications e.g. backup or distributed processing. The self-healing and repairing nature of the mesh makes it idea for this type of application. The final application type relates to the convergence of multiple types of wireless devices, permitting intercommunication amongst them. An example of this is the interconnection of 3G and WLAN networks to create a 4G network. Any future research into new applications needs to harness the unique advantages of WMN.

7.3.6 Standards Activities

At the time of writing, the IEEE 802.11 working groups which are working towards improving the data-rates within wireless products (IEEE 802.11n), there are also several working groups within IEEE 802 who are working towards standardising WMNs.

IEEE 802.11s is working towards defining MAC and PHY layers for mesh networks which will promote a scalable and redundant architecture with automatic topology learning and dynamic path configuration. Typically wireless base stations work in one of two modes, either ad hoc or infrastructure based. In ad hoc mode, each base station uses the wireless infrastructure to relay data behalf of other nodes. Whereas with infrastructure based networks, inter-node communication is carried via a wired network connection, reducing the bandwidth load on the wireless infrastructure. IEEE 802.11s aims to create a hybrid mode, by extending the MAC and PHY layers, to permit the use of an infrastructure using both a wired and wireless medium.

IEEE 802.15 aims to provide mesh networks within a personal area network (PAN) scale, providing mesh networks between devices such as mobile phones and portable audio devices. This standards group is broken down into several smaller working groups, aimed at various application scenarios. IEEE 802.15.3a [401] aims to provide high bandwidth (up to 1.3 Gbps) over a small area (10 meters or less) with wireless extensions for USB and IEEE 1394 devices. IEEE 802.15.4 are aimed at telemetry based Mesh Networks, which aim for high battery life and low device cost mesh networks, typically at lower data rates. The ZigBee [806] alliance is developing higher-level protocols that will run on 802.15.4 MAC and PHY layers.

Finally IEEE 802.16 aims to provide a metropolitan area network which supports point to multi-point communication. This standard extends beyond the traditional range of IEE 802.11 products to be a true last-mile solution to extend traditional fibre infrastructures. The commercialisation of the product (WiMAX) is aiming to provide high speed internet connections to areas which were typically beyond the reach of DSL and other wired technologies. There are however some notable problems which the current standard has yet to address. IEEE 802.16 provides limited scalability which each network capable of supporting up to 100 users due to scheduling and message structures. Due to the connectionless MAC layer it is also feared that QoS may be an issue [227].

7.3.7 Mesh Management

WMNs are seen to be autonomic and capable of self-management, ultimately requiring software to manage and control the behaviour both of the mesh network and the users connected to it. Mesh networks can be seen as either a standard mesh network, or a community one. This difference relates to the style of management which takes place on the network. Community Mesh Networks are typically run by people with little or no technical expertise and so rely on the software to manage and control the network on their behalf. Whereas standard WMNs can be configured for a specific purpose and can be highly customised to match certain specifications. Management within WMNs covers several areas a selection of which is covered below:

Configuration. Configuration within WMNs provides the ability to customise a mesh network to work towards a particular goal or task. Within traditional WMNs, the network is configured for a specific purpose by an individual with technical experience. However, as mesh networks move into community environments, they are being run by individuals without that experience, producing the requirement to map high level requirements understandable by a user to that of a lower level setting.

Gateway Selection. One of the primary functions of WMNs is that of providing connectivity to the wider Internet. This requires a mesh network to peer with the Internet or another private network. In the presence of multiple gateways there are a number of factors which must be considered in order to choose the correct gateway.

- Distance from a client to a gateway in terms of wireless hops and the saturation of those links between client and gateway.
- Type of connection which the uplink point provides and the bandwidth/QoS provisions associated with it.
- If a client is accessing content from a particular service provider, is one gateway a direct connection to that network provider or closer than other gateways.
- Is it suitable to use multi-home the connection to increase the speed and reliability of the connection.

In addition to the gateway specific requirements there are also other factors to consider, such as the stability of the gateways, policy controls (e.g. from the network provider) and cost of using a particular gateway.

User Management and Billing. A key aspect to Community WMNs relates to the users, their management and how their interactions with the network are controlled. Typically unauthorised users will be restricted through an access control list (ACLs), WEP/WPA encryption key, or username/password combination. Users which are permitted to connect to the network may also be restricted either by available bandwidth or types of services they can utilise on the network. Users may also be subject to charging and billing either through a contract service (e.g. per month) or through the amount of bandwidth or types or services used.

7.3.8 (Transmission) Power Management

Unlike wireless sensor networks WMNs usually have a fixed power supply and so do not have power constraints. When power management is addressed in WMNs it relates to the transmission power of the wireless network device within a mesh device. If a single channel is used within a network, then the transmission power of a single node impacts the spatial-reuse factor (the number of nodes sharing the same physical "air space"). If transmission power is reduced then interference decreases, however, this is at the cost of possibly introducing hidden nodes into the network. Additional material on power management is discussed in [479].

7.4 Testbeds

Over the past five years there have been several large scale notable research projects and associated testbeds which have created considerable interest in the WMN research community. This section highlights a number of these projects alongside a discussion on their benefits and shortfalls to the research community. Divided into two sections, the academic testbeds provide insight into research environments and deployments on the cutting edge of the technology. Secondly the commercial testbeds provide practical deployments with a commercial objective in mind.

7.4.1 Academic Testbeds

This section discusses some of the most prominent academic testbeds over the past few years.

MIT RoofNet. Research into unplanned deployment of WMNs by MIT produced a series of papers on MITs RoofNet [7, 87]. RoofNet is a multi-hop network constructed in an ad hoc and unplanned fashion, consisting of 37 nodes interconnecting to internet gateways using a bespoke DSR like routing protocol Srcr, derived from ETX in [217]. One of the interesting aspects to this research project is the establishment of a WMN without prior planning. The aim was to determine the plausibility of one such a network outside of a laboratory environment and away from simulations. Ultimately a success with throughput speeds of up to 627 Kbit/s/s, there a number of notable problems and concerns with the testbed.

The goal of the Srcr routing protocol used within RoofNet is to determine a route to the internet gateway which can carry the most throughput. The route metric does not consider hop count or signal quality in its analysis. While this is successful in delivering bandwidth to end users, there are no significant details on the latency between nodes on the network, especially when under load. Additionally there are no details on how the network performs when one of the links on the network are saturated. It is highly likely that when under load pathways with poor signal strength fail resulting in reduced capacity or failure on the network. An interesting comparison, not mentioned by the authors of the papers would be to see a comparison against a planned deployment. They stipulate that the network is unplanned. However, a large number of the nodes on the network are run by technically competent members of MIT. Ultimately this adds a level of technical expertise to the project which may not be present in normal Community Wireless Mesh Networks. As the aim of the network was not to test the Community aspects of WMN, this can be overlooked. On the whole the research provides evidence that unplanned deployments are plausible and that the Srcr routing protocol is robust, even in unplanned ad hoc networks.

Wray Broadband Project. Lancaster University investigated the challenges of deploying a WMN into a rural community setting, both to see the feasibility of the Wireless Mesh Network technology and how it operates outside of a laboratory environment, with users who were previously without an Internet connection. The WMN deployed within Wray village consists primarily of LocustWorld Mesh devices located strategically throughout the village. The back-bone of the mesh operates using IEEE 802.11b network technology, on top of which AODV provides the routing to the network backhaul, a 5.8 GHz wireless link accessed from the local school. Clients connect to the Internet wirelessly via one of the Mesh devices, using off-the-shelf IEEE 802.11b network cards. Wireless Mesh devices were sited in strategic locations within the village, determined partly by geography, but also based on areas expected to have high utilisation. Individual Mesh devices used an externally mounted omni-directional antenna to distribute the signal locally as well as providing connectivity to the village school for the Internet uplink.

The research carried out within [406] highlights while they were successful in deploying the network and supporting a relatively large number of users there is still a significant amount of work before WMNs can become a mainstream product. One of the main issues to arise related to the self-management characteristics of the technology, while they were capable of some form of self-management, component interaction was poor resulting in expected results. There was also lack of any statistics gathering platform to portray the current status of the network. On a positive note, there were a number of unexpected social benefits which arose from the deployment of the network including improved faced-to-face communication and community awareness within the village (to solve IT problems) and a significant boost for rural businesses to match their urban counterparts.

Berlin RoofNet. Community Wireless Mesh Networking (CWMN) and the area of self-organization are researched under the Berlin RoofNet project [763]. The main focus of their CWMN is to improve the community aspects of the network, removing the need for operator assistance in running the network. This functionality revolves around a hash table to provide distributed DHCP, DNS and ARP. This DHT runs a-top of a modified version of the DSR routing protocol and a bespoke "Software Distribution Platform" (DSP).

Each mesh device on a given network provides their own services such as DHCP, with the distributed hash table providing the ability to share information on events such as the allocation of an IP to a client. This provides benefits to the automation of the network, including robust failover should a mesh device fail, and "seamless roaming" between access points. The network therefore looks the same to every client from any access point on the network. This allows the network to be modified without impacting a client significantly.

In [763] they conduct simulations using the network simulator ns–2 on the benefits of DHT for ARP and DHCP over traditional flooding mechanisms. They also study the behaviour of the network in a testbed environment. The simulations show that the DHT method of synchronising data is more efficient, both in terms of response time and data overhead. Their testbed is implemented across two buildings, although the location and purpose of the buildings are unclear. They indicate that their DHT provides an efficient way of providing auto configuration with CWMN and their results would seem to reinforce this.

While a DHT presents one possible way of providing distributed service provision (over and above flooding), there are a number of factors which may lead to problems on the mesh network. There is a lack of information on how it will behave should the network split and the same IP address be allocated to two clients. Furthermore they have only tied the DHT into low level network aspects of the mesh network. Any additional mechanisms requiring distributed state, e.g. traffic shaping or statistics collection would required re-engineering of the DHT, making this approach is therefore inflexible to change.

Additional Research Testbeds. The Champaign-Urbana Community Wireless Network (CUWiN) project [208] is a non-profit organization aiming to automate the process of installing CWMN. The CUWiN system prioritises routes among mesh devices based on the software developed for MIT's RoofNet. CUWiNs also utilises the WMN to build a local intranet as well as providing for Internet-connectivity. A village using CUWiNs system is also creating a community-wide local area network over which collaborative services can be provided (e.g. centralised printing).

The Broadband and Wireless Network (BWN) Lab at Georgia Institute of Technology [119], is made of 15 IEEE 802.11-based mesh devices. With some nodes acting as gateways providing access to the internet. The purpose of the testbed is to evaluate the feasibility of protocols applicable to heterogeneous wireless networks.

The WMN testbed presented by Raniwala [687] known as Hyacinth is a multi-channel WMN built using IEEE 802.11 technology. The testbed aims to solve the problems with interference caused by too many Wireless devices operating on the same channel. They utilise a channel assignment algorithm to balance network connectivity and bandwidth.

The UCSB Mesh Testbed [535] is an experimental wireless mesh network deployed on the campus of UC Santa Barbara. The network consists of 25 nodes equipped with multiple IEEE 802.11a/b/g wireless radios distributed over a five floors campus building. Each node is composed by two Linksys WRT54G wireless devices. One devices act as mesh device running the AODV routing protocol. The second device is used for out-of-band management of the AODV mesh device.

Early work in [395] presents the MeshDVNet testbed, which uses the auto-configuration properties of IPv6 to provide automation and self-configuration. This is advantageous as the protocol is standardised allowing interaction between non-mesh networks.

There are a number of testbeds which stray from Community WMN into other application environments. Recent work in [488] highlights a WMN designed to provide coverage to traffic lights, in a traffic control network. While the way in which the testbed is implemented is not unique, the application and use of the network (to provide traffic control and monitoring) is. A similar example of such a testbed is that presented in [110]. This demonstrates a WMN being used to relay a video feed from inside of a burning building. Finally, there are a number of testbeds which aim to provide network connectivity to low or zero population areas. For example, [872] details the deployment of a testbed to over 2000 acres of park land.

7.4.2 Commercial Testbeds

The testbeds presented in the previous section originate from an academic research stand point, these typically are designed without any commercial requirements. This section discusses a number of commercial avenues which highlight both research and purchasable systems.

Microsoft Research have created a loadable windows driver called the Mesh Connectivity Layer (MCL) [570] which permits the creation of ad hoc networks using a modified version of DSR. The aim of this system is to provide "Self-Organizing Neighbourhood Wireless Mesh Networks". The majority of research carried out is to provide network layer connectivity between devices on the network. The MCL sits between the network and data link layers of the network, acting as a layer 2.5 protocol providing both pathway selection and device support. Their published research focuses mainly on lower level aspects to WMN, including Models of Wireless Interference [900], Channel Assignment [466] and MAC layer behaviour [542]. One feature, Multi-Radio Unification (MUP) [5] details the use of both non-overlapping channels and different radio technologies to provide increased bandwidth and coverage. The non-overlapping channels increase the throughput available to the network. While using lower frequency radios permits extended coverage in sparsely populated networks, with the option of swapping to higher frequency radios as the number of nodes on the network expands.

Firetide Inc. [281] is a wireless mesh technology company which develops Wireless Mesh Network equipment aimed at providing high performance, yet easy to deploy networks. They are bespoke solution is aimed at a number of environments including Internet access HotZones, airports, hotels and other locations where wiring is difficult. The key architecture behind FireTide's broadband radio mesh HotPoint Network is the Reverse-Path Forwarding (TBRPF) routing protocol to manage the unique and dynamic environment of the mesh network. A patented technology which allow the HotPoint routers to find each other automatically.

MeshNetworks recently acquired by Motorola [567] aims at providing mobile broadband internet access by means of supporting high speed mobile users through the use of QDMA (quad-division multiple access) protocol, a proprietary radio technology developed and used by the military. Another important feature of MeshNetworks mesh networking technologies is the building of its proprietary MeshNetworks Scalable Routing (MSR) protocol. MSR technology utilises an optimized ad hoc routing algorithm that combines both proactive and reactive routing techniques. This enables support for high speed mobility, unmatched scalability and low messaging overhead.

Tropos Networks MetroMesh architecture [813] including the Tropos MetroMesh OS and Predictive Wireless Routing Protocol (PWRP) is a purpose-built outdoor mesh network operating at 2.4 GHz. PWRP aims to provide scalable routing while trying to negate the effects of radio frequency interference. Tropos claims that their network scales to thousands of nodes.

BelAir Networks [74] approach to wireless mesh networking utilises lower level components to provide the mesh networking, e.g. physical layer feedback and radioaware routing algorithms. Typically BelAir WMN utilise multiple antennas allowing dynamic selection of which antenna to utilise, reducing link congestion and failure.

MeshDynamics [566] provides multi-interface mesh solutions based on 802.11 devices. This architecture provides separate backhaul and service functionalities and dynamically manages channels of all of the radios so that all radios are on noninterfering channels.

7.5 Summary and Outlook

This chapter has introduced Wireless Mesh Networking and described how it has emerged as an important technology for building next-generation networks. The main body of the chapter was focused on introducing some of the key research areas behind the technology and some practical examples of this in action. While a significant amount of research has already been conducted, the testbed deployments highlighted that work is still required.

There are numerous challenges which are still faced in the deployment of the technology, these include: The advancement in bandwidth available on wired infrastructures has not been matched by its wireless counterparts, caused by incompatibilities amongst components such as MAC and Transport layer clashes. A second challenge faced is that of managing WMNs. The current aim for research is in investigating the underlying technology and not in how it is managed at a higher level. Managing WMN can be divided into component, network and user management. Component management reflects the ability to maintain and replace the functional components which build up to create a mesh network, permitting WMNs to operate dynamically in flexible environments. Network management provides the ability to control the behaviour of the network side of a WMN, dealing with aspects such as routing, gateway selection and QoS. The final management challenge relates to user management, this includes user security, access controls, billing in addition to Quality of Service. An additional challenge relates to how the components which make up a WMN are configured. At present WMNs are created based on a static operational environment with each component configured to operate in this environment. The challenge faced is to permit a WMN setup to be dynamically altered between environments and setup conditions.

Part III
Middleware Issues

The third part covers different problems that typically solved ate middleware level, such as efficient broadcast, structured message dissemination, event dissemination and subscription, distributed coordination, security, and adaptation. Namely:

- R. Friedman, A.-M. Kermarrec, H. Miranda and L. Rodrigues address in Chap. 8 the use of gossip-based protocols in mobile ad hoc networks. Gossip-based protocols have the advantage of requiring little or no structure to operate, making them particularly suitable for dynamic systems such as wireless self-organizing networks.
- In Chap. 9, M. Allani, B. Garbinato and F. Pedone address the problem of structured dissemination of data, which is usually solved via some application layer multicast. In particular, the authors show how it is possible to efficiently disseminate information without forcing nodes to have global knowledge of their networking environment, typically by relying on some form of overlay creation and maintenance.
- Chapter 10 by R. Baldoni, L. Querzoni, S. Tarkoma, and A. Virgillito addresses the issue of achieving scalable information dissemination using the publish/subscribe paradigm. This paradigm is of particular interest for mobile ad hoc networks, as it offers both anonymity and asynchrony to communicating nodes.
- Chapter 11 addresses the implementation of tuple spaces, an abstraction that supports data sharing and coordination among components of a distributed system. P. Costa, L. Mottola, A. Murphy, and G. Picco concisely present the state-of-the-art concerning middleware platforms based on the tuple space abstraction and expressly designed for wireless scenarios.
- Mobile device applications can be highly security- or privacy-sensitive, which is often difficult to ensure due to the inherent limitations of mobile device platforms. In Chap. 12, B. De Win, T. Goovaerts, W. Joosen, P. Philippaerts, F. Piessens, and Y. Younan elaborate on possible threats and applicable security solutions for mobile devices.
- A fundamental characteristic of mobile systems is the variability of their deployment and execution environments. In this context, dynamic adaptation is an essential technique when it comes to ensure that these systems continually provide the required level of service, in spite of continuous changes. In Chap. 13, P. Grace investigates the software techniques for performing adaptation, as well as the adaptive middleware technologies that have been used to develop dynamic mobile applications.

Chapter 8
Gossip-Based Dissemination

Roy Friedman, *Technion, Israel*
Anne-Marie Kermarrec, *INRIA, France*
Hugo Miranda, *FCUL, Portugal*
Luís Rodrigues, *INESC-ID/Instituto Superior Técnico, Portugal*

8.1 Introduction

Gossip-based networking has emerged as a viable approach to disseminate information reliably and efficiently in large-scale systems. Initially introduced for database replication [222], the applicability of the approach extends much further now. For example, it has been applied for data aggregation [415], peer sampling [416] and publish/subscribe systems [845]. Gossip-based protocols rely on a periodic peer-wise exchange of information in wired systems. By changing the way each peer is selected for the gossip communication, and which data are exchanged and processed [451], gossip systems can be used to perform different distributed tasks, such as, among others: overlay maintenance, distributed computation, and information dissemination (a collection of papers on gossip can be found in [451]). In a wired setting, the peer sampling service, allowing for a random or specific peer selection, is often provided as an independent service, able to operate independently from other gossip-based services [416].

While most gossip-based protocols have been deployed in wired networks, their robustness (namely in the face of dynamic topologies) makes them an appealing solution for Mobile Ad hoc NETworks (MANETs). In this case, the epidemic dissemination is supported by the broadcast communication nature of wireless nodes. Interestingly, there are several inherent differences between MANET and Internet settings. These include, for example, the fact that MANETs operate on a radio broadcast medium, where links are unreliable and may suffer from a high percentage of message losses. Additionally, in wireless networks, remote nodes can only communicate with one another with the help of intermediate nodes, while in the Internet, all the routing is handled by dedicated routers, which are typically separate entities from the devices on which applications (and middleware) run. Moreover, in mobile ad hoc networks, the network topology is constantly changing, which makes routing extremely expensive. As much as mobility can be a burden as it introduces high dynamics, it also provides opportunities that can be exploited by gossip protocols. For instance, the peer sampling component may benefit from the mobility

---

Algorithm 8.1: Push Based Gossip in the Internet

```
broadcast(m)
begin
    gossip(m);
end

upon receiving( m for the first time)
begin
    deliver m to the application;
    gossip(m) ;
end

gossip(m)
begin
    for i = 1 to k do
        for j = 1 to ℓ do
            p ← pick_random_node() ;
            send m to p ;
        end
        wait(Δ) ;
    end
end
```

---

of peers. Thus, gossip-based protocols for MANETs have to take into account the specific nature of these networks.

This chapter addresses the use of gossip-based protocols to implement data dissemination and data placement in MANETs. First we show how gossip-based protocols can be used to broadcast information in the network in an efficient manner. Then we show how the same principles can be adapted to support information dissemination in a setting where each node is interested in just a subset of the information being disseminated. Finally, we show how gossip-based dissemination can be combined with autonomous data-placement strategies to keep frequently accessed data in the close vicinity of its clients.

8.2 Gossip-Based Broadcast Protocols

Gossip based broadcasting protocols can be *push* based or *pull* based. Generally speaking, in push based protocols, a node that receives a message for the first time forwards the message to other nodes with a certain probability. On the other hand, in pull based protocols, nodes exchange gossip about which messages they have received. When a node p learns through this gossip exchange that it is missing a message m, then p sends an explicit request for the message m to q, and q is supposed to reply by forwarding m to p. Algorithms 8.1, 8.2, 8.3, 8.4 capture the generic structure of gossip protocols using the dichotomies "push based vs. pull based" and "Internet based vs. MANET based" protocols.

Algorithm 8.2: Pull Based Gossip in the Internet

```
broadcast(m)
begin
    list_of_messages += m;
end


upon receiving( (Message, m) for the first time)
begin
    deliver m to the application;
    list_of_messages += m;
end


periodically
begin
    remove from list_of_messages all messages received more than Δ seconds ago;
    for i = 1 to l do
        p ← pick_random_node();
        hdrs ← headers of all messages in list_of_messages;
        send (Ihave, hdrs) to p;
    end
end


upon receiving( (Ihave, hdrs) from p)
begin
    if hdrs contains the header h of a message m ∉ list_of_messages then
        send (Request, h) to p;
    end
end


upon receiving( (Request, h) from p)
begin
    if list_of_messages contains a message m whose header is h then
        send (Message, m) to p;
    end
end
```

Specifically, Algorithm 8.1 presents the push based gossip code for Internet based systems. As can be seen, a message is gossiped k times to l random nodes. The value l is a parameter of the protocol called the *fan-out*. The value of k is also a parameter which controls how many gossip iterations the protocol will perform, each separated by Δ seconds. Together, these three parameters control the *latency* of the protocol, its *reliability*, *contention level*, and *communication overhead*. Latency is the time from the invocation of the broadcast method by the application until all nodes receive the message. Reliability is the percentage of nodes that receive the message out of the entire network. Notice that due to the probabilistic nature of gossip, some nodes may never receive a given message even if there are no failures. However, this probability can be made arbitrarily small by controlling the values of k, l, and the random selection function. The contention level of the protocol is the number of messages that are sent concurrently by nodes, whereas the communication overhead is defined as the total number of messages generated by the protocol for each broad-

---

Algorithm 8.3: Push Based Gossip in a MANET

```
broadcast (m)
begin
    data_link_broadcast (m) ;
end

upon receiving ( m for the first time)
begin
    deliver m to the application ;
    optionally: probabilistic_gossip (m);
    optionally: counter_based_gossip (m);
end

probabilistic_gossip (m)
begin
    invoke data_link_broadcast (m) with probability P;
end

counter_based_gossip (m)
begin
    timeout ← random_period()
    wait (timeout)
    if the total number of copies of m received so far is less than c then
        data_link_broadcast (m);
    end
end
```

---

cast invocation. Clearly, when the messages are small, it is possible to incorporate multiple gossip messages inside a single network level message [295].

Algorithm 8.2 presents the pull based gossip code for Internet based systems. Here, in each iteration, a node gossips about the headers of messages it is aware of with l randomly selected nodes (l is a parameter). The parameter Δ controls how long nodes remember the messages they received both in order to gossip about their headers and in order to forward theses messages if asked. This parameter is used in order to limit the memory space needed for storing received messages and the size of gossip messages. Together, Δ and l control the reliability level of the protocol. The latency of broadcast is dictated by the value of l (the larger the l the faster the information spreads) and the period by which gossip occurs (a shorter period decreases the latency but increases the load on the network).

8.2.1 Gossip in MANET

Applying Internet gossip protocols in MANETs as is has two drawbacks. First, addressing a random node in MANETs involves routing, which is expensive. This is especially severe in light of the fact that by targeting a random node at each step, one prevents route reuse. Hence, the cost of route discovery cannot be amortized

---

Algorithm 8.4: Pull Based Gossip in a MANET

```
broadcast (m)
begin
    list_of_messages += m;
end


upon receiving ( (Message, m) for the first time )
begin
    deliver m to the application;
    list_of_messages += m;
end


periodically
begin
    remove from list_of_messages all messages received more than Δ seconds ago;
    hdrs := headers of all messages in list_of_messages;
    data_link_broadcast (Ihave, hdrs)
end


upon receiving ( (Ihave, hdrs) from p )
begin
    if hdrs contains the header h of a message m ∉ list_of_messages then
        send (Request, h) to p;
    end
end


upon receiving ( (Request, h) from p )
begin
    if list_of_messages contains a message m whose header is h then
        send (Message, m) to p;
    end
end
```

---

over multiple invocations. Second, wireless communication occurs over a broadcast medium. In other words, each message sent by a node p is received, at least at the network interface card level, by every node within the transmission range of p. Hence, the preferred embodiment of gossip protocols in wireless ad hoc networks apply what can be thought of as "broadcast gossip" [296]. That is, rather than having the sender pick the receivers at random, the receivers are picked by the topology of the network. The receivers then apply some probabilistic mechanism to determine who should react to the gossip message, as listed in Algorithms 8.3 and 8.4.

In the push based protocol of Algorithm 8.3, a receiver probabilistically decides whether to rebroadcast the message using two optional mechanisms. The first is listed in the method `probabilistic_gossip()`. Here, the decision whether to broadcast is determined by a probability $\mathcal{P}$. Various gossip protocols differ in the way they set this probability, and we discuss some of them below.

The specific instance of push based gossip in which the retransmission probability $\mathcal{P}$ at all nodes is set to 1 is called *flooding*, whereas when $\mathcal{P}$ is less than 1, the protocol is often referred to as *probabilistic flooding*. Notice that it is also possible to give different retransmission probabilities to different nodes, e.g., based on their locally observed density [236], or distance from the originator of the message [140, 353, 727, 808]. For example, in RAPID, the retransmission probability is set with inverse proportion to the node's density [236]. On the other hand, in the GOSSIP2 protocol of Haas et al. [353], the protocol behaves like flooding up to k hops from the originator of the message. Beyond that, the retransmission probability is reduced to some value $\mathcal{P} < 1$.

An alternative to the above setting is to set the retransmission probability indirectly as appears in the counter_based_gossip() method of Algorithm 8.3. This approach, known as *counter-based* forwarding, was initially proposed by Tseng et al. [818, 817]. Specifically, in the counter-based approach, a node p that receives a message m for the first time picks a random period to wait before sending the message. If during this period p hears more than $c(n)$ retransmissions of m, it cancels its own retransmission. Here, $c(n)$ is the counter threshold, which may depend on the number of neighbors n a node has. Otherwise, if fewer than $c(n)$ transmissions of m are received, then p rebroadcasts m at the end of the waiting period.

The *distance-based* approach is an extension of the counter-based mechanism [140, 583, 817]. In the distance based approach, the random waiting period is biased so that the waiting period is inversely proportional to the distance between p and the node q from which p received m for the first time. The rationale here is that if p and q are close to each other, a retransmission by p will hardly reach any new nodes. On the other hand, if they are far apart, then it is likely that a retransmission by p will reach many new nodes. Measuring the exact distance between two nodes is often hard. Hence, the Pampa protocol approximates the distance using the power level of the received message [583].

Finally, it is possible to combine both the implicit probabilistic scheme with the counter-based approach and use them in tandem, as been done in the RAPID protocol [236]. The rationale for the various options of setting the retransmission probability and for using the above approaches is described in the following sections.

8.2.2 On the Benefits and Limitations of Gossip Protocols

As hinted above, broadcast protocols have three main goals, which are somewhat conflicting. These goals are low latency, high efficiency, and high reliability. Latency measures how fast the message is disseminated to all its recipients, efficiency measures the communication burden that the protocol imposes on the network, whereas reliability measures how many nodes are likely to receive a broadcast message.

Consider for example flooding. This protocol enjoys low latency, since messages are propagated along all possible paths and, moreover, every message is likely to reach every node along the fastest route. Moreover, as long as the network is not overloaded, this protocol is likely to deliver messages to all nodes, due to its high level of redundancy. Yet, this protocol generates a great amount of traffic, a phe-

No. of neighbors	1	2	3	4	5	6	7	8	9	10	11	12	More
Counter value	2	3	4	5	5	4	4	4	3	3	3	2	2

Fig. 8.1 Preferred counter values as a function of the number of neighbors according to [818]

nomenon called *broadcast storm*. Interestingly, this phenomenon indeed occurred in an experimental deployment of the OLPC project in Mongolia [648].

A probabilistic flooding protocol (with $\mathcal{P} < 1$) can also maintain relatively low latency, since messages are forwarded immediately. Yet, in lightly loaded networks, the latency can be slightly higher than flooding since not all paths are being used. On the other hand, lowering the broadcasting probability can reduce the reliability of the protocol. This is because when $\mathcal{P} < 1$, there is a non-zero probability that none of the receivers of a message will decide to rebroadcast it.

The work of Haas et al. has investigated empirically the impact of various strategies of setting the rebroadcasting probability on the reliability of the protocol in MANETs [353]. The best approach among the ones they tried was the one nick-named GOSSIP3, in which the retransmission probability is set to 0.65, yet there is also a counter based phase that requires hearing at least one retransmission after a constant timeout.[1] Moreover, both the works of Haas et al. [353] and Sasson et al. [727] have investigated the behavior of reliability vs. probability when the re-broadcasting probability is the same at all nodes. They have found phase transition phenomena, where below a certain probability, very few nodes receive the messages, yet above this probability most nodes receive the message.

Alternatively, the work of Tseng et al. has investigated the relation between the value of the counter, in the pure counter-based approach, and the reliability of the protocol [818]. This study was based on an analytical model of the intersection between coverage areas of multiple retransmissions in the same neighborhood coupled with simulation based measurements. They have found that the minimal value of the counter needed to ensure high reliability is a function of the local network density. The preferred values of the counter as a function of the size of the local neighborhood of a given node are reported in Fig. 8.1. These numbers were obtained for networks in which nodes are uniformly distributed.

Drabkin et al. have investigated the reliability of a broadcast scheme vs. the number of retransmissions in each neighborhood, assuming such retransmission are uniformly distributed in the neighborhood [236]. The results of their formal probabilistic analysis are summarized in Fig. 8.2. Specifically, it can be seen that the graph has a concave shape. In order to obtain relatively high reliability of 85–90%, it is enough to have a few retransmissions in each neighborhood (between 2.5–3.5). However, in order to obtain even higher reliability of near 100%, a much larger number of retransmissions is needed. Hence, the conclusions of [236] are as follows: (*i*) the retransmission probability should be set inversely proportional to the

---

[1] The probability of retransmission is a parameter in GOSSIP3. In the experiments reported in [353], for uniformly distributed networks whose density is sufficient to ensure connectivity [344] and the number of nodes is up to a few hundreds, the value of 0.65 was deemed the best.

Fig. 8.2 # of transmissions in each neighborhood vs. reliability

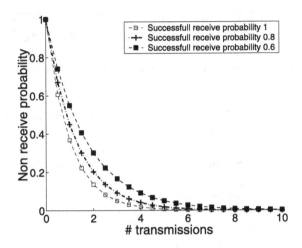

number of neighbors a node has, (*ii*) this probability should ensure roughly 2.5–3.5 retransmissions per neighborhood, and (*iii*) the protocol should be complemented with additional measures to ensure near 100% reliability. Notice that when nodes are uniformly distributed, if the retransmission probably is set inversely proportional to the number of neighbors, then the expected number of nodes that decide to rebroadcast a message in each neighborhood is roughly the same, regardless of the size of each neighborhood. Consequently, the RAPID protocol combines a probabilistic gossip phase with a counter based gossip phase as well as a pull based gossip phase along the lines of Algorithm 8.4. RAPID sets the retransmission probability of node i to $\mathcal{P} = 3.5/N_i$, where N_i is the neighborhood size of i, with the counter threshold of the counter-based phase set to 2.

When comparing probabilistic gossip to counter based gossip, the following tradeoff appears. The counter based scheme can obtain a better reliability to communication overhead ratio than pure probabilistic ones. However, probabilistic protocols can obtain better latency. This is because in counter based protocols nodes wait for a while before retransmitting a message. While in each hop the delay may be small, along a multiple hop path, these delays accumulate.

8.2.3 Combining Pull Based Gossip with Other Approaches

As noticed in [236], a protocol that only has probabilistic gossip and counter-based mechanisms *cannot* obtain complete reliability. Specifically, consider a network topology as depicted in Fig. 8.3. In this example, node p and all n_i nodes are within the transmission range of s, but q is not. Node q is also within the transmission range of p, but outside the transmission range of all other nodes. In a pure probabilistic protocol, it is possible that p will decide not to rebroadcast the message. Additionally, with a counter based mechanism, given any counter threshold number k, it is

Fig. 8.3 The necessity of pull based gossip

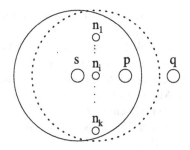

possible that the nodes $n_1, \ldots, n_k$ will retransmit a given message before p, and therefore p will avoid retransmitting this message.

The above mentioned situation can be overcome using pull based gossip, along the lines of Algorithm 8.4. Hence, it is possible to increase the reliability of any push based gossip protocol by adding to it a pull based phase, as was done in RAPID [236]. It is possible to combine a pull based gossip phase with any deterministic dissemination protocol in order to recover from omissions at the data link level. This was originally proposed for Internet based networks in the seminal PBcast work [88]. Yet, in MANETs, it is best to use the broadcast gossip scheme of Algorithm 8.4 rather than random point-to-point gossip as one can leverage on the broadcast nature of wireless communication. An example of such a protocol is BDP [235].

Finally, it is important to realize that while pull based gossip is helpful in terms of reliability, it also increases the protocol's overhead. When the application generates messages frequently, this is negligible, since gossip messages are only sent periodically. However, if the application rarely sends messages, the periodic gossip creates constant useless communication. In particular, in a sensor network environment in which battery power is scarce, it may be advisable to either avoid it altogether, or develop optimizations in which the rate of gossip is greatly reduced during periods in which the application is quiescent.

8.3 Gossip-Based Publish/Subscribe in Mobile Ad Hoc Networks

In the previous section we have addressed the broadcast service. In this service, all nodes are interested in all messages. Quite often, only a subset of the nodes is interested in a given broadcast. A paradigm that matches data producers with data consumers is the publish/subscribe paradigm. Publish/Subscribe systems have recently received increased attention. Such a communication service is of specific relevance in the MANET context [387], where event notification is an extremely useful mechanism for many applications such as location-based applications [278], context-aware applications [293] or vehicular applications, for example.

Although a large amount of research has been conducted on publish/subscribe systems for the wired networks (see Chap. 10), there is a pressing need to extend publish/subscribe to MANETs. In this context, gossip-based protocols emerge as relevant candidates to ensure an efficient implementation, given their ability to cope with high dynamics. In addition, the broadcast nature of the communication in MANETs has an intrinsic flavor of epidemic dissemination. In this section, we first briefly enumerate the key characteristics of publish/subscribe systems. We then review a few works in the area of gossip-based publish/subscribe protocols for MANETs.

8.3.1 The Publish/Subscribe Paradigm

Publish/Subscribe is a paradigm for distributed applications where *subscribers* may register their interest in an event (or a pattern of events) in order to be asynchronously notified of any event matching their interest, published by *publishers* [265]. The strong decoupling of subscribers and publishers in time, space and synchronization matches the loosely coupled nature of contemporary distributed applications extremely well. Publishers generate events without being aware of the subscribers. Similarly, subscribers have the ability to express their interest in events regardless of the publishers.

While the publish/subscribe paradigm is clearly useful in wired systems, the level of decoupling it provides is very appealing for mobile ad hoc networks subject to frequent topological changes. *Time decoupling* refers to the fact that publishers and subscribers do not need to actively participate in the interaction at the same time. More specifically, the publisher might publish events when the subscribers are disconnected. This is of particular relevance in the context of MANETs where nodes might be frequently disconnected. *Space decoupling* refers to the fact that publishers and subscribers do not need to be aware of each other. The publish/subscribe system solely takes care of the event-matching. Again, in many MANET applications, nodes do not know each other. Even if they do, there is most of the time no routing infrastructure available. Finally, *synchronization decoupling* refers to the asynchronous nature of the publish/subscribe paradigm. Neither the publishers nor the subscribers need to be blocked during the event matching. Again, due to the frequent disconnections and the mobile nature of nodes in MANETs, this is a particularly relevant characteristic. This decoupling removes all explicit dependencies between publishers and subscribers and provides a flexible and powerful paradigm, particularly well suited to MANETs.

Publish/Subscribe systems can be classified in two main classes: *topic-* and *content-* based systems (again, see Chap. 10 for a more detailed description of these classes). Although content-based system account for much more expressiveness, implementing content-based publish/subscribe systems in a fully decentralized manner remains a very complex task. As we will see later, there are still only a few fully decentralized approaches to content-based systems in wired networks [345, 845]. This also remains mostly an open area in the context of MANETs.

8.3.2 Gossip-Based Approaches to Publish/Subscribe in MANETs

Many implementations of publish/subscribe systems, namely topic-based publish/-subscribe, rely on maintaining a specific data dissemination structure, such as a tree. Unfortunately, these structures are clearly difficult to maintain in a mobile network, due to the highly dynamic nature of the system. Therefore, the most promising approaches for MANETs are structure-less, where the event notification is usually ensured by spreading messages to interested subscribers, leveraging on the mobility patterns, and common interests or characteristics among nodes (also called node affinity).

In the rest of this section, we focus on unstructured approaches, briefly describing four concrete solutions. This description is not meant to provide all details that can be found in the related publications. We purposely leave out of this section broker-based or infrastructure-based approaches such as [19, 138], for example. Ensuring a reasonable level of reliability without flooding the network is challenging. Likewise, resources (bandwidth, memory, etc.) in a MANET may be limited, so the state to maintain as well as event buffering should be limited. The approaches described below share the aims of providing a reliable event notification scheme, while limiting the number of redundant messages nodes receive and the amount of state information required to keep track of the subscription patterns of other nodes.

Although approaches that we will describe in the next section have been proposed for different scenarios, they share several common characteristics. There is a natural trade-off between the efficiency of a system (measured typically by the latency and reliability of event delivery), and the overhead and *spam* generated. The overhead is measured by the load imposed on each node during the dissemination. The spam refers to the number of messages delivered to a node that do not match that node's interests. In order to limit the overhead and the spam, all the approaches below assume that some *affinity* between nodes can be detected and exploited. In publish/subscribe systems, the affinity usually refers to the overlap in interests in events. In this case, affinity can be exploited during event dissemination by favoring nodes with similar interests to limit the spam. Also, the design of the presented approaches relies on the fact that mobility in wireless ad hoc networks can be exploited, rather that suffered from, to opportunistically disseminate events. Leveraging both mobility and affinity could be considered as basic fundamental strategy to implement publish/subscribe in MANETs.

8.3.3 Frugal Event Dissemination

Baehni *et al.* [40] propose an event dissemination scheme for topic-based publish/-subscribe. The algorithm reduces the number of redundant messages (duplicate and non-subscribed events received by subscribers) by leveraging on the mobility of both publishers and subscribers to achieve a high level of dissemination. Each node has a unique identifier and a limited amount of memory, may be subject to movement or failures, and may subscribe to one or several topics. Topics are organized hierarchically. Each event published has an associated validity period. The algorithm has three phases: (*i*) neighborhood detection, (*ii*) events dissemination, and (*iii*) garbage collection.

- During neighborhood detection, each node periodically exchanges heartbeat messages containing its id, the list of current subscriptions and its current physical speed. This information is used for each node to build a one-hop neighbor table. Each entry contains the node identifier and a list of topics to which it has subscribed. A neighbor q is added to the one-hop neighbor table of node p only if the subscriptions of q and p overlap. Events are kept in an event table, constantly garbage collected according to the validity period of each event. When nodes get in contact with one another they check the events they are interested in against the event identifiers of their neighbors. Only if a nodes finds that it is interested in an event carried by another node, the event data is actually transferred.
- During the dissemination phase, when a node detects that some of its neighbors need an event, it broadcasts the event to its neighborhood. Each node then carries on the dissemination if it knows of any other process interested in the event. If a process receives an event it is not interested in, it simply drops the event. Although this algorithm does not completely avoid redundant messages due to the broadcast communication paradigm of MANETs, it limits it by not involving non interested nodes in the dissemination process.
  The garbage collection phase consists in removing events when the buffer is full according to its validity period.

The algorithm has been evaluated through simulations of two mobility models: Random Waypoint [429] and city section [216] (the later captures the movement of humans in a city). Results have shown that in the Random Waypoint model, the reliability of event dissemination is highly dependent of the number of subscribers and of the movement speed. This is due to the fact that nodes forward events only if they have subscribed to the associated topic and when they have neighbors interested in the same events. The efficiency of the protocol arises from the dissemination of events identifiers rather than the events themselves. The garbage collection scheme captures the fact that there is no need to keep forwarding messages that are no longer valid. The experiments also show that increasing the heartbeat rate does not always improve reliability as it increases the risk of collisions in a MANET. The experiments on the city model exhibit better results in terms of reliability. This is due to the fact that the affinity between nodes is captured: nodes subscribing to similar topics are more likely to encounter one another in a mobility model reflecting social links than in a Random Waypoint Model.

In this work, each node keeps affinity information for nodes in its one hop neighborhood (as opposed to the whole network). Results demonstrated that the mobility pattern of nodes clearly affect the reliability of the event dissemination. This work is interesting because it is based on a simple and efficient dissemination scheme; for instance, authors did not consider sophisticated distance computation between nodes interests to avoid complex computation to be conducted at each node. On the other hand, the approaches presented in the next paragraphs do rely on the computation of some distance between nodes.

8.3.4 Autonomous Gossiping

Datta *et al.* [213] present an epidemic protocol for selective event dissemination for MANETs, denoted autonomous gossip (AG). Each node has a limited memory and to each node is associated a profile that captures its subscription patterns. The subscriptions (characterized by categories in the system model) may correspond to topics or more elaborate content description. Each node hosts some data items, each of which has also an associated profile described in a similar manner as the mobile hosts profile, as well as an associated *utility*. Data items are also associated to a given geographical location. Such a model is particularly relevant in context-aware or location aware applications. At every node, a weight is associated to other nodes for each topic reflecting the overlap in interest, and therefore the affinity between nodes. Similarly, data items have a given affinity with respect to a particular topic. Affinity is computed using standard information retrieval techniques [41].

In this work, the nodes communicate through wireless links in a given radius. Their movement follows a hybrid version of city walk and random walk. This is intended to simulate a population mobility in a city. The originality of this approach is to implement several migration and replication policies to move or replicate data items from one node to another depending on some parameters about the comparisons of similarities between data item and node profiles. Data are pushed to nodes exhibiting the most similar profile.

This stateless protocol's objective is to provide a best effort dissemination scheme where efficiency and overhead are traded against coverage. The protocol can be applied in any dynamic arbitrary environment. As in [40], instead of flooding, nodes forward data to 1-hop neighbors interested in the same set of events while avoiding non interested subscribers.

The simulation results show that the protocol achieves an effective content-based dissemination scheme, but that the reliability[2] is impacted by the strictness of the matching criteria: the stricter the criteria, the less accurate the protocol. The accu-

---

[2] The reliability of a message dissemination in a publish/subscribe systems also refers as the completeness of the dissemination: a dissemination is complete when all the nodes subscribing to an event receive it.

racy is expressed as the difference between the events that a node receives and the ones that it should receive. This refers to the *missed* messages, the ones that were not received while they should have been and the *false positive* messages, the ones received that shouldn't. These results are similar to the ones of [40] where a decrease in the number of subscribers, decreases the number of potential information carriers and therefore degrades the completeness of the protocol.

8.3.5 Socially-Aware Publish/Subscribe Systems

Finally, although the approach is not explicitly characterized as a gossip-based approach,[3] we present a recent work in this area: SocialCast [192]. SocialCast is a routing framework for topic-based publish/subscribe systems that exploits social interactions to identify the best nodes that could forward a message. SocialCast relies on the fact that in a human operated MANET, node movements are dictated by social behaviors. While the network might be disconnected, the connectivity of related subscribers might be reflected by the social links between nodes, or more specifically, between users operating the nodes. Such social links may be leveraged to deliver messages opportunistically to interested subscribers. The basic idea is that socially related users are likely to subscribe to the same events, regularly meet, and therefore those links should be preferably used to disseminate and carry events. Note that this has also been exploited in wired systems [652].

In order to take into account the social links between nodes, observations about social connections are exploited to predict future movement using Kalman filter forecasting techniques. These predictions are used to identify the relevant information carriers for specific events. This work is done in the context of topic-based publish/subscribe systems in intermittent networks. For example, when a message has to be routed to a set of interested subscribers, it should be carried by a node which has a high probability to meet those interested subscribers in the future. Such movements are predicted using forecasting techniques based on recent history. The basic assumption behind this model is that nodes subscribing to a same topic have a high probability to be co-located from time to time. Even if they are temporarily disconnected, their movements will bring them back close to each other so that events can be propagated.

SocialCast also relies on the notion of *node utility* with respect to a given interest. This is used in the routing protocol to identify the *good* message carriers in the routing process. The utility of a node is derived from its co-location with other nodes as well as its movements. SocialCast is divided in three phases: (*i*) interest dissemination; (*ii*) carrier selection and (*iii*) message dissemination. Interest dissemination consists for each node to disseminate to its one hop neighbors its list of interests. This information is used by the carrier selection phase to compute the utility of each node. This phase enables to identify the most useful information carriers for events

---

[3] The way interests are broadcast and messages disseminated is extremely similar to the so-called gossip-based approaches presented above.

of a given interest. Finally, during the message dissemination phase, messages are forwarded based on the new subscriptions received and on the computed utilities. Typically, the message related to interest i is sent to all neighbors whose subscriptions contain i.

The originality of this work is to leverage social links to carry events in temporarily disconnected networks and in particular to predict future behavior. More specifically, the probability of a user to be co-located with another user sharing some interest is the most important attribute taken into consideration.

SocialCast has been evaluated by simulation using the community-based mobility model. The synthetic model used in this work has been validated against real traces. The simulation results show that SocialCast achieves a better reliability than a simple dissemination scheme where no prediction is used and achieves better delay, better resource optimization and low overhead.

8.3.6 Socio-Aware Overlay for Publish/Subscribe in DTNs

We now describe an approach that identifies communities of nodes through gossip in a Delay Tolerant Network (DTN) packet switched network in order to implement an overlay of brokers; brokers are defined as the most central nodes in each community and are in charge on ensuring the dissemination within each community [886]. Beyond the broker-based architecture which has some drawbacks that we will come back to later, the originality of this work lies in the way the communities are detected. In addition, this work has been evaluated against real traces from MIT, Cambridge and UCSD.

In a mobile wireless network, each individual node maintains a *familiar set*, which represents the direct neighbors from a social standpoint, and a local community set, representing the neighbors from a geographical standpoint. This informations is exchanged whenever two nodes meet each other through a gossip-based protocol. The gossip protocol enables nodes to build up these sets and also to potentially merge communities. A novel community merging algorithm is presented to make communities evolve over time. When a node p comes across node q, they need to figure out if (i) they should be integrated in each others communities and (ii) their communities should be merged. Node p is integrated in q's community if the overlap between the familiar set of p and the community set of q is greater than a given threshold. Similarly, two communities are merged if the overlap between the communities exceeds a given threshold. This algorithm builds communities in a fully decentralized way. Simulation results show that the community views computed locally are very close to the global view (provided by an oracle, for example).

The publish/subscribe system is then built as follows. Based on these community computations, one or several brokers are extracted from each community to compose the broker overlay. Brokers are chosen according to their *centrality*, which reflects the *connectivity* of a node with other nodes of the community. The brokers are in charge of disseminating published events by matching events against the subscription of the nodes in their community, which they store.

This work provides a novel and efficient algorithm to compute social communities in a decentralized way, which could be used in many other approaches. The broker approach might be an issue with respect to scalability for several reasons. The load is not evenly spread as only brokers are in charge of disseminating events. Therefore they might be exhausted before other nodes. In addition, a fair amount of state needs to be maintained at each broker node and, more generally, at every node for community computation. Yet this is an interesting approach exploiting social connections between nodes to opportunistically disseminate events in a delay tolerant network.

8.4 Gossip-Based Data Distribution

Quite often, nodes in a MANET need to reliably share data items. In a wired setting, two basic approaches may be used to implement data sharing: data-pull and data-push. In a data-pull approach, data is kept in one central server (maybe with a limited number of replicas). However, due to the high failure and disconnection rate of mobile devices, this solution is not satisfactory for MANETs, as the data may easily become unavailable. In a data-push approach, data dissemination protocols such as the broadcast or publish/subscribe protocols described in the previous sections may be used to disseminate the data updates to all interested nodes. However, a pure data-push approach requires the interested nodes to store the data updates for later use. If the number of data items is large, nodes may quickly exhaust their limited local resources.

In this section we discuss a third interesting alternative that consists in combining data-push, data-pull and data replication to build a reliable distributed data-store, where all the nodes in the MANET contribute a portion of their resources to ensure that a copy of the data items can always be retrieved from the vicinity of any node.

In this alternative, data is initially pushed, creating a small number of replicas. This first stage aims at making at least one copy of the data easily reachable from any node. Later, nodes can pull the data from one of the nodes storing a replica. In the background, a gossip mechanism monitors the state of the cache of each node to ensure that the number and location of the copies remains adequate.

The location of the hosts storing each replica is a crucial decision for this approach. In some applications, for example, when there is geographical proximity of data producers and consumers or when data is not frequently requested, it may be preferable to store the replicas in the vicinity of the data source; in other applications it may be preferable to keep replicas close to the nodes that are more likely to access the data. This is the case when network partitions are frequent or when data is frequently accessed. The remainder of this section focuses on gossiping algorithms providing a uniform distribution of the replicas over the MANET.

8.4.1 Uniform Geographical Replication

Non-uniform dissemination is adequate if the relevance of the information decays with its distance to the source. However, this is not the case in the majority of the applications. In general, dissemination should uniformly distribute the replicas over the region covered by the MANET. There are two classes of reasons to promote a uniform distribution of the replicas:

Improved resilience to failures.   If replicas are geographically distributed, the probability of having all of them being affected by some localized malfunction of the network, such as interference, is reduced. If the network becomes partitioned, geographical distribution may also increase the probability of having replicas available in the different partitions of the network.

Decreased cost of data retrieval.   Assuming that there is no relation between the location of the nodes and the data they query, geographical distribution of the replicas reduces both the cost (in number of messages) and the time to access the data, because the probability of finding the data item in some close neighbor of the client increases.

PADIS [584] aims at the uniform dissemination of data assuming that either: *i*) there is no predictable (biased) access pattern to the objects or; *ii*) that this access pattern cannot be derived a priori or even during the lifetime of the system. We present in Fig. 8.4 an illustrative example of one run of PADIS. The figure depicts in black the nodes that store a replica of the item and in gray nodes that forward the message.

The dissemination begins with the broadcast of a message by the producer of the data item. The item is stored at the producer and included in the message (Fig. 8.4a). The messages also carries a *Time From Storage* (TFS) field which records the distance (in number of hops) from the node transmitting the message to the known closest copy of the item. The TFS that would be placed in a message forwarded by the node as part of the propagation algorithm is depicted at its center. Notice that this value may change with the reception of other copies.

Figures 8.4b and 8.4c show the progress of the dissemination. Nodes use the Pampa distance-based broadcast protocol (see Sect. 8.2.1) to propagate the data. Pampa provides two important contributions for the efficiency of the algorithm: *i*) a small number of nodes are required to retransmit the message; and *ii*) it selects for retransmission the nodes that are geographically more distant from the previous transmitters. Pampa imposes a small delay on all nodes before they retransmit a message. During this delay nodes count the retransmissions they hear and compute the minimum value of the TFS field of the retransmissions they receive. The minimum TFS observed is kept in a variable named *mTFS*. When forwarding the message, a node sets the TFS field to *mTFS*+1, accounting with the additional hop needed to reach the closest copy of the item.

A constant *Distance Between Copies* (DbC) dictates the maximum value of the TFS field and, implicitly, the degree of replication of the items. This example uses *DbC*=2. Figure 8.4d shows that a node with *mTFS*=2 at the end of the hold period

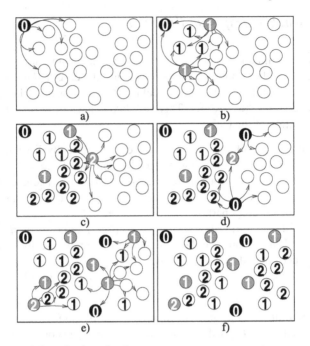

Fig. 8.4 Example of dissemination of an item

stores a copy of the item and retransmits the message. The TFS of the message is reset to 0 to let other nodes learn about the newly stored copy and update their *mTFS* variables accordingly (see, for example Fig. 8.4e).

The final state of the system after the dissemination of the item is depicted in Fig. 8.4f. Although only a small number of nodes have stored the item, a replica is stored at no more than *DbC* hops away from any of the nodes.

8.4.2 Effects of Movement in Data Placement

Intuition suggests that the movement of nodes could create scenarios where items are unevenly distributed, even if initially replicas were placed using a perfect algorithm. To assess if this intuition proves to be correct, we prepared a simple simulation where 100 nodes are initially deployed according to one of the Random Waypoint (RWP) or Manhattan (Man) movement models in a 1500 m × 500 m region. Nodes have a transmission range of 250 m and store a replica of each data item (*DbC*) every 5 hops. The configuration requires a small number of copies thus permitting to emphasize the impact of node movement on their distribution.

Initially, nodes do not move and use PADIS to distribute 100 items. The values depicted at time 0 in Fig. 8.5 are the average and the standard deviation of the

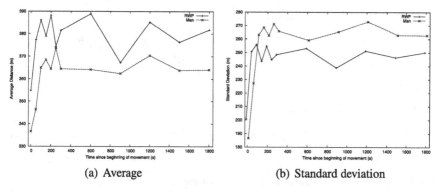

(a) Average   (b) Standard deviation

Fig. 8.5 Effects of movement in the distribution of the replicas

distance (in meters) of each node to the closest copy of each item at the end of
the dissemination period. As it can be observed, PADIS places a copy of each data
item on average at 336.64 m (resp. 354.81 m) of each node, in the Manhattan (resp.
Random Waypoint) movement model. The ratio of the standard deviation to the
average is at this stage of 55.4% (resp. 56.6%).

After the placement, nodes moved according to the predefined movement models
for 1800 s, without pausing and with a speed of approximately 10 m/s. The distance
of each node to the closest node storing a copy of each item was measured periodically. Figure 8.5 shows that the average distance increases rapidly as soon as nodes
begin to move and remains significantly higher than the initial value for the rest of
the simulation.

Relevant for our study is the property of the standard deviation estimator stating
that when comparing two values for the same average, the higher one will include
samples with higher distances to the average. The large increase (up to 50%) in the
standard deviation and the increase of the ratio between the average and the standard
deviation (which with node movement go up to 73.7% in Manhattan and 66.3% in
Random waypoint) shows that there is a much more irregular distribution of the
distance of the nodes to the closest copy of each item.

8.4.3 Shuffling Algorithms

The *Probabilistic Shuffling* algorithm [579] extends PADIS to continuously adjust
which data replicas are stored by each node, to address the uneven distribution resulting from node movement. Probabilistic shuffling does not require the transmission of dedicated control messages. Instead, nodes gossip fragments of the state of
their storage space piggybacked in the messages broadcast to retrieve data items.

The algorithm fills query messages with data items randomly selected from the
storage space up to the Maximum Transmission Unit (MTU) of the wireless media.
Data items are broadcast with two complementary goals: *i*) to inform nodes in the

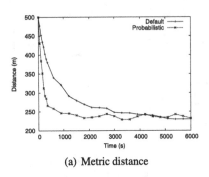 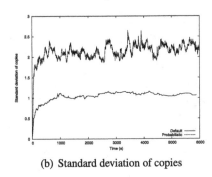

(a) Metric distance   (b) Standard deviation of copies

Fig. 8.6 Performance of probabilistic shuffling in a worst case scenario

vicinity about its cached items, so that duplicates can be discarded by neighbors; and *ii*) to create replicas of the discarded items in distant nodes.

A new replica will be created in a node that receives the message if all the following conditions are met:

1. the record is signaled in the message as being discarded;
2. the node is more than DbC hops from the source;
3. the node has storage space available;
4. a random number generator selects a certain number.

The randomness of the algorithm, which is fundamental to increase the diversity of records in each neighborhood, is increased by permitting forwarding nodes to change the content of the piggybacked payload (by replacing items added by other nodes with their own data items).

8.4.4 Illustration

To evaluate the benefits of the probabilistic shuffling algorithm, a simulation scenario called "worst-case" was prepared. In the simulation, 100 nodes were randomly deployed in a 1500 m × 500 m region. Each node produced one data item. In this scenario, four copies of each data item were deployed: one in the node that produced it and one in each of the remaining three nodes that were geographically closest. After the deployment of the replicas, 3000 queries were performed, distributed over 6000 s. Figure 8.6 compares the probabilistic shuffling algorithm with a baseline "Default" algorithm. In the default algorithm, a node decides to store a replica of a data item *i* if its query for item *i* received a reply from a node more than DbC hops away, randomly replacing some other item.

Figure 8.6a evaluates the average metric distance of each node to the closest copy of each data item. As it can be seen, the probabilistic algorithm rapidly redistributes the replicas, placing them, on average, much closer of every node in the network. A second aspect considered in this evaluation addressed the impact of the

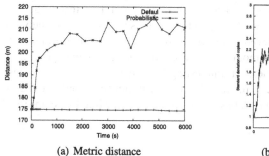

(a) Metric distance

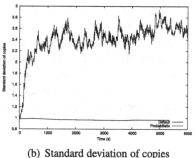

(b) Standard deviation of copies

Fig. 8.7 Performance of probabilistic shuffling in a standard scenario

algorithms in the number of copies. The results presented in Fig. 8.6b show the standard deviation of the number of copies, calculated after every update to the storage space of any node.[4] As expected, the probabilistic algorithm performs a more uneven distribution of the number of replicas. However, there are two relevant aspects that deserve some attention: *i*) like in the distance evaluation, the metric seems to stabilize. This suggests that the algorithm does not continuously benefit some data items by increasing its number of copies and; *ii*) a comparison between Figs. 8.6a and 8.6b show that the rapid convergence of the metric distance is achieved with a biased distribution of the number of replicas, improving the geographical distribution by reducing the number of replicas of those items that are placed in more central locations.

In another scenario, called "early", the same number of items were initially distributed using PADIS, thus providing an early geographical distribution of the replicas. Performance evaluation results depicted in Fig. 8.7 show that starting from a good distribution, the shuffling algorithm tends to slightly worsen the average distance. Interestingly, it was shown elsewhere [584] that PADIS creates more replicas than what could be predicted. The performance of the probabilistic shuffling algorithm is attributed to its relocation of some of the copies, thus effectively creating space for the storage of additional items.

8.5 Summary and Outlook

The characteristics of MANETs are such that designers must face several challenges, namely: (*i*) avoiding the need of rigid structures such as a tree, which are extremely difficult to maintain in an environment subject to frequent changes; (*ii*) limiting the amount of state information that should be maintained at each node as nodes in some MANET applications may have limited resources; (*iii*) implementing the trade-off between reliable dissemination (completeness) and flooding.

---

[4] The standard deviation is the relevant metric because the space for storing data items is constant.

An interesting aspect in gossip-based approaches for MANETs is that mobility is both a challenge and an asset: while it leads to frequent disconnections and network partitioning, it can actually be leveraged to opportunistically spread information in an epidemic manner. Leveraging mobility in this context is extremely dependent of the mobility patterns of the nodes. An interesting open research area is to devise techniques that are able to capture the mobility model in run-time, and to dynamically adapt both dissemination and data sharing protocols according to the observed mobility patterns.

Acknowledgments

The authors and indebted to Paul Grace and Eric Ruppert for their helpful comments on earlier version of this chapter. Work described in this chapter has been partially supported by project REDICO, PTDC/EIA/71752/2006, through FCT and FEDER.

Chapter 9
Application Layer Multicast

Mouna Allani, *University of Lausanne, Switzerland*
Benoît Garbinato, *University of Lausanne, Switzerland*
Fernando Pedone, *University of Lugano, Switzerland*

9.1 Introduction

An increasing number of Peer-to-Peer (P2P) Internet applications rely today on data dissemination as their cornerstone, e.g., audio or video streaming, multi-party games. These applications typically depend on some support for multicast communication, where peers interested in a given data stream can join a corresponding *multicast group*. As a consequence, the efficiency, scalability, and reliability guarantees of these applications are tightly coupled with that of the underlying multicast mechanism.

At the network level, IP Multicast offers quite good efficiency and scalability but best-effort reliability only. Moreover, the deployment of IP Multicast requires all routers to be appropriately configured, which makes it quite impractical in large scale or open settings, i.e., where one does not have full control over the networking environment. For this reason, several research directions have focused on trying to offer some form of multicast communication at a higher level, typically via an application-oriented middleware. Such a higher level data dissemination support is often referred to as *Application Layer Multicast* (ALM). Of course, ALM-based solutions to data dissemination usually rely on the lower level networking mechanisms, as suggested in Fig. 9.1.

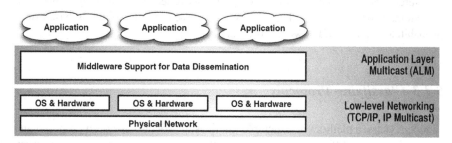

Fig. 9.1 Multicast support for data dissemination

9.1.1 From Peer-to-Peer to MANETs

Recently, mobile ad hoc networks (MANETs) have gained remarkable interest and popularity. By definition, a MANET represents a collection of wireless mobile nodes dynamically formed with the motivation that users can benefit from collaborating with each other. P2P and MANET applications share similar requirements; at the heart of these requirements, one usually find some form of reliable multicast.

While largely developed independently of each other, P2P and MANET applications share many properties: (1) Both P2P and MANET are self-organized and decentralized networks, relying on distributed components with no specific role. (2) They are both exposed to high *churn*, due to nodes frequently joining and leaving the network and, in the case of MANETs, to the mobility of the nodes. Given the similarities of both environments, their dedicated multicast protocols face similar challenges (e.g., coping with frequently changing topologies). We argue that due to their nature, multicast solutions designed to MANETs could in general be mapped to P2P. As we will discuss, the reverse is less straightforward as MANET components are usually limited on resources compared to P2P Internet components.

9.1.2 Probabilistic vs. Structured Approaches

Roughly speaking, data dissemination solutions can be categorized as following either a *probabilistic* approach or a *structured* approach. The probabilistic approach usually relies on a gossiping protocol, which consists in having each peer forward the data it receives to a set of randomly chosen neighbors. As a consequence, the path followed by the disseminated data is not deterministic. By contrast, the structured approach consists in first organizing the network peers into some overlay network and in routing disseminated data through this virtual topology.

The tradeoff between these two approaches is basically *scalability* vs. *resource consumption*. Structured dissemination uses fewer messages than probabilistic dissemination to ensure the same reliability. The reliability of a probabilistic approach hinges on the redundancy of sent messages; its scalability results from the fact that no overlay creation or maintenance is required. In this chapter, we focus on the structured approach, i.e., on multicast protocols based on an overlay network to disseminate data, whereas Chap. 8 discusses dissemination solutions based on a probabilistic approach, known as gossiping.

9.1.3 Classifying Protocols

In recent years, a plethora of structured multicast protocols have been proposed and there exists many different ways to classify them. Some of the protocols we survey were designed for MANETs and most of them were originally designed for Internet

P2P applications. It turns out that some P2P protocols could be used in MANETs; we discuss how applications in MANETs can benefit from them. In doing so, we characterize the proposed protocols using a small set of dimensions we consider the most relevant. To structure our discussion, we group these dimensions into three *viewpoints*, depending on their *applicative* focus, their underlying *architecture* or their *Quality of Service*.

9.2 Applicative Viewpoint

In this section, we discuss multicast solutions in terms of their *applicative* dimensions, i.e., those characterizing the applicative context in which structured dissemination is to be used. Intuitively, discussing these dimensions boils down to answering the three questions listed below.

- What is the usage focus of the multicast solution?
- What is the communication model it considers?
- Can MANET applications benefit from its use?

9.2.1 Usage Focus

Existing multicast solutions to achieving structured dissemination pursue a wide variety of goals, which are directly related to their main *usage focus*. Roughly speaking, these solutions typically focus either on some specific applicative usage (e.g., asynchronous or on-demand streaming), context-aware data dissemination, or on some system-level contraints. In turn, these contraints are usually expressed in terms of a Quality of Service (QoS) to achieve. Examples of QoS-oriented goals include minimizing latency or maximizing throughput, achieving reliability or scalability, coping with real-time constraints.

9.2.2 Communication Model

Another key applicative dimension to characterize multicast solutions is their communication model. Here we mainly distinguish between the *specific-source* communication model and the *multi-source* communication model. In a multi-source solution, the same overlay network can be used by *any peer*, while in a specific-source solution, only *one peer* can use a given overlay network.

When considering tree-based overlays for instance,[1] specific-source solutions create a separate and dedicated tree for each peer that needs to disseminate data.

---

[1] Sect. 9.3.1 discusses tree-based overlays in detail.

Therefore, these solutions tend to offer a more efficient multicast protocol than multi-source solutions. As the number of sources increases however, the overhead induced by creating and maintaining multiple trees grows. So multi-source solutions tend to make a better use of the available resources, be it memory, network bandwidth, or computing power.

9.2.3 MANET Suitability

As already suggested in Sect. 9.1, a large number of multicast solutions were devised to support P2P interactions on the Internet, out of which many, yet not all, can be transposed to MANETs. This is no surprise since interactions in a MANET are inherently peer-to-peer due to the absence of an infrastructure. For this reason, we characterize the multicast solutions discussed in this chapter using the notion of *MANET suitability* to indicate whether a multicast solution that was designed for P2P can be adapted to MANETs. In doing so, we restrict our discussion to how messages are routed. The question of how members join or leave a given multicast group is out of the scope of this survey.

Indeed, in spite of their similarities, P2P and MANET contexts tend to impose very different constraints when it comes to joining groups.[2] For example, several Internet-based P2P multicast solutions rely on a predefined set of fixed peers, known as *rendezvous points*, in order to join a multicast group. In the context of MANETs, this approach is simply impossible since by definition a MANET has no predefined set of members. With respect to how messages are routed, we rely on three criteria to decide whether a routing protocol can be used in the context of MANETs. If one or more of these criteria are not satisfied, we consider the multicast solution as not well suited for MANETs. These criteria are described hereafter.

Fully Peer-to-Peer. This criterion states that the routing protocol must strictly assume a *peer-to-peer communication model*, i.e., it cannot assume the existence of dedicated routing nodes. Furthermore, any routing solution that relies on logical IDs associated to nodes, e.g., obtained by hashing their IP addresses, is not suitable for MANETs. Intuitively, this criterion reflects the fact that MANETs do not rely on a predefined networking infrastructure and thus cannot assume the existence of a predefined set of nodes with known IDs. The notion of node identification, with its logical and physical variants, are further discussed in Sect. 9.3.1.

Dynamically Adaptable. This criterion implies that the protocol must be *self-configurable* and *self-improving*, i.e., it must be able to cope with a dynamically changing environment. In particular, the overlay structure it uses to route messages must be readjusted continuously as the network changes. The notions of self-configurable and self-improving networks are further discussed in Sect. 9.4.1.

---

[2] Discovering and joining multicast groups is a particular case of a more general problem known as *service discovery*, which is extensively discussed in Chap. 15.

Resource Friendly. This criterion captures the fact that mobile nodes are usually *resource constrained*, i.e., they have limited memory, processing and battery power and must use low bandwidth networks. For instance, protocols that minimize message forwarding help better share the limited wireless bandwidth of MANETs.

9.3 Architectural Dimensions

In this section, we discuss multicast solutions supporting structured dissemination in terms of their *architectural* dimensions. Intuitively, discussing these dimensions boils down to answering the four questions listed below.

- What is the topology of the overlay network?
- How is the overlay network created and maintained?
- Does the proposed solution rely on some infrastructure support?
- How is the networking environment modeled by the proposed solution?

9.3.1 Overlay Structure

As suggested in Sect. 9.1, structured dissemination usually implies the existence of some overlay network in charge of routing and delivering the data. The creation and maintenance of such overlays basically depends on two aspects: (1) how nodes are identified and (2) what is the topology of the overlay.

Node Identification

To define an overlay including all nodes in a multicast group, some approaches designate nodes using *physical IDs*, whereas other approaches rely on *logical IDs*.

Physical IDs. An overlay network based on physical IDs tends to match the underlying physical network and to better account for the resources and contraints of peers and links. With this approach, an *explicit overlay network* must be created prior to performing any data dissemination, usually adopting either a tree-based or a mesh-based topology. Multicast solutions based on physical IDs can often be adapted to MANETs precisely because they do not try to hide the physical contraints imposed by the underlying network.

Logical IDs. A logical ID may reflect some physical properties of its corresponding node (e.g., nodes proximity) or on the contrary it may be completely agnostic of the underlying physical network. The actual structure of the overlay network then depends on the motivation to use logical IDs. When the main goal is to achieve scalability and to minimize the maintenance overhead, the overlay structure is undefined and implicitly formed by the routing mechanism performed on logical IDs. Routing schemes based on distributed hash tables are an example of such implicit overlay structures. That is, logical IDs implicitly embed a notion of logical neighborhood, which is then used to route data from one peer to another. As a consequence, the multicast protocol does not need to explicitly create or maintain an overlay network, which tends to make it more scalable. When the goal is to balance the load of forwarding data, on the other hand, the overlay structure is explicit and usually follows a well-known topology (e.g., a tree, built on top of logical IDs).

When it comes to MANETs, multicast solutions based on logical IDs are usually harder to apply because they tend to hide important contraints of the underlying physical network. For example, two peers that happen to be located in the same neighborhood at the physical level might be quite distant in terms of logical neighborhood, and vice versa. As a result, latency is difficult to control, since data might have to be routed across many peers, some of which might be slow, and across high-latency links.

Overlay Topology

While a graph topology overlay is often associated to physical ID approaches, some Logical ID approaches build a graph on top of the node abstraction aiming to routing load balancing. The widely used overlay topology is the tree.

Tree-Based Topologies.   Tree-based overlay networks connect any two peers via a unique path and are acyclic, which greatly simplifies data routing. In addition, the way such networks are constructed tends to make them match the underlying physical network. For example, when the outbound bandwidth of a peer is limited, the number of children of that peer in the tree tends to be proportionally limited as well.[3] On the other hand, tree topologies are very sensitive to failures or partitioning: as soon as a non-leaf peer leaves or crashes, the tree breaks, which prevents members of the partitioned subtree from receiving the disseminated data.

Interestingly, tree-based overlay networks can be used to disseminate data both from a single source or from multiple sources. In addition, some multicast solutions rely on hierarchical clusters of peers to create and maintain various source-specific delivery trees.

Mesh-Based Topologies.   Various multicast solutions consist in first building a mesh-based overlay network and then deriving a data dissemination tree on top of that mesh. The mesh-based overlay network is usually constructed using some knowledge about the underlying network, so as to improve overall performance overlay. Note that it is possible to take advantage of standard routing protocols to construct the data dissemination tree, in which case the routing protocol automatically handles potential loop problem.

Creating and maintaining an intermediate mesh-based overlay has several advantages over directly building a data dissemination tree. For example, a mesh-based

---

[3] As we shall see in Sect. 9.5, many different tree construction and maintenance techniques have been proposed, resulting in overlay networks exhibiting various interesting properties.

overlay better tolerates peer failures than a tree structure because it contains multiple paths from the data source(s) of a multicast group to its members. In addition, the existence of multiple paths in a mesh makes it possible to optimize the overlay, e.g., for partition recovery, load-balancing.

9.3.2 Overlay Creation and Maintenance

The overlay creation and maintenance process can either be *centralized*, in which case it is the responsibility of a single peer, or *distributed*, in which case the responsibility is shared among several peers. Distributed creation and maintenance offers several advantages over a centralized approach. For example, it is more resilient to failures, since the crash of one peer will not prevent the construction and maintenance of the overlay network. It is also more scalable, since this process relies on several peers and thus tends to avoid bottleneck issues.

Building an overlay in a distributed manner, however, induces a potential loss of simplicity and optimality, because no peer has a global view on the constraints of the underlying physical network. Some multicast solutions propose an hybrid approach, based on a controller peer to manage the multicast group, e.g., for joining or leaving the group, while the overlay construction and maintenance is still performed cooperatively.

Some protocols rely on a distributed and *layered* creation and maintenance process. In such a layered architecture, the multicast group is divided into a hierarchy of clusters, each containing a set of peers and one being the leader. This layered architecture is hierarchical in the sense that clusters are grouped into layers, where all leaders of some layer are also (simple) members of a cluster in the higher-level layer. A leader has then the responsibilities to organize members of its cluster and to represent them in the whole system. On the one hand, a layered architecture is scalable and has rather low overhead in terms of network overlay management, since this is done concurrently in each cluster. On the other hand, cluster management induces an additional (but local) overhead. Furthermore, all optimizations are done locally to each cluster, so peers in different clusters are unable to establish overlay links. This may lead to a suboptimal overlay network.

9.3.3 Infrastructure Support

The majority of multicast protocols do not rely on any infrastructure, i.e., they are pure *peer-to-peer* solutions. Nevertheless, some solutions rely on strategically deployed *overlay proxies*, such proxies being typically dedicated to message routing, overlay creation and maintenance, etc.

The advantage of using proxies lies in the fact that they can be deployed on strategic hosts, those that are more reliable or have more resources. One can also make them aware of both low-level and high-level contraints. In the case of on-demand data streaming for example, proxies can arbitrate the various applicative demands, based on their knowledge of the underlying networking constraints. However, proxies may become bottlenecks or single points on failure.

9.3.4 Networking Environment Modeling

In order to create and maintain an overlay network, peers must somehow model their physical networking environments, e.g., in terms of their neighborhood, of the message loss probability for their outgoing links.[4] Usually, members of a multicast group have only a *local knowledge* of their environment, and more seldom a *global knowledge*. Typically, this knowledge is built up using techniques based on some probing interactions between peers.

For multicast solutions, the choice between global knowledge and local knowledge translates into a tension between efficiency and scalability. Indeed, the more a group of peers knows about its networking environment, the more efficient the multicast protocol will be. However, modeling a very large networking environment potentially requires storing a large amount of information, which challenges the scalability of solutions based on global knowledge.

9.4 QoS Viewpoint

In this section, we discuss multicast solutions supporting structured dissemination of data in terms of their *Quality of Service* (QoS). The notion of QoS being very broad, as well as the range of QoS guarantees offered by existing multicast solutions, it is not our intention to provide an exhaustive characterization of those solutions in terms of QoS. Rather, we focus on a limited set of QoS dimensions that we believe are the most relevant to the subject, namely *adaptivity, scalability* and *reliability*. Intuitively, discussing these dimensions implies answering the three questions listed below.

- How does a multicast solution adapt to changes in its environment?
- How does it scale when the number of peers increases?
- How reliable is the network overlay it creates?

---

[4] Of course, this observation does not apply to multicast solutions based on *logical ID* overlays, discussed in Sect. 9.3.1. Indeed, the very nature of *logical ID* overlays is to be decoupled from the underlying physical network.

9.4.1 Adaptivity

Adaptivity captures the degree to which a multicast solution adapts to a changing environment, e.g., changes in the underlying physical network, changes in the membership of the multicast group. In terms of adaptability, we distinguish between three degrees of multicast solutions: *static, self-configurable* and *self-improving*.

A static solution typically builds its dissemination overlay once, assuming that all members of the multicast group are known and will not change in the future. On the contrary, a self-configurable solution builds its dissemination overlay incrementally, based on a continuous stream of peers joining and leaving the multicast group. Finally, a self-improving solution continuously improves its dissemination overlay, as ways to make a better usage of the underlying physical network are discovered. Depending on its usage focus, a given self-improving solution will aim at adapting to one or more properties of the physical network, e.g., network latency or bandwidth, links reliability.

In contrast, some solutions relying on a Logical ID overlay network can never exhibit a self-improving behavior, as it does not take the underlying physical network as its starting point. Such solutions are sometimes said to be *network agnostic*.

9.4.2 Scalability

The scalability of a solution depends on several factors, such as the resources required to create and maintain its overlay network, the scope of knowledge each peer needs to maintain in order to model its environment, etc. These factors are almost always directly correlated with the number of peers in the multicast group. For this reason, this survey characterizes the scalability of a multicast solution via the maximum size of multicast groups it can support. Such sizes can either be *small* (tens of members), *medium* (hundreds of member) or *large* (thousands of members).

9.4.3 Reliability

When reliability is considered, the majority of multicast solutions offer a reactive mechanism to recover from peer or link failures. Fewer solutions also rely on a more proactive strategy, which aims at preventing such failures. A typical prevention mechanism consists in creating a reliable and possibly redundant overlay, which includes the most reliable network components (peers and links). The reliability of components is then usually computed by probing, as discussed in Sect. 9.3.4.

In this survey, we classify the reliability of multicast solutions as *reliable* if the solution provides an explicit failure prevention, as *fault-tolerant* if it simply proposes a mechanism for recovering from peer or link failures, and as *best-effort* if it only relies on the peer-to-peer reliability offered by low-level networking layers (e.g., TCP/IP).

9.5 A Plethora of Protocols

In the following, we illustrate the dimensions discussed in the previous sections by detailing some multicast protocols for structured dissemination. Since they are quite numerous, we structure the presentation by classifying them according to the topology of their overlay network, as discussed in Sect. 9.3.1.

9.5.1 Tree-Based Overlay

ALMA. The Application Level Multicast Architecture (ALMA) [883] targets video streaming across heterogeneous networks. It has three components: a centralized Directory Server (DS), a Web Server (WS), and the client nodes. Data streaming is undertaken over the overlay tree formed by the clients. The DS is responsible for maintaining the tree and does not participate in the actual data delivery, while the WS provides a global user interface for sources to advertise their sessions and for clients to discover the sessions available. Different dedicated trees are created for different sources. A client who wishes to join an advertised session first contacts the WS and chooses the session to join. Next, it issues a join request to the DS which in response sends back a list of potential parents. The client selects the best parent according to some metric such as the round trip time (RTT), measured at the application level. A new branch is then added to the propagation tree and the parent begins streaming data to the new client.

All clients in the tree overlay are organized into levels. Level 1 clients are between the source and the rest of the clients, and therefore are seen as synonymous to the source; clients at higher levels have joined the session at a later time. Every client on the tree monitors RTT and gossips with potential parents from the list initially sent by the DS. A client can decide to switch to another parent if this improves the overall performance. The gossip list of a client is dynamic: if possible parents become incapacitated and the list diminishes below a certain threshold, the client requests a new list from the DS. In order to improve robustness, the tree structure is enhanced with a simple spiral mechanism that withstands node failures and overlay partitions—as long as consecutive nodes of the same branch do not fail. The spiral mechanism allows a client to maintain a connection with its grandparents in the overlay, making it possible for the client to reconnect to its grandparent if the parent fails, instead of requesting a new list of potential parents from the DS. If consecutive nodes fail, the gossip mechanism will be used by affected clients to switch parents.

ALMI. The Application Level Multicast Infrastructure (ALMI) [659] is a multicast for interconnected groups of relatively small size (e.g., tens of members). Potential applications include video conferencing, multi-party games, private chat rooms, web cache replication and database/directory replication. End hosts are connected using unicast connections according to a minimum spanning tree built using application-specific performance metrics, such as the round-trip delay between group members.

Every ALMI session consists of a centralized component, the session controller, and session members. The session controller handles member registration and maintains the multicast tree; it is assumed to be accessible by all members. The session controller performs two functions. First, it ensures connectivity of the spanning tree when members join and leave the session or in the presence of network or host failures. Second, it is responsible for periodically recalculating the tree efficiency based on the measurements received from session members. To avoid single points of failure, the protocol can be extended to support multiple back-up session controllers, known to all session members, that operate in stand-by mode and periodically receive the state from the primary controller.

Although the multicast tree is calculated by a central component, and so it should be loop free, tree update messages are disseminated to all session members independently, and can be lost or received out-of-order by different group members, resulting in loops and tree partitions. To prevent such problems, the session controller assigns version numbers to trees. When generating a packet a source includes the latest tree incarnation number in the packet header. Each communication participant caches recent multicast trees. After a packet is received, the receiving node checks whether the tree version is in its cache. If so, it forwards the packet following this tree version; otherwise the packet is discarded. If a member receives a packet with a newer tree version, it contacts the controller to receive a new tree.

BTP. The Banana Tree Protocol (BTP) [365] is a simple multicast for networks of a few hundreds of nodes. BTP is unreliable; it assumes that reliability can be provided by the protocol implemented on top of it. The protocol assumes a bootstrap mechanism that allows multicast nodes to find out about some of the nodes in the system and to connect to them. The first node to join the multicast becomes the tree root. Later nodes join the group by choosing one node to connect to (e.g., the root node). A node is allowed to switch its parent to optimize the tree cost, where the cost of a link is the unicast latency, i.e., the distance between two nodes on the physical network. To avoid loops, a conservative policy limits the switch of a node only to its sibling. Moreover, a node rejects all switching attempts if it is itself executing a parent switching. Tree partitioning is resolved by having a node whose parent has left the group reconnect to the root node. Thus, BTP is more eligible for systems where nodes do not leave or fail frequently. Two different protocols were built on top of BTP. The Banana Tree Simple Multicast Protocol (BTSMP) is a wrapper around BTP that provides many-to-many group communication for sending and receiving packets. The Banana Tree File Transfer Protocol (BTFTP) provides reliable one-to-many file transfer.

HMTP. The Host Multicast Tree Protocol (HMTP) [897] takes advantage of IP multicast infrastructure to build an overlay multicast layer. Its key design goals are: (1) to be deployable in the Internet; (2) to be compatible with IP Multicast; and (3) to be scalable. To achieve these goals, the proposed protocol automates the interconnection of IP multicast islands, i.e., designated members self-organize into a multicast tree that connects the various IP multicast islands. The tree structure is revisited periodically to deal with changes in the network topology and the group membership.

The environment modeling in HMTP is rather static, as it assumes the existence of IP multicast islands and the protocol basically consists in bridging these islands. Members of a multicast group can join and leave the group, which triggers an algorithm aimed at repairing the tree. Furthermore, so called designated members, responsible for tunneling requests from one island to another, can also dynamically decide to quit their tunneling role. HMTP periodically reruns the join procedure to accommodate changes in the network and group membership.

MAODV. The Multicast Ad hoc On-Demand Distance Vector (MAODV) [719] is a multicast solution based on the AODV routing protocol designed for MANETs. It builds a multicast tree on-demand, connecting multicast group members. Such a multicast route is discovered on-demand using a broadcast route discovery mechanism. When a node needs a route to a destination, it broadcasts a *Route Request* (RREQ). A node with a current route to that destination or the destination itself unicast a *Route Reply* (RREP) back to the source node. Information gathered through RREQ and RREP messages defines the multicast route.

In MAODV, each process maintains two tables: The *Route Table* is used for recording the next hop for routes to other nodes in the network. This table is filled by adding a new entry at the reception of an RREQ or an RREP message for the first time. Mainly, an entry includes the destination's IP address, its sequence number, and the next hop. The *Multicast Route Table* contains entries to the multicast group the node is the router. New entries are added to this table by the reception of a *Multicast Activation* message (MACT) from the source node that previously asked for a route for a multicast group. A node that broadcasts an RREQ for a multicast group often receives more than one RREP. The source keeps the received route with the greatest sequence number and the shortest hop count to the nearest member of the multicast tree. Then, it unicasts the MACT to this selected next hop. If this node is not a member of the multicast tree, it forwards the MACT to the best next hop from which it received a RREP and enables the corresponding entry in its multicast route table. This process continues until reaching a tree member that already generated a RREP. The *Multicast Activation* message builds a loop-free multicast tree. Consequently, nodes only forward data packets along activated routes in their multicast route tables.

NICE. NICE [60] is designed to support low-bandwidth data streaming applications with large receiver sets, such as real-time stock updates and quotes. An average member has limited knowledge about other members in the group, i.e., it maintains state for a constant number of other members. NICE creates a hierarchical arrangement of members by assigning them to distinct layers numbered sequentially with the lowest being layer zero. Hosts in each layer are grouped into a set of clusters according to some metric. This hierarchical arrangement is used to define overlay structures for control messages and data delivery paths. The control overlay is a cyclic structure where a member has connections to all peers in the layers it belongs to. The neighbors in the control topology periodically exchange soft state refresh messages. A cyclic structure with high connectivity makes the protocol converge quickly. The data delivery path on the other hand needs to be loop-free to avoid packet duplications.

NICE chooses the data delivery path to be a source-specific tree where in each cluster the topology is a star. The forwarding rule used in NICE enables a source to send a data packet to all its peers in all the clusters it belongs to. A received package will be forwarded further only if the receiving host is the cluster leader. To keep the size of each cluster bounded, whenever it exceeds the upper size limit, the cluster is split into two clusters of the same size; or if the size gets below the bottom limit a merge operation is performed. By having all members send periodic probe messages carrying new information according to the metric used, cluster refinements are enabled. It is possible for members to have distinct inconsistent views of the cluster membership due to update transition periods and for transient cycles to appear on the data path. These cycles are eliminated once the protocol restores the hierarchy invariants and reconciles the cluster view for all members. Finally, the protocol assumes a Rendezvous Point (RP) that all members must be aware of. The RP can be a member at the top layer of the hierarchy or a dedicated host that knows about the member that is the leader of the topmost layer.

OMNI. The Overlay Multicast Network Infrastructure (OMNI) [61] targets real-time applications such as media streaming. OMNI consists of a set of service nodes named Multicast Service Nodes (MSNs) and a set of clients, i.e., end-hosts interested in receiving multicast data. MSNs are deployed on strategic positions in the network. They organize themselves in a distributed fashion to form a spanning tree that represents the multicast data delivery backbone. An important property of OMNI is its ability to adapt the overlay tree to changing network conditions and the changing distribution of clients at the different MSNs. The later is conducted by managing the MSNs' position in the overlay depending on their relative importance represented by the number of clients they are responsible for. OMNI strives to minimize the latency of the overlay structure and to bound the out-degrees of the multicast service nodes.

The multicast source is connected to a MSN called the root MSN. The protocol requires each MSN to maintain a limited state information for all its tree neighbors and ancestors including unicast latency measurements and the overlay path from the root to itself. The initialization of the overlay structure is performed off-line, prior to the data distribution. During this initial phase every participating MSN measures the unicast latency between itself and the root MSN and sends this information to the root. After gathering all the messages from different MSNs, the root orders MSNs according to the latencies. Furthermore, it uses a centralized algorithm for constructing the initial data delivery tree respecting the maximum out-degree requests for each MSN, and distributes it to the MSNs. Constructing the tree that will represent the optimal solution to the minimum overall latency problem with an additional out-degree constraint belongs to the class of NP-hard problems.

Assuming that the out-degree of all MSNs is d, the root first chooses d closest MSNs in terms of the latency as its direct children. The next d closest MSNs are then assigned to be children of the closest child to the root, then the next d closest MSNs are assigned to the second closest child of the root, and so forth. In order to cope with dynamic environments where network conditions may change during time, OMNI proposes periodic transformations to the overlay involving nearby MSNs within two levels of each other. These transformations allow MSNs or clients to swap among themselves to improve tree properties in terms of the overall latency.

oStream. oStream [205] targets the unpredictability of user requests in on-demand media distribution. In particular it addresses *asynchrony*, where users may request the same media object at different times; *non-sequentiality*, where users do not request data sequentially (from beginning to end); and *burstiness*, where the request rate for a certain media object is highly unstable over time. The idea is to establish a minimum spanning tree and use media buffering at the hosts to aid in the distribution of asynchronous service requests for the same streaming media. To do so, oStream defines a *Temporal Dependency Model*, which determines whether user requests overlap, and can therefore be reused; the model also determines the order in which requests should be retrieved in order to maximize reuse.

oStream builds a Media Distribution Graph (MDG) that includes requests as nodes connected with weighted directed links. The weight of a link represents the number of hops between the two end hosts carrying the connected requests. MDG changes dynamically to insert new requests and remove terminated ones. Based on MDG, oStream constructs the Media Distribution Tree (MDT), which is the minimal spanning tree on the overall transmission cost of media distribution. MDT is constructed and maintained incrementally and in a distributed manner. A node (representing a request) that leaves the MDT, first deletes itself and then notifies its children to find a new parent. A node joining the MDT first finds a parent whose transmission cost is minimal and then notifies its successors to consider it as their new parent if it offers a lower cost.

Overcast. Overcast [412] is a multicast solution that builds the overlay tree incrementally. The tree includes the source as the root and dedicated nodes called overcast nodes. Using bandwidth estimation measurements, Overcast organizes the dedicated nodes with the goal of creating high bandwidth channels from the source. To allow a quick join of clients, Overcast maintains a global status at the root, built from an information table maintained by each node about all nodes lower than itself in the hierarchy and a log of all changes in the table.

The source stores content and schedules it for delivery to the Overcast nodes. A multicast group is represented as an HTTP URL: the hostname portion names the root of an overcast network and the path represents a particular group in the network. Typically, once the content is delivered, the publisher at the source generates a web page announcing its availability. The final consumers of the content in an Overcast network are HTTP clients. When a client clicks on the URL for published content, Overcast redirects the request to a nearby Overcast node that serves the content. By representing a multicast group as an HTTP URL, Overcast allows an unmodified HTTP clients to join the multicast group. It also permits the archival of content so that the client can specify a starting time when joining an archived group, such as the beginning of the content.

The overlay tree is continuously reevaluated. Each node periodically reassesses its position in the tree by measuring the bandwidth to its current siblings, parent, and grandparent. As a result, nodes constantly reevaluate their position in the tree and an overcast network is inherently tolerant of nodes failures.

RanSub. RanSub [471] distributes a random subset of the participants to each node; these subsets change periodically and with a uniform representation of all participants. The execution proceeds in epochs, where an epoch consists of a distribution phase, during which data is distributed down an overlay tree, and a collection phase, during which data is aggregated up the tree. In the collection phase each participant propagates to its parent a randomly selected *collect set* containing nodes in the subtree it roots. In the distribution phase each node sends to its children a *distribute set* composed of randomly chosen participants from the collect sets gathered during the previous collect phase and the received distribute set. Once the root receives all collect sets from its children, it launches the distribution phase.

SARO. The Scalable Adaptive Randomized Overlay (SARO) builds on RanSub to construct adaptive overlays constrained by degree, delay and bandwidth. During each epoch, nodes measure the delay and bandwidth between themselves and all members of their distribute set. Each SARO node A performs probes to members of the distribute set that RanSub transmits to it to determine if a remote node B would deliver better delay or bandwidth to its descendants. If so, A attempts to move under B by issueing the add request to B and waiting for the response. If the request is accepted, A notifies its old parent, communicates its new delay from the root to all its children, and notifies the new parent B of its farthest descendant. The use of RanSub for SARO is circular: SARO uses RanSub to probe peers and locate neighbors that meet performance targets, and at the same time, RanSub uses the SARO overlay for efficient distribution of collection and distribution messages. SARO requires no global coordination to perform overlay transformations. A special instance of RanSub ensures a total ordering among all participants such that no two simultaneous transformations can introduce loops into the overlay.

RMX. RMX [165] proposes to partition the heterogeneous receiver set into a number of small homogeneous data groups. Each data group is assigned a Reliable Multicast proXy (RMX). RMXs represent strategically located agents organized into an overlay network. Placing RMXs strategically to form an overlay network is a hard problem. RMX uses manual configuration of receivers into independent data groups that are bridged by a manually constructed spanning tree of RMXs.

A network of application-aware RMXs uses detailed knowledge of application semantics to adapt to heterogeneity constraints. As data flows through an RMX, it dynamically alters the content of the data or adapts the rate and ordering of data objects. Each receiver can define its own level of reliability and decide how and to what extent individual data objects can be transformed and compressed. A source or a receiver joins the specific RMX session by subscribing to its local data group. RMXs spread across the data groups organize themselves into a spanning tree via unicast interconnections. To communicate with other RMXs across wide-area links, an RMX uses a reliable unicast communication protocol such as TCP; within each data group, RMX relies on multicast forwarding. Data dissemination is performed in the following manner: a source generates data and distributes it to its local group and the corresponding RMX. The RMX forwards data from the local group towards the participants in the other groups. Whenever an RMX receives data on a link, it forwards it to all of its remaining links using spanning-tree flooding.

The recovery of lost data is performed in such a way that in the first attempt the receiver tries to retrieve the data form the local group. If this fails, the group's RMX forwards the recovery request upward along the RMX hierarchy towards the source of the data. Intermediate RMXs first attempt to recover the data from their local group, and if that fails, forward the request toward the source. Each RMX maintains a table that is used to determine the path toward sources from that RMX.

TAG/STAG. Topology-Aware Grouping (TAG) [486] is a single-source multicast protocol that exploits network topology information in the overlay construction. It assumes that it is possible to obtain the underlying path and bandwidth information to a designated member by using one of the existing route discovery mechanisms. The shortest path information that IP routers maintain is exploited when performing the path matching algorithm. This algorithm allows a TAG member to select as a parent a member whose shortest path from the source has the longest common prefix, in terms of the number of hops, with its own shortest path from the source. Alternatively, a partial matching algorithm can be used, thereby a node may join a parent that provides enough bandwidth, even if that does not follow the longest prefix matching scheme. After joining a multicast group the member maintains a family table (FT) where parent-child relationships for this node are stored.

The multicast tree management protocol encompasses three main actions: member join, member leave, and member failure. A new member that wishes to join a session sends a JOIN message to the source (the root of the tree). Upon the receipt of the JOIN message, the source computes the underlying paths to the new member and executes the path matching algorithm. If the parent of the new member is the source itself, the FT of the source is updated and the joining procedure is completed; otherwise, the source sends a FIND message to its child that shares the longest prefix with the new member's path. The procedure continues until the appropriate parent for the new member is found. To leave a session, a member sends a LEAVE message to its parent together with its FT. Upon receipt of the LEAVE message, the parent removes the leaving member from its FT and takes over its children by adding their FT entries to its own FT. To detect failures, session members periodically exchange reachability messages with their children. The failure of a child is detected as the absence of exchange data, in which case the parent simply discards the child from its FT. When the failure of a parent is detected, the child must rejoin the session.

STAG. The Subnet Topology-Aware Grouping (STAG) proposes two variations to the join procedure in TAG: reverse control and delay control. With reverse-control, in parallel to sending the JOIN message to the source node, a new node broadcasts the JOIN message in the subnet. As a result, some subnet nodes may reply back with a HELLO message. The new node accepts as a parent the node whose response arrives first and sends an OK message with the new parent's address to the other nodes. Since the join request is sent to the source as well, there is no time loss in cases where the node does not find a parent in its subnet. However, if the parent is found within the subnet, a STOP message is sent up the tree to stop the continuation of the join process. With delay-control, the two actions are separated and performed sequentially. The new node first broadcasts a JOIN message within the subnet. If it does not receive any HELLO message in response for a defined time interval, it will send the JOIN message to the source.

TBCP. In the Tree Building Control Protocol (TBCP) [558] participating members have a partial knowledge of the network topology and of the group membership before the spanning tree is computed. Every host in a TBCP tree is allowed to constrain its out degree by setting up the maximum number of children it is willing to support. In doing this, the host can control the amount of traffic it will be responsible for. When a new node wishes to join the tree and become a member of a multicast group, it uses the root of the spanning tree as a rendezvous point, assumed to be known by the node. It then sends a HELLO message to the root, which replies back with the list of its children and starts a timer waiting for the response from the new node. The root is not allowed to accept any further HELLO messages before the join procedure is completed or the timer expires. The new node estimates the distance (based on host-to-host measurements) to every child of the root and sends this information in a JOIN message to the root. If the timeout has not expired, the root evaluates all possible configurations based on the measurements received from the new node. If it judges that becoming the parent of the new node is the best configuration, it replies with a WELCOME message, the new node acknowledges it, and the joining procedure is completed. Alternatively, the root node may decide to redirect the new node to one of its children, thereby the join procedure starts again with the root's child playing the role of the new root. If on the other hand the timeout expires, the root sends a RESET message to the new node signalizing that it must restart the join procedure. Additional rules for the overlay tree construction can be used to improve the efficiency and the shape of the overlay tree. Such rules assume a hierarchical organization of the receiver nodes in a certain number of domains, so that receivers belonging to the same domain are grouped in the same sub-tree. TBCP considers the distance metric used for estimating the score of a configuration to be an application specific.

TOMA. Two-tier Overlay Multicast Architecture (TOMA) [490] requires infrastructural support for providing the multicast service. It implements an aggregated multicast approach in which multiple groups share one delivery tree. Although this reduces the tree management overhead as well as the amount of multicast state information stored in routers, it may result in some hosts receiving information they have not subscribed for, leading to bandwidth waste.

Special service nodes referred to as member proxies (edge nodes) and host proxies (internal nodes) are connected in a tree structure. This overlay proxy placement is an optimization problem defined as follows: given the number of group members for all routers and the shortest distance between any two routers, find at most K routers as proxies such that the weighted sum of distances from each router to its nearest proxy is minimized. End hosts, both sources and receivers, are grouped in clusters according to a distance metric and connected to a tree structure called P2P multicast tree.

Each TOMA group has a unique URL-like identifier. End users subscribe to the specific group by sending a request to the group membership server, which decides about the member access and maintains the group membership information. A list of IP addresses of advertised member proxies is obtained from a DNS server, and one is selected from the list, taking into account latency and load. The selected member proxy creates the P2P multicast tree using a centralized approach.

When an end host wishes to multicast a message, it communicates with other participants and disseminates data. The message is transmitted to all other users in the same cluster and to the member proxy in charge of that group. This member proxy will disseminate the message to other member, and they will send the data further on to the group members in their clusters.

AEMD/GASR. AEMD (Adaptive Algorithm for Efficient Message Diffusion) [305] and GASR (Gambling Algorithm for Scalable Resource-Aware Streaming) [16] aim at reliable information diffusion. Both algorithms use a modular approach in two layers: while the lower layer is responsible for the environment modeling, the upper layer builds a spanning tree based on the environment modeling to reach all destinations.

To measure the reliability of links and nodes, nodes probe their direct neighbors. The environment approximation solution permits each node to have a continuously updated view of the system, including the components' reliability and the topology. The node's view is a subgraph of the system's topology, built incrementally. In AEMD this view is complete, in which case nodes try to approximate the entire topology; in GASR it is partial, defined by the maximum diameter of each node's view.

AEMD uses an overlay named Maximum Reliability Tree (MRT) built in a centralized manner by the source when a message should be disseminated. The MRT is built in two steps: First, the source selects the most reliable components to include in the MRT, using the information from the environment modeling layer. Second, an optimize function assigns to each of the included links in the MRT the number of retransmission necessary to reach a given success probability.

GASR builds a Maximum Probability Tree (MPT) to stream data. The MPT matches the physical properties of the system and is built in a distributed manner. Some nodes, called *Incrementing Nodes*, contribute to the construction of MPT using theirs respective partial knowledge about the system. The process is incremental, starting from the source, by aggregating subtrees covering different partial views. The goal is to maximize the probability to reach all members in the neighborhood of a node given a fixed quota of messages that can be used per node.

9.5.2 Mesh-Based Overlay

Narada. Narada [178] proposes an architecture where multicast is implemented assuming only unicast IP service. The protocol is robust to failure of end systems and to dynamic changes in group membership.

Narada builds an overlay structure in a self-organized and fully distributed manner. End systems gather information of the network characteristics using passive monitoring and active measurements. An overlay structure representing a spanning tree is constructed in two steps. First, Narada builds a mesh with desirable performance properties such as the quality of the path between any pair of members, using for example delay or bandwidth, and a limited number of neighbors each member has in the mesh. For doing so, Narada requires all the nodes to have a global knowledge, i.e., to maintain a list of all other members in the group. The protocol continuously refines the overlay structure as new information is available.

An out-of-band bootstrap mechanism is assumed to exist that allows a member to retrieve a list of group members. This list needs neither to be complete nor accurate, and must contain at least one currently active group member. Dynamic changes are taken care of by allowing the protocol to incrementally improve the mesh quality by adding or dropping some of the overlay links. Members probe each other at random and gather information about utility of links. The decision on adding or dropping a link is based on these measurements. A good quality mesh must ensure that for any pair of members there exist paths along the mesh which can provide performance comparable to the performance of the unicast path between the members. The precise utility function depends on the performance metrics that the overlay is optimized for.

ODMRP. ODMRP [499] is a mesh-based multicast routing protocol. It particularly aims at mobile ad hoc wireless networks as it reduces the link/storage overhead. It is based on the concept of a *forwarding group*, which represents a set of nodes responsible for forwarding multicast data on shortest paths between any member pairs, building consequently a *forward mesh* for each multicast group. ODMRP has two phases: a request phase and a reply phase. In the request phase, the multicast source periodically sends to the entire network an advertising packet, named *Join Request*. When a node receives a *Join Request* for the first time, it stores the upstream node ID and rebroadcasts the packet. When the *Join Request* reaches a multicast receiver, the receiver creates or updates the source entry in its member table. At the reply phase, each node sends periodically a *Join Table* to its neighbors. When a node receives a *Join Table*, it checks if one entry corresponds to its own ID. If so, that node realizes that it is on the shortest path to the source and thus is part of the *forwarding group*. Forwarded by the *forwarding group* member, the *Join Table* reaches the source via the shortest path. This implicitly constructs and maintains a mesh with the *forwarding group* ensuring the shortest path from the sources to the receivers.

OverStream. OverStream [524] uses a heuristic approach to construct both a delay- and degree-bounded application-level multicast tree in a distributed fashion. The algorithm constructs an application-level multicast tree (ALMT) intended to provide an efficient solution to live audio and video data streaming in a single source scenario. The solution takes into consideration the essential characteristics of live streaming applications, such as delay sensitivity, real-time constraints and the amount of concurrent continuous user requests. This results in an NP-hard optimization problem referred to as Maximum Sum of Nodes Multicast Tree (MSNMT) subject to delay and out degree bound constraints. The ALMT is constructed on top of the overlay network such that the tree has the source node as its root. In ALMT there is an overlay path from the source to every other node. The delay constraint requires that the overlay path delay of any node in the tree be less than a predefined delay bound. The degree constraint requires that the out-degree of any node in ALMT be less than a maximum out-degree.

OverStream works as follows: initially, each node continuously sends probe messages to other nodes in order to approximate the overlay. Nodes approximate only a portion of the network and construct an incomplete directed overlay. Afterwards, the ALMT is constructed in a fully distributed manner using a heuristic named the Powerful Propagating Ability First (PPAF). When a new node wishes to join the ALMT, it first constructs a sub-tree having itself as a root and calculates PPAF for every node in its probed node set. PPAF estimates the sum of nodes that a child node brings to a parent, so that a child can choose a node with the maximal PPAF as its parent. If the parent node is saturated and incapable of supporting more children, the child with the minimal PPAF is chosen from the set of all parent's children, the new node takes its place, and the old child is requested to perform a rejoin procedure. To leave the ALMT a node sends a leave message to its parent node and sends rejoin messages to all of its children. If a node detects missing control messages it infers the failure of its parent, and as a reaction, it performs a node join procedure and sends messages to its siblings requiring them to perform rejoin as well.

Yoid. Yoid [290] employs a single shared tree of all members of the group and creates a mesh topology to recover from tree partitions. The first member to join a tree becomes its root. Each subsequent joining member contacts a centralized rendezvous point (RP) and obtains a list of members. These members represent the candidate parents of the joining member. The parent selection depends on a performance metric, e.g., loss rate, latency. To optimize the tree, each node periodically queries RP for updates on their initial lists of candidate parents. A node is allowed to switch to a new parent if this one provides a better latency or packet loss performance. To guarantee a non-partitioned mesh topology, each member M establishes a small number of other members as mesh neighbors. These members are randomly selected, with the exceptions that they must not include members that are tree neighbors, and must not include members that have already established a mesh link to M. The reason for this latter restriction is to prevent trivial cliques, where three or four members all use each other as mesh neighbors, thus partitioning themselves from the rest of the mesh topology.

9.5.3 Logical ID Overlay

CAN-multicast. The Content Addressable Networks (CAN) multicast [690] is designed to scale to large groups without restricting the service to a single source. It makes use of the Content-Addressable-Networks (CAN) [689] architecture to provide an application level structured peer-to-peer overlay network whose constituent nodes form a virtual d-dimensional Cartesian coordinate space and each member owns its individual distinct zone in this space. CAN-multicast can be deployed for large group sizes where each member needs to maintain knowledge about only a small subset representing its neighbors. Neighborhood relationships are dictated by addresses assigned to hosts rather than performance. This technique gains in simplicity and scalability by not having to run routing algorithms to construct and maintain delivery trees.

To join CAN, a new node first looks up the CAN domain name in DNS to retrieve a bootstrap IP address. The bootstrap node supplies the IP address of several randomly chosen nodes currently in the system. The new node then randomly choses a point (x, y) in the space and sends a JOIN request destined to point (x, y). The current occupant node then splits its zone in half and assigns one half to the new node. In CAN-multicast, if all the nodes in CAN are members of a given multicast group, then the multicast data topology is implicitly defined by performing directed flooding on the control topology. If only a subset of the CAN nodes are members of a particular group, then the members of this group first form a mini CAN and then the multicast is achieved by flooding over the mini CAN. CAN-multicast's flooding solution aims at reducing message duplication; it is less robust than the naive flooding algorithm because the loss of a single message results in the breakdown of message delivery to several subsequent nodes.

DT. The *Delaunay Triangulation* (DT) multicast protocol [518] is based on an overlay network composed of logical coordinates assigned to the members of the underlying network topology (see also Chap. 4). The DT protocol can build and maintain very large overlay networks with relatively low overhead, at the cost of suboptimal resource utilization due to a possibly poor match of the overlay network to the network-layer infrastructure. Hence the DT protocol trades off economy of scale for increased scalability. The use of DT as an overlay has several advantages. For example, a DT has a set of alternate non-overlapping routes between any pair of vertices that represent an alternate path when nodes fail. Moreover, a DT is established and maintained in a distributed manner.

A Delaunay Triangulation for a set of vertices A is a triangulation graph with the defining property that for each circumscribing circle of a triangle formed by three vertices in A, no vertex of A is in the interior of the circle. In order to establish a Delaunay triangulation overlay, each node is associated with a vertex in the plane with given (x, y) coordinates. The coordinates are assigned via some external mechanisms (e.g., GPS or user input) and can be selected to reflect the geographical locations of nodes. Multicast forwarding in the Delaunay triangulation is done along the edges of a spanning tree that is embedded in the Delaunay triangulation overlay, and that has the sender as the root of the tree. Each node can locally determine its child nodes with respect to a given tree using its own coordinates, the coordinates of its neighbors, and the coordinates of the sender. Local forwarding decisions at nodes are done using compass routing [477]. Although compass routing in general planar graphs may result in routing loops [443], they can not happen in Delaunay triangulations [477].

Bayeux. Bayeux's multicast [911] is based on Tapestry [903], an application-level routing protocol. Nodes in Tapestry have names independent of their location and semantic properties, in the form of random fixed-length bit-sequences represented by a common base. Tapestry uses local routing maps stored at each node named neighbor maps. These maps are used to incrementally route overlay messages to the destination ID digit by digit.

Bayeux uses a source-specific model for data dissemination. Every multicast session is identified by a unique tuple. Tapestry's data location services are used to advertise Bayeux multicast sessions in the following manner: A unique session identifier is first securely mapped into a 160-bit identifier using a one-way hash function; then a file named as the identifier is created and placed on the multicast session's root node. Using Tapestry location services, the session's root (source) server advertises that document into the network. Clients willing to join a session must know the unique tuple that identifies the session. After joining a session, clients can then perform the same operations to generate the file name, and query for it using Tapestry.

A session root represents a single point of failure that can compromise the entire group's ability to receive data. It also represents a scalability bottleneck since all messages have to go through the root that has to maintain global knowledge by keeping a list of all group members. To overcome these issues, Bayeux allows the creation of multiple root nodes and partitions receivers into disjoint membership sets, each containing receivers closest to a local root in network distance. Every root contains the object to be advertised, and a new member uses Tapestry location service as before, and becomes a member of the nearest root's receiver set.

SCRIBE. SCRIBE [146] is a multi-source decentralized multicast infrastructure built on top of Pastry [718]. Pastry leverages a fully decentralized peer-to-peer model according to which a scalable and self-organizing overlay network of nodes is built. Nodes are assigned 128-bit uniformly distributed identifiers, i.e., nodeIds. Reliable message dissemination in Pastry is performed as follows: given a message and a key, Pastry routes the message to the Pastry node that has the nodeId numerically closest to the key, among all nodes. Each node maintains a routing table with entries that map the nodeId to the node's IP address. Every entry in the routing table can refer to potentially many nodes with the appropriate prefix. Among these nodes with the appropriate prefix, only the closest one to the present node is chosen according to some metric, e.g. a proximity metric such as the round trip time. Besides a routing table, each node maintains a leaf set, which represents sets of nodes with the numerically closest larger and numerically closest smaller nodeIds.

When a node wishes to join a multicast group, it chooses a nodeId and issues the request to a nearby node to route a special message using that nodeId. The message is routed by Pastry to the node that has numerically closest nodeId to the given one. The new node then initializes its state by obtaining the leaf set from the closest node. To detect node failures, neighboring nodes periodically exchange keep-alive messages. A node that does not send a response within a certain defined period of time, is assumed to be failed. The nodes from the leaf set of the failed node are then informed to update their leaf sets.

Every multicast group in SCRIBE has a unique groupId and is associated with a rendezvous point. A node that has a nodeId numerically closest to the groupId is chosen to act as the rendezvous point. The overlay tree is built upon Pastry by joining the Pastry routes from each group member to this rendezvous point that represents the root of the tree. Each forwarder, i.e., a node that is a part of a group's multicast tree, maintains a children table containing an entry for each of its children in the tree. When a SCRIBE node wishes to join a group, it asks Pastry to route a JOIN message with the groupId as the key. If on the route a node is not a forwarder, it creates an entry for the group and adds the source node as a child in its children table. The process is repeated until the JOIN message reaches the rendezvous point. SCRIBE manages group joining operations in a fully distributed manner and supports large and dynamic membership. When a SCRIBE node wishes to leave, it sends a LEAVE message to its parent initiating the appropriate structures to be updated.

In the presence of a node failure, a child node that detects the heartbeat message loss will suspect its father has failed and call Pastry to route a JOIN message which will lead to a new parent. In order to tolerate the rendezvous point's failure, SCRIBE allows its state to be replicated in a certain number of the neighboring nodes from the leaf set. SCRIBE provides best-effort delivery of messages, but leaves an open framework for applications to implement stronger reliability guarantees.

SplitStream. Based on Scribe and Pastry [718, 146] to build and maintain a tree-based application-level multicast, SplitStream [147] defines several tree structures forming a forest to duplicate and forward content. The key idea in SplitStream is to split the content into k stripes and to multicast each stripe using a separate tree. Peers join as many trees as the stripes they wish to receive and specify an upper bound on the number of stripes they are willing to forward. The resulting forest of multicast trees distributes the forwarding load subject to the bandwidth constraints of the participating nodes.

SplitStream does not only adapt the outbound bandwidth but also the inbound bandwidth contraints. Assuming that the content has a bandwidth requirement of B and the content is split into k stripes. Participating peers may receive a subset of the stripes, thus controlling their inbound bandwidth requirement in increments of B/k. Similarly, peers may control their outbound bandwidth requirement in increments of b/k by limiting the number of children they adopt.

PeerCast. PeerCast [899] is a reliable multicast service for dynamic high-churn environments of possibly unreliable peers that may enter and depart the system at arbitrary points in time. It is based on a decentralized mechanism that utilizes peer network proximity measurements in order to efficiently cluster end hosts in the Peer-Cast P2P network. The protocol further uses these clusters for building efficient multicast trees that minimize latency. PeerCast accounts for the diversity of computing capacities and network bandwidth of large networks by organizing hosts into an overlay which balances host multicast workloads. PeerCast takes advantage of Internet landmark-based techniques to guarantee that peer identifiers are allocated such that peers near each other in terms of their locations in the physical network have numerically close identifiers.

Load imbalance due to node heterogeneity is addressed in PeerCast using virtual nodes. The idea is used to allow any peer to be assigned one or more virtual nodes based upon the resource availability at the peer. The resource availability of a peer is designed to represent a weighted sum of three components, namely bandwidth availability, CPU availability, and memory availability. The resource availabilities are estimated using some of the existing techniques in the literature, while the weight assignment is application specific and given by the specification of the publisher of the content. Peers periodically monitor the resource availabilities and perform a procedure called virtual node promotion that ensures corresponding multicast tree reorganization if necessary.

In order to cope with high churn-rates and avoid multiple re-subscription of many nodes in cases where a node gracefully leaves or fails, PeerCast proposes a passive service replication scheme. Each peer maintains a Replication List of other peers that can take over the role of a parent and proceed with sending data to the downstream nodes in case of a failure or a peer leave. The replication scheme is installed right after the group information is established on a peer; the selected replicas are kept consistent as hosts join or leave the end-system multicast group.

9.6 Summary and Outlook

Tables 9.1, 9.2 and 9.3 show a recapitulative view of the main aspect characterising the multicast solutions described before. With general properties (e.g. goals), this table includes architecture, service and performance properties.

Most multicast protocols surveyed focus on minimizing latency and improving scalability. These goals are justified by the nature of the environment targeted by typical application-level multicast protocols, namely loosely coupled large-area networks, such as the Internet. Indeed, most solutions aim at large systems with hundreds or thousands of nodes. In these contexts, protocols do not seem to significantly favor a design for a specific source or for multi sources, in which the same overlay network is reused by any peer willing to disseminate information.

While most of the protocols surveyed were originally designed for P2P Internet environments, some could be used in MANET settings. Yet, most of them are not

Table 9.1 Summary of the properties of the various techniques (applicative viewpoint)

Solution	Usage focus	Communication model	MANET Fit
AEMD	Reliability	Specific source	No
ALMA	Min. Latency RTT metric	Specific source	Yes
ALMI	Min. Latency	Multi sources	No
Bayeux	Fault tolerance	Specific source	No
BTP	Min. Tree cost	Multi sources	No
CAN-multicast	Scalability	Multi sources	No
DT	Scalability Low overhead	Multi sources	No
GASR	Resources adaptive	Specific source	No
HMTP	Min. Tree cost	Multi sources	No
MAODV	Low overhead	Multi sources	Yes
Narada	Min. Latency	Specific source	Yes
NICE	Scalability Low overhead	Specific source	Yes
ODMRP	Min Latency	Multi sources	Yes
OMNI	Min. Latency	Specific source	No
oStream	Asynchronous streaming	Specific source	No
Overcast	Adapt. to bandwidth	Specific source	No
OverStream	Min. Latency Bounded out-degree	Specific source	No
PeerCast	Min. Latency Resource adaptive	Specific source	No
RanSub	Min. Latency Max. throughput	Specific source	No
RMX	Reliability	Multi sources	No
SCRIBE	Scalability Min. Latency	Multi sources	No
SplitStream	Load balancing Adapt to bandwidth	Specific source	No
TAG	Min. Latency	Specific source	Yes
TBCP	Bounded out-degree	Multi sources	Yes
TOMA	Scalability	Multi sources	No
Yoid	Min. Latency	Multi sources	No

suitable to MANETs. At first glance, this may look surprising due to the similarities between P2P networks and MANETs. In reality, however, MANET components are more resource-constrained then P2P Internet peers. Moreover, some P2P protocols rely on assumptions that are not feasible in MANETs (e.g., the existence of rendezvous points).

With respect to overlay structure, the large majority of protocols uses physical node IDs, as opposed to logical IDs. Moreover, by far and large most protocols build tree overlays, and the few ones based on meshes use it mainly to control information, not to disseminate data. Disseminating information over a mesh would

Table 9.2 Summary of the properties of the various techniques (architecture viewpoint)

Solution	Node ID / Topology	Creation & Maintenance	Environment modeling	Infrastructure support
AEMD	Physical ID/Tree	Centralized	GK	No
ALMA	Physical ID/Tree	Distributed	LK at clients GK at DS	No
ALMI	Physical ID/Tree	Centralized	GK	No
Bayeux	Logical ID/ Tree on Tapestry	Distributed	LK at nodes GK at root	No
BTP	Physical ID/Tree	Distributed	GK	No
CAN-multicast	Logical ID/ undefined	Distributed	LK	No
DT	Logical ID undefined	Distributed	LK	No
GASR	Physical ID/Tree	Distributed	LK	No
HMTP	Physical ID/Tree	Distributed	LK at clients GK at RP	Yes
MAODV	Physical ID/Tree	Distributed	LK	No
Narada	Physical ID Mesh-tree	Distributed	GK	No
NICE	Physical ID Mesh-tree	Layered	LK	No
ODMRP	Physical ID/Tree	Distributed	LK	No
OMNI	Physical ID/Tree	Distributed	LK	Yes
oStream	Physical ID/Tree	Distributed	LK	Yes
Overcast	Physical ID/Tree	Distributed	LK at Proxies GK at root	Yes
OverStream	Physical ID/Tree	Distributed	LK	No
PeerCast	Logical ID/ Tree on DHT	Distributed	LK	No
RanSub	Physical ID/Tree	Distributed	LK	No
RMX	Physical ID/Tree	Distributed	GK	Yes
SCRIBE	Logical ID/ Tree on Pastry	Distributed	LK	No
SplitStream	Logical ID/ Tree on Scribe	Distributed	LK	No
TAG	Physical ID/Tree	Distributed	GK	No
TBCP	Physical ID/Tree	Distributed	LK	No
TOMA	Physical ID/ Aggregated Trees	Distributed	LK GK for proxies	Yes
Yoid	Physical ID/ Tree-mesh	Distributed	LK	No

create redundant data paths and waste bandwidth in the absence of failures. While this would provide higher reliability, as we will see next, protocols tend to resort to different mechanisms to handle peer or link failures.

Table 9.3 Summary of the properties of the various techniques (QoS viewpoint)

Solution	Adaptivity	Scalability	Reliability
AEMD	Static	Small	Reliable
ALMA	Self-configurable/Self-improving	Medium	Fault tolerant
ALMI	Self-configurable	Small	Fault tolerant
Bayeux	Self-configurable	Large	Fault tolerant
BTP	Self-configurable/Self-improving	Medium	Fault tolerant
CAN-multicast	Self-configurable	Large	Best effort
DT	Self-configurable	Large	Best effort
GASR	Self-improving	Large	Reliable
HMTP	Self-configurable	Medium	Fault tolerant
MAODV	Self-configurable	Medium	Best effort
Narada	Self-configurable/Self-improving	Medium	Best effort
NICE	Self-configurable/Self-improving	Large	Best effort
ODMRP	Self-configurable	Medium	Best effort
OMNI	Self-configurable/Self-improving	Large	Best effort
oStream	Self-configurable	Medium	Best effort
Overcast	Self-configurable/Self-improving	Large	Fault tolerant
OverStream	Self-configurable/Self-improving	Large	Fault tolerant
PeerCast	Self-configurable/Self-improving	Large	Reliable
RanSub	Self-configurable/Self-improving	Medium	Fault tolerant
RMX	Self-configurable	Large	Reliable
SCRIBE	Self-configurable	Large	Best effort
SplitStream	Self-configurable/Self-improving	Large	Fault tolerant
TAG	Self-configurable	Medium	Fault tolerant
TBCP	Local/Self-improving	Medium	Best effort
TOMA	Self-configurable	Large	Best effort
Yoid	Self-configurable/Self-improving	Medium	Fault tolerant

It is perhaps less surprising that most solutions implement a distributed overlay creation and maintenance, where the responsibility is shared among several peers, as opposed to a centralized approach in which one peer is responsible for building the overlay. One consequence of the centralized technique is that one node must have global knowledge about the system. While most multicast protocols surveyed strive to avoid global knowledge, a few adopt a hybrid approach in which some nodes must maintain a complete view of the system. This is typically the case with rendezvous points, used by nodes to join the system.

Infrastructure support is not described in most of the techniques we surveyed. In part this is due to the fact that some of the papers considered focus on algorithmic aspects instead of systems considerations. This also has an impact on the way some of the algorithms are evaluated, often by simulation.

As for adaptivity, while most techniques are self-configurable—that is, the overlay is built incrementally, based on peers joining and leaving the system—some are also self-improving, meaning that they adapt to the physical network characteristics. Interestingly, although almost all protocols provide some support to handle failures, most prefer to recover from failures, instead of preventing them from happening. One possible reason for this is that preventing failures in the occurrence of component errors usually involves some level of redundancy, and a consequent increase in the cost of the protocol (e.g., in terms of bandwidth).

Summing up, the prototypical ALM protocol, one that would encompass most of the features of existing systems, would be designed to *minimize latency*, probably most appropriate for *P2P* networks, based on *physical IDs* structured as an *overlay tree* and built for a *specific source*. The overlay would be created and maintained in a *distributed* manner, using *local-knowledge*. Finally, the protocol would be *self-configurable* and perhaps also *self-improving*, adapted to *medium and large networks*, and would probably be *unreliable* (i.e., reacting to failures or best effort). It turns out that five protocols in our survey match the prototypical ALM protocol: ALMA, Narada, OMNI, OverStream, and RanSub.

Given the large number of existing proposals, covering different aspects and characteristics, it is difficult to speculate on the future of structured information dissemination research. Nevertheless, one could consider ideas that have been fruitful in related areas of data dissemination (e.g., group communication for smaller-scale systems). In this context, application semantics have been used to design protocols tailor-made for specific applications. The idea is that applications can specify requirement constraints, and only pay the price (e.g., in terms of bandwidth or response time) for the requirements chosen. We have identified one protocol that builds on similar ideas (i.e., RMX) but we could expect to see others in the future.

As seen before, adaptivity is currently accounted for using mostly self-configurable and self-improving techniques. It turns out that systems can also improve reliability by adapting to the changing characteristics of the environment in which they execute (e.g., node crashes). One class of such solutions is self-stabilizing algorithms [231]. We have not identified any such a protocol in the context of structured dissemination. Among its advantages, besides handling environment changes naturally, it could provide an adequate framework for designing provable system properties.

Acknowledgements

The authors are partially supported by the Swiss National Science Foundation, in the context of the Colloc project (SFNS 200020-120188). We would also like to thank Marija Stamenkovic for helping us prepare a preliminary version of this chapter.

Chapter 10
Distributed Event Routing in Publish/Subscribe Systems

Roberto Baldoni, *Sapienza Università di Roma, Italy*
Leonardo Querzoni, *Sapienza Università di Roma, Italy*
Sasu Tarkoma, *Helsinki University of Technology and Nokia NRC, Finland*
Antonino Virgillito, *Sapienza Università di Roma, Italy**

10.1 Introduction

Since the early nineties, anonymous and asynchronous dissemination of information has been a basic building block for typical distributed applications such as stock exchanges, news tickers and air-traffic control. With the advent of ubiquitous computing and of the ambient intelligence, information dissemination solutions have to face challenges such as the exchange of huge amounts of information, large and dynamic number of participants possibly deployed over a large network (e.g. peer-to-peer systems), mobility and scarcity of resources (e.g. mobile ad hoc and sensor networks) [53].

Publish/subscribe (pub/sub) systems are a key technology for information dissemination. Each participant in a pub/sub communication system can take on the role of a publisher or a subscriber of information. Publishers produce information in form of *events*, which is consumed by subscribers issuing *subscriptions* representing their interest only in specific events. The main semantic characterization of pub/sub is in the way events flow from senders to receivers: receivers are not directly targeted from publisher, but rather they are indirectly addressed according to the content of events. Thanks to this anonymity, publishers and subscribers exchange information without directly knowing each other, this enabling the possibility for the system to seamlessly expand to massive, Internet-scale size.

Interaction between publishers and subscribers is actually mediated by the pub/sub system, that in general is constituted by a set of distributed nodes that coordinate among themselves in order to dispatch published events to all (and possibly only) interested subscribers. A distributed pub/sub system for scalable information dissemination can be decomposed in three functional layers: namely the overlay infrastructure, the event routing and the algorithm for matching events against subscriptions. The overlay infrastructure represents the organization of the various entities that compose the system, (e.g., overlay network of dedicated servers, peer-to-peer structured overlay, etc.) while event routing is the mechanism for dispatching

---

* Antonino Virgillito is currently at ISTAT, Italy.

information from publishers to subscribers. Event routing has to effectively exploit the overlay infrastructure and enhance it with routing information in order to achieve scalable event dispatching. Several research contributions have appeared in the last years proposing pub/sub solutions in which these functionalities are not sharply separated and their dependencies have not been clearly pointed out. This makes all the different proposals difficult to fully understand and compare among each other.

This chapter is aimed at giving the reader a structured overview of current solutions for event routing in distributed pub/sub systems. It firstly introduces a general pub/sub architectural model for scalable information dissemination, that decomposes a generic pub/sub system into the three layers identified above. Then the chapter focuses on the solutions proposed in the literature for event routing and discuss their relations with the overlay network level solutions and possible network deployments. As a result any specific pub/sub system can be easily characterized as a stack of solutions available at each layer. Specifically the chapter categorizes event routing algorithms into six classes, each of which corresponds to a basic general dispatching method. These classes are discussed according to their underlying assumptions in terms of aspects such as the induced message overhead, routing information required at each node, dependency from the subscription language, adaptivity to dynamic changes of the underlying network. Moreover we specify how algorithms in each class relate to the overlay network layer, in particular pointing out which overlay infrastructure is more suitable for a specific event routing solution and why. In the final part of the chapter we discuss specific issues related to the deployment of pub/sub systems in mobile environments.

Being focused on scalable event routing and its relation with the underlying overlay network, this chapter complements other surveys related to publish/subscribe systems such as [265, 756, 528, 90]. The main aim of [265] has been indeed to position the pub/sub paradigm with respect to other communication paradigms, whereas [756] concentrated on software engineering aspects of a pub/sub system. Liu and Plale in [528] propose a survey of pub/sub systems considering overlay topology, matching algorithms and aspects such as reliability and security.

10.2 Elements of a Publish/Subscribe System

A generic pub/sub communication system (often referred to in the literature as *Event Service* or *Notification Service*) is composed of a set of nodes distributed over a communication network. Clients to the systems are divided according to their role into publishers, which act as producers of information, and subscribers, which act as consumers of information. Clients are not required to communicate directly among themselves but they are rather *decoupled*: the interaction takes place through the nodes of the pub/sub system. This decoupling is a desirable characteristic for a communication system because applications can be made more independent from

Fig. 10.1 High-level interaction model of a publish/subscribe system with its clients (*p* and *s* indicate a generic publisher and a generic subscriber respectively)

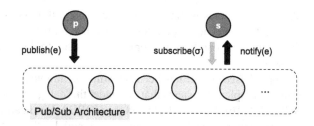

the communication issues, avoiding to deal with aspects such as synchronization or direct addressing of subscribers from publishers.[1]

Operationally, the interaction between client nodes and the pub/sub system takes place through a set of basic operations that can be executed by clients on the pub/sub system and vice versa (Fig. 10.1). A publisher submits a piece of information *e* (i.e., an event) to the pub/sub system by executing the publish(*e*) operation. Commonly, an event is structured as a set of attribute-value pairs. Each attribute has a *name*, a simple character string, and a *type*. The type is generally one of the common primitive data types defined in programming languages or query languages (e.g. integer, real, string, etc.). On the subscribers' side, interest in specific events is expressed through *subscriptions*. A subscription σ, is a filter over a portion of the event content (or the whole of it), expressed through a set of constraints that depend on the subscription language. A subscriber installs and removes a subscription σ from the pub/sub system by executing the subscribe(σ) and unsubscribe(σ) operations respectively.

We say a notification *e matches* a subscription σ if it satisfies all the declared constraints on the corresponding attributes. The task of verifying whenever a notification *e* matches a filter *f* is called *matching* ($e \sqsubseteq f$).

10.3 Subscription Models

Various ways for specifying the events of interest have led to identifying distinct variants of the pub/sub paradigm. The subscription models that appeared in the literature are characterized by their expressive power: highly expressive models offer to subscribers the possibility to precisely match their interest, i.e. receiving only the events they are interested in. In this section we briefly review the most popular pub/sub subscription models.

Topic-Based Model. Notifications are grouped in topics i.e., a subscriber declares its interest for a particular topic and will receive all events related to that topic. Each topic corresponds to a logical channel ideally connecting each possible publisher to all interested subscribers. For the sake of completeness, the difference between

---

[1] The interested reader can refer to [265] for a more deep discussion on the publish/subscribe paradigm and subscription models.

channel and topics is that topics are carried within an event as a special attribute. Thanks to this coarse grain correspondence, either network multicast facitilies or diffusion trees, one for each topic, can be used to disseminate events to interested subscribers.

Topic-based model has been the solution adopted in all early pub/sub incarnations. Examples of systems that fall under this category are TIB/RV [639], iBus [18], SCRIBE [716], Bayeux [911] and the CORBA Notification Service [635].

The main drawback of the topic-based model is the very limited expressiveness it offers to subscribers. A subscriber interested in a subset of events related to a specific topic receives also all the other events that belong to the same topic. To address problems related to low expressiveness of topics, several solutions are exploited in pub/sub implementations. For example, the topic-based model is often extended to provide hierarchical organization of the topic space, instead of a simple flat structure (such as in [39, 639]). A topic B can be then defined as a sub-topic of an existing topic A. Events matching B will be received by all clients subscribed to both A and B. Implementations also often include convenience operators, such as wildcard characters, for subscribing to more than one topic with a single subscription. For the sake of completeness, we point out that the word *subject* can be used to refer to hierarchical topics instead of being simply a synonymous for topic. Analogously, *channel-based* is sometimes [635] used to refer to a flat topic model where the topic name is not explicitly included in the event.

Content-Based Model. Subscribers express their interest by specifying conditions over the content of notifications they want to receive. In other words, a filter in a subscription is a query formed by a set of constraints over the values of attributes of the notification composed through disjunction or conjunction operators. Possible constraints depend on the attribute type and on the subscription language. Most subscription languages comprise equality and comparison operators as well as regular expressions [143, 742, 271]. The complexity of the subscription language obviously influences the complexity of matching operation. Due to this fact, it is uncommon to have subscription languages allowing queries more complex than those in conjunctive form (examples are [131, 91]). A complete specification of content-based subscription models can be found in [593]. Examples of systems that fall under the content-based category are Gryphon [336], SIENA [789], JEDI [203], LeSubscribe [682], Ready [335], Hermes [673], Elvin [741].

In content-based publish/subscribe, events are not classified according to some predefined criterion (i.e., topic name), but rather according to properties of the events themselves. As a consequence, the correspondence between publishers and subscribers is on an event basis. Then, the higher expressive power of content-based pub/sub comes at the price of the higher resource consumption needed to calculate for each event the set of interested subscribers [142, 264]. It is straightforward to see that a topic-based scheme can be emulated through a content-based one, simply considering filters comprising a single equality constraint.

Type-Based. The type-based [263, 260] pub/sub variant events are actually objects belonging to a specific type, which can thus encapsulate attributes as well as methods. With respect to simple, unstructured models, Types represent a more robust data model for application developer, enforcing type-safety at the pub/sub system, rather than inside the application [262]. In a type-based subscription the declaration of a desired type is the main discriminating attribute. That is, with respect to the aforementioned models, type-based pub/sub sits itself somehow in the middle, by giving a coarse-grained structure on events (like in topic-based) on which fine-grained constraints can be expressed over attributes (like in content-based) or over methods (as a consequence of the object-oriented approach).

Concept-Based. The underlying implicit assumptions within all the above-mentioned subscription models is that participants have to be aware of the structure of produced events, both under a syntactic (i.e., the number, name and type of attributes) and a semantic (i.e., the meaning of each attribute) point of view. Concept-based addressing [180] allows to describe event schema at a higher level of abstraction by using ontologies, that provide a knowledge base for an unambiguous interpretation of the event structure, by using metadata and mapping functions.

XML. Some research works [158, 159, 761] describe pub/sub systems supporting a semistructured data model, typically based on XML documents. XML is not merely a matter of representation but differs in the fact that introduces the possibility of hierarchies in the language, thus differentiating from a flat content-based model in terms of an added flexibility. Moreover, it provides natural advantages such as interoperability, independence from implementation and extensibility. As a main drawback, matching algorithms for XML-based language require heavier processing.

Location-Awareness. Publish/subscribe systems used in mobile environments typically require the support for location-aware subscriptions. For example, a mobile subscriber can query the system for receiving notifications when it is in the proximity of a specific location or service. Works describing various forms of location-aware subscriptions are [560, 278, 201, 761]. The implementation of location-aware subscriptions requires the pub/sub system the ability to monitor the mobility of clients.

10.4 Architectural Model

In this section we describe the reference architectural model we use in our presentation. The architectural model is depicted in Fig. 10.2, including four logical layers, namely *Network Protocols*, *Overlay Infrastructure*, *Event Routing* and *Matching*. We present in the following the functionality associated to each layer as well as the different possible solutions for its realization (also illustrated in the figure).

In Table 10.1 we report how different pub/sub systems present in the current literature implement the proposed architectural model. For each system we tried to abstract its main characteristic in order to let them fit our model. Note, however, that this table is not meant to be complete as it only includes a limited subset of

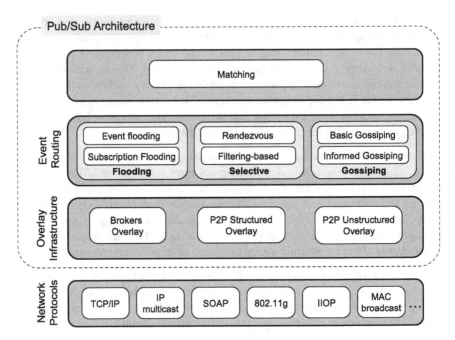

Fig. 10.2 Publish/subscribe architectural model

all existing pub/sub solutions, and for those presented here, it avoids to detail many important technical aspects that characterize and differentiate each single system.

10.4.1 Network Protocols

Network protocols anchor a pub/sub system to the underlying network by allowing transmission of data among pub/sub system components. Due to the fact that a pub-/sub system could span over heterogeneous networks (e.g., LANs, WANs, mobile networks, etc.), it could employ more than a single network protocol either to cope with different software/hardware conditions that could be found in a given part of the network or to maximize performance. For example a pub/sub system deployed over a WAN could use MAC broadcast inside a LAN to reach in one shot all recipients of an events while sending evens between two LANs using TCP connections.

Transport Level. Pub/sub systems are usually built exploiting the functionality of common transport-level protocols. That is, nodes in the overlay infrastructure communicate directly through TCP or UDP sockets or using specific TCP-based middleware protocols (like IIOP or SOAP). This choice allows the greater flexibility and ease of deployment, though in such situations, the deployment over a wide-area

Table 10.1 Pub/Sub Systems

System	Subscription Model	Network Protocol	Overlay Infrastructure	Event Routing
TIB/RV [639]	Topic	TCP & MAC bcast	Brokers	Filtering
Scribe [146], CAN [690]	Topic	TCP	P2P Structured	Rendezvous
Siena [143], Gryphon [59], Rebeca [801]	Content	TCP/UDP	Brokers	Filtering
Hermes [673]	Content	TCP	Brokers	RVs/Filtering
Meghdoot [345]	Content	TCP	P2P Structured	Rendezvous
DADI (Kyra [132])	Content	TCP	Brokers	RVs/Filtering
DADI (MEDYM [133])	Content	TCP & IP mcast	Brokers	Sub. Flooding
GREEN (WAN) [761]	Various	TCP	P2P Structured	Rendezvous
GREEN (Mobile) [761]	Various	802.11g	Unstructured	Gossiping
Sub-2-Sub [845]	Content	UDP	P2P Structured & Unstructured	Informed Gossiping
Spidercast [176]	Topic	UDP	P2P Unstructured	Gossiping
TERA [57]	Topic	UDP	P2P Unstructured	Informed Gossiping

network can be limited by the presence of network firewalls or private networks, requiring the intervention of an administrator for configuration.

Network-Level Multicast. Directly exploiting local-area or wide-area multicast and broadcast network primitives is an efficient way to realize many-to-many diffusion experiencing low latencies and high throughput, thanks to the small delays introduced by implementing the protocols exclusively involving routers and switches. For example, IP Multicasting can be directly used in wide-area topic-based systems, as each topic corresponds exactly to one multicast group. Using IP multicast for content-based systems is not as straightforward because subscribers cannot be directly mapped to multicast groups. This has inspired some research work targeting at organizing subscribers in clusters, where subscribers in the same cluster contain most of the subscriptions in common [645, 706, 707, 340]. The main drawback of IP multicast is in its lack of a widespread deployment [230, 751]. Hence, network-level multicasting cannot in general be considered as a feasible solution for applications deployed over a WAN (for example TIB/RV or the CORBA Notification Service uses network multicast only for diffusing notifications inside a local area network).

Mobile Networks. Several works on pub/sub systems address the mobile network scenario, in two different fashions. One group of works consider each node in the infrastructure having the possibility to move, e.g. constituting a mobile ad hoc network (MANET) [387, 19, 761, 560]. In other works, part of the nodes form a fixed infrastructure and only clients can roam, being always one-hop away from the fixed infrastructure [278, 135, 606]. In both cases the network interfaces for mobile nodes can be either a transport protocol or a data link protocol (such as 802.11b). Obviously mobility induces specific resource constraints over the overall architecture design such as battery drain, limited bandwidth etc. Also phenomenon such as temporary disconnections and node unavailability should be considered common and have to be dealt with.

Section 10.7 specifically focuses on issues related to mobility in pub/sub systems.

Sensor Networks. A sensor network is composed by small devices capable of taking various measurements from the environment, and transmitting them toward applications hosted into specific base stations. Sensors communicate (among each other and with base stations) through broadcast-based facilities over wireless protocols such as 802.15.4. It is evident that the pub/sub paradigm fits naturally this context: sensors publish measurements, that are received by subscribers placed in base stations. With respect to a general pub/sub system, a further assumption can be made in this context: the number of subscribing nodes is very low, as sensors are exclusively publishers and they are predominant in number with respect to base stations.

From the architectural point of view, sensor networks are similar to MANETs, regarding aspects such as the topology determined by the devices transmission range and the limited power supply. Obviously, the fact that a sensor network is a fixed network reduces the complexity related to dynamic topology changes, though dynamicity has still to be taken into account, because devices are in general failure-prone and they frequently rely on stand-by periods for power saving. Works presenting pub/sub solutions suited to sensor networks are [355, 189].

10.4.2 Overlay Infrastructure

A pub/sub system generally builds upon an application-level overlay network. In the following we discuss the possible pub/sub overlays, characterized by the organization of the nodes, the role of each node (pure server or also acting as client) and the overall functionality on which the event-routing algorithm rely on. The discussion includes the conditions under which each infrastructure is more feasible and the constraints it imposes.

Broker Overlay. The support for distributed applications spanning a wide-area, Internet-size network requires the pub/sub system to be implemented as a set of independent, communicating servers. In this context, each single server is called a *broker*. Brokers form an application-level overlay and typically communicate through an underlying transport protocol. Clients can access the system through any broker and in general each broker stores only a subset of all the subscriptions in the system. The particular case of systems composed by a single broker (centralized architecture) is often considered in the literature [271, 878].[2]

The broker network is implemented as an application-level overlay: connections are pure abstractions as links are not required to represent permanent, long-lived connections, so that the neighborhood in the network is determined purely by a knowledge relation. The topology is assumed to be managed by an administrator, based on technical or administrative constraints. For this reason, a broker overlay is inherently static: topology changes are considered to be rare, mainly to face events such as addition of new brokers or repairing after a failure.

The broker network is the most common choice in actual pub/sub implementations, being used by system such as TIB/RV [639], Gryphon [336], SIENA [789], JEDI [203] and REDS [202], as well as in several event routing algorithms proposed in the literature [132, 851]. Apart from the routing protocols, that we analyze in Sect. 10.5, the main aspect to be clarified in this type of infrastructure is the topology formed by the brokers themselves. There are basically two solutions, hierarchical or flat. In a hierarchical topology, brokers are organized in tree structures, where subscribers' access points lie at the bottom and publishers' access points are roots (or vice versa). Many contributions [86, 851] rely on this topology, thanks to the simplifications it can allow since notifications are diffused only in one direction. In a flat topology, a broker can be connected with any other broker, with no restrictions. [143] shows the more effective load-balance obtained with a flat topology with respect to a hierarchical one, due to the fact that brokers belonging to upper levels of the hierarchy experience a higher load than ones at lower levels.

Peer-to-Peer Structured Overlay. A peer-to-peer structured overlay infrastructure is a self-organized application-level network composed by a set of nodes forming a structured graph over a virtual key space where each key of the virtual space is mapped to a node. The structure imposed to the graph permits efficient discovery of data items and this, in turns, allows to realize efficient unicast or multicast communication facility among the nodes. A structured overlay infrastructure ensures that a correspondence always exists between any address and an active node in the system despite churn (the continuous process of arrivals and departures of nodes of the overlay) and node failures. Differently from a broker overlay infrastructure, a structured overlay allows to better handle dynamic aspects of the systems such as faults and node joins. Then, it is more suited in unmanaged environments (for example, large-scale decentralized networks) characterized by high dynamicity, where human administration interventions cannot be considered a feasible solution.

As a consequence of the popularity of structured overlays, many techniques to build structured overlays have been proposed: we cite among the others Pastry [718], Chord [773], Tapestry [904] (unicast diffusion) or CAN [690], I3 [774] and Astrolabe [838] (multicast diffusion). Structuring a pub/sub system over an overlay network infrastructure means leveraging the self-organization capabilities of the in-

---

[2] Though centralized architectures are of practical interest being particularly suitable for small-scale deployments, they are evidently out of the focus of our work and will not be considered in the following.

frastructure, by building a pub/sub interface over it. The event routing algorithm is realized only exploiting the communication primitives provided by the underlying overlay, therefore these systems are usually considered more suited to implement pub/sub systems with exact-match interfaces (like topic-based systems) rather than those with range-based subscriptions. Examples of systems using this solution are Bayeux [911] and Scribe [146], for what concerns topic-based systems, and Meghdoot, Hermes [674], Rebeca [801] and [6], for what concerns content-based systems. Finally, we cite SelectCast [114], a multicast system built on top of Astrolabe providing a SQL-like syntax for expressing subscriptions.

Peer-to-Peer Unstructured Overlays. Unstructured overlays strive to organize nodes in one flat or hierarchical small diameter network (like a random graph) despite churn and node failures [649]. Differently from broker overlays, nodes in these overlays are not necessarily supposed to be dedicated server but can include workstations, laptops, mobile devices and so on, acting both as clients and as part of the pub/sub system. Moreover, the topology of the overlay is obviously unmanaged (that is, it does not rely on a human administrator).

Unstructured overlays use flooding, gossiping or random walks [669, 845, 176, 57] on the overlay graph to diffuse and to retrieve information associated with the nodes. This is due to the difficulties involved in maintaining deterministic data structures for event routing in a setting characterized by the dynamic behaviour of participants (see Sect. 10.5.2 for details). On the other hand, unstructured overlays are widely used for file sharing applications for their simplicity in handling joins and leaves of nodes (with respect to their structured counterparts) and for the fact that, in such applications, there is no need for precise searches. So unstructured overlay are probabilistic in nature as there is non-zero possibility that some item present in the network is not found during a search.

Overlays for Mobile Networks. In a mobile setting, topology changes in the overlay are due to the mobility pattern as well as churn and node failures. Mobility determines the topology of the network and makes impossible to make optimizations as in the peer-to-peer unstructured overlays, such as keeping small diameter networks. Moreover, specific algorithms are required for creating and maintaining the conditions under which the event routing algorithm can work, such as connectivity or consistency of the event routing data structures [387, 19, 671]. We detail this aspect in Sect. 10.5.4. Running algorithms for keeping tree-topology over a set of mobile nodes can be expensive in terms of resources by blocking the computation till a tree is formed. This, as well as the scarce computational resources normally available to mobile nodes, makes the broker overlay and the structured overlay infrastructures not suited to this environment. Hence, typical solutions to event routing are based on an unstructured overlay and rely directly on the MAC layer (e.g. 802.11b) by exploiting its beaconing system and its broadcast characteristics [54]. These approaches have similarities to data management in the Bayou system [666] and can be easily ported over a unstructured peer-to-peer overlay network.

Overlays for Sensor Networks. Differently to mobile networks, in sensor networks it is less difficult to maintain some form of structured overlay, though the limited reliability and computational capability of devices pose some limits to its realization. Broker overlays are obviously unfeasible, while structured overlays can be realized, including accurate design solutions that allow to cope with frequent failures. Still, unstructured overlays suits more seamlessly to the communication model used by the sensor devices.

10.4.3 Event Routing

The core mechanism behind a distributed pub/sub system is event routing. Informally, event routing is the process of delivering an event to all the subscribers that issued a matching subscription before the publication. This involves a visit of the nodes in the Notification Service in order to find, for any published event, all the clients whose registered subscription is present in the system at publication time.

The impossibility of defining a global temporal ordering between a subscription and a publication that occurred at two different nodes makes this definition of routing rather ambiguous. A discussion on this point as well as formal specifications of the event routing problem can be found in [55].

The main issue with an event routing algorithm is scalability. That is, an increase of the number of brokers, subscriptions and publications should not cause a serious (e.g., exponential) degradation of performance. This requires on one hand controlling the publication process, in order to possibly involve in propagation of events only those brokers hosting matching subscriptions. On the other, reducing the amount of routing information to be maintained at brokers, in order to support and flexibly allow subscription changes. These two aspects are evidently conflicting and reaching a balance between them is the main aim of a pub/sub system's designer.

We have identified and classified the approaches presented in literature for event routing. Routing approaches are oblivious to the particular architectural solution in the sense that a same routing algorithm can be used in different infrastructures, though each approach can be more suitable for a specific architecture. Section 10.5 is entirely devoted to describing and comparing routing algorithms, as well as identifying which type of infrastructure is more suitable for each solution.

10.4.4 Matching

Matching is the process of checking an event against a subscription. Matching is performed by the pub/sub system in order to determine whether dispatching the event to a subscriber or not. As we show in the following section, event routing algorithms often also require a matching phase to support the routing choices. As the context of interest is that of large-scale systems, we expect on one side the overall number of subscriptions in the system to be very high, and on the other a high rate of events

to be processed. Then, in general the matching operation has to be performed often
and on massive data sizes. While obviously this poses no issue in a topic-based system, where matching reduces to a simple table lookup, it is a fundamental issue for
the overall performance of a content-based system. The trivial solution of testing
sequentially each subscription against the event to be matched often results in poor
performance. Techniques for efficiently performing the matching operation are then
one important research issue related in the pub/sub field. They can be grouped in two
main categories [710], namely predicate indexing algorithms and testing network algorithms. Predicate indexing algorithms are structured in two phases: the first phase
is used to decompose subscriptions into elementary constraints and determine which
constraints are satisfied by the notification; in the second phase the results of the
first phase are used to determine the filters in which all constraints match the event.
Matching algorithms falling into the predicate indexing family are [877, 660, 271,
141]. Testing network algorithms ([8, 322, 131]) are based on a pre-processing of
the set of subscriptions that builds a data structure (a tree in [8] and [322] or a binary
decision diagram in [131]) composed by nodes representing the constraints in each
filter. The structure is traversed in a second phase of the algorithm, by matching
the event against each constraint. An event matches a filter when the data structure
is completely traversed by it. This quick overview is not intended to cover all the
works proposed in the literature and was introduced here mainly for the sake of
completeness, since the focus of our work is on distributed event routing. A formal
complexity analysis and comparison of matching algorithms can be found in [437].

10.5 Event Routing

In this section, we investigate the general solutions for event routing to achieve scalable information dissemination. Three categories are identified, flooding algorithms
(event flooding and subscription flooding), selective algorithms (rendezvous-based
and filter-based) and event gossiping algorithms (basic gossiping and informed gossiping). Roughly, flooding algorithms are based on a complete deterministic dissemination of event or subscriptions to the entire system. Selective algorithms aims
at reducing this dissemination thanks to a deterministic routing structure built upon
subscriptions, that aids in the routing process. Event gossiping are probabilistic algorithms with no routing structure, suitable for highly dynamic contexts such as mobile ad hoc networks. The general characteristics of the algorithms are summarized
in Table 10.2, reporting for each algorithm the type of routing decisions (probabilistic or deterministic), the nodes that perform the filtering (producers, consumers
or intermediate nodes on the path from publishers to subscribers) and the nodes to
which events and subscriptions are sent (none, all or a subset).[3]

---

[3] In the following we refer as node to a generic node of the pub/sub system, let this be a broker in a
broker overlay or a peer in a structured/unstructured overlay. Clients in broker-based architectures
are not considered and their behavior is completely handled by nodes in the pub/sub system.

Table 10.2 Classification of event routing algorithms

		Routing	Filtering	Nodes storing Subs	Nodes handling Events
Flooding	Event flooding	Det.	Subscribers	None	All
	Subs Flooding	Det.	Publishers	All	None
Selective	Filter-Based	Det.	Intermediaries	Subset	Subset
	Rendezvous-based	Det.	Intermediaries	Subset	Subset
Gossiping	Basic gossiping	Prob.	Subscribers	None	All
	Informed gossiping	Prob.	Intermediaries	Subset	Subset

In the remainder of this section we give a detailed description of all routing solutions, stating the relationship between each algorithm and the various overlay infrastructures and identifying their trade-offs in terms of the following dimensions[4]:

- Message overhead: the overhead induced on the network by sending both publication and subscription messages. It is normally measured in terms of overlay hops, that is the number of nodes that are traversed by an event along propagation. Ideally, an event routing algorithm should reach all subscribers in a single hop. All the further messages besides these are considered as overhead.
- Memory overhead: the amount of information stored at each process. Related to subscription replication, which is the number of copies of each subscription that are present in the system.
- Subscription language limitations: the routing mechanism may induce limitations on the supported subscriptions, for example regarding the type of constraints.

Besides these aspects, one has finally to consider that event routing algorithms are subject to two types of dynamic changes: i) the behavior of users dynamically changing their subscriptions and ii) the changes in the composition of the system due to the process of arrival, departure and failure of nodes (that is, churn). While all event routing algorithms (except event flooding) are equally subject to the first type of dynamicity, some are more sensitive than others to the latter type, according to the overlay infrastructure they are deployed over. We also highlight this issue in the presentation of the algorithms and discuss it in detail in Sect. 10.5.2.

10.5.1 Event/Subscription Flooding

The trivial solution for event routing consists in propagating each event from the publisher to all the nodes in the system (*event flooding*, Fig. 10.3(a)). This algorithm can be simply implemented in all the architectures: a network-based solution consists in broadcasting each event in the whole network, while with any form of overlay it suffices for a node to forward each event to all the known processes. The

---

[4] A simulation study of event routing algorithms has recently appeared in [90].

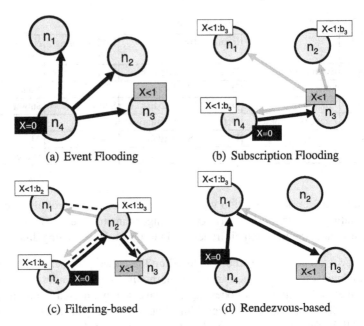

(a) Event Flooding   (b) Subscription Flooding

(c) Filtering-based   (d) Rendezvous-based

Fig. 10.3 Example of event routing algorithms. Black boxes and arrows represent published and sent events, gray boxes and arrows represent stored and sent subscriptions and white boxes represent stored routing information

obvious drawback is that this routing mechanism does not scale in terms of message overhead. However, event flooding presents minimal memory overhead (no routing information needs to be stored at a node) and there are no language limitations.

On the other side of the spectrum of routing solutions with respect to the routing information stored at each node lies the *subscription flooding* approach: each subscription is sent to all the nodes together with the identifier of the subscriber. That is, each node has the complete knowledge of the entire system, thus recipients can be reached in a single hop (the ideal value) and non-interesting events can be immediately filtered out at producers (Fig. 10.3(b)). However, both simulations studies ([594, 143]) and practical experiences ([742]) report that subscription flooding can rarely be considered a feasible solution if subscriptions change at a high rate, as each node has to send all the changes to all other nodes (in other words, the overlay is completely connected). For example, the complete flooding of subscriptions was a characterizing feature (referred to as "quenching") of an older version of Elvin [741], which was removed in a successive version ([742]), since it proved to be very costly. A recent work presenting a subscription flooding approach is ME-DYM [133], an algorithm part of the DADI framework.

10.5.2 Selective Event Routing

The principle behind *Selective Event Routing* algorithms is to reduce the message overhead of event flooding by letting only a subset of the nodes in the system store each subscription and a subset of the nodes be visited by each event (both subsets possibly spanning the whole system). Selective routing algorithms allow to save network resources particularly when an event has to be transmitted only to a restricted portion of subscribers. When most events are of interest for a large number of subscribers, flooding can be considered an option [144, 595] since it avoids the overhead due to the storage and update of event routing information.

Filtering-Based Routing

In *Filtering-based routing* [143] events are forwarded only to nodes that lie on an overlay path leading to interested subscribers. Message overhead is reduced by identifying as soon as possible events that are not interesting for any subscriber and arrest their forwarding. This approach has been largely studied and used in the literature [595, 90].

The construction of diffusion paths requires routing information to be stored and maintained on the nodes. Routing information at a node is associated to each of its neighbors in the overlay and consists in the set of subscriptions that are reachable through that broker. This allows to build reverse paths to subscribers followed by events. In practice, copies of all the subscriptions have to be diffused toward all possible publishers, and in the general case when all nodes may act as publishers for any subscription, this means again flooding all subscriptions. However, differently from the subscription flooding approach, a node communicates directly only with its neighbors, thus reducing the local message overhead due to subscription update. Subscription diffusion can also be limited in this approach by exploiting subscription containment, as done in SIENA and REBECA.

The pseudo-code of filtering-based routing at a broker is presented in Algorithm 10.1. A broker can handle *publish* or *subscribe* messages, respectively sent by a client or by another broker. Each broker maintains three structures: a *neighbors* list, a *routing* table and a *subscription* list. The *routing* table associates a neighbor with an entry representing a set of subscriptions. The *subscription* list associates a node to its subscription. The *match* function matches an event against either the subscription list or a routing table entry and returns a list with all the matching nodes. An example of Filtering-based routing is depicted in Fig. 10.3(c), where the dashed lines represent connections at overlay level.

The natural architecture for this kind of solution is the brokers' network, usually structured in an acyclic topology (tree or graph). Actually, the presence of cycles requires duplicate detection while diffusing both event and subscriptions and thus is usually avoided in implemented systems. The addressing scheme of a structured overlay does not represent a useful feature in this type of solution, except for the fact that it can keep the consistent association between a node and its position in

---

Algorithm 10.1: Pseudo code of Filtering-based routing

```
upon receiving (publish(event e) from node x )
begin
      matchlist ← match(e,subscriptions) ;
      send notify(e) to matchlist ;
      fwdlist ← match(e,routing) ;
      send publish(e) to fwdlist − x ;
end

upon receiving (subscribe(subscription s) from node x )
begin
      if x is client then
            subscriptions += s ;
      else
            routing += (x,s) ;
            send s to neighbors − x ;
      end
end
```

---

the overlay, allowing easy overlay repair upon failure. However, the consistency of information in the routing tables has still to be provided by specific event routing-level algorithms. This type of solution is considered in Hermes [674] and in [801, 187]. The use of Filtering-based routing over unstructured overlays suffers mainly from the dynamicity of the network, that requires frequent updates of the routing information. Moreover, it is not possible to assume an acyclic topology.

The performance of filtering-based routing is obviously influenced by the topology of the overlay network. In particular, the diameter of the topology is related to the length of the overlay paths traveled by events, thus affecting notifications latency. Obviously, increasing the number of neighbors of a node lowers the diameter of the network, but also the amount of routing information kept by nodes (memory overhead) increases. This is the reason why the efficiency of the matching algorithm also impacts on delivery latency.

Finally, filtering-based routing does not impose any limitation on the subscription language. Indeed, the only point in the algorithm dependent on the language is the *match*() function, that can be implemented easily for any data type.

Rendezvous-Based Routing

Rendezvous-based event routing is based on two functions, namely SN and EN, used to associate respectively subscriptions and events to nodes in the pub/sub system. In particular, given a subscription σ, $SN(\sigma)$ returns a set of nodes, named *rendezvous nodes of* σ, which are responsible for storing σ and forwarding events matching σ to all the subscribers of σ. $EN(e)$ complements SN by returning the *rendezvous nodes of* e, which are the nodes responsible for matching e against sub-

Algorithm 10.2: Rendezvous-based routing

```
upon receiving (publish(event e) from node x at node i)
begin
    rvlist ← EN(e) ;
    if i ∈ rvlist then
        matchlist ← match (e,subscriptions) ;
        send notify(e) to matchlist ;
    else
        send (e) to rvlist ;
    end
end

upon receiving (subscribe(subscription s) from node x at node i)
begin
    rvlist ← SN(s) ;
    if i ∈ rvlist then
        subscriptions += s ;
    else
        send (s) to rvlist;
    end
end
```

scriptions registered in the system. Upon issuing a subscription σ, a subscriber sends σ to the nodes in $SN(\sigma)$, which store σ and the subscribers' identifier.

Then, rendezvous-based event routing is a two phase process: a publisher sends their events to nodes in $EN(e)$, which match e against the subscriptions they host. For each subscription matched by e, e is forwarded to the corresponding subscriber. In order for the matching scheme to work and forward e to the consumers, it is necessary that the rendezvous nodes of e collectively store all the subscriptions matched by e, i.e., if $e \in \sigma$ for any subscription σ, then $EN(e) \cap SN(\sigma) \neq \emptyset$. We refer to this property as the *mapping intersection rule* [56]. The pseudo-code of rendezvous-based routing is presented in Algorithm 10.2 (*subscriptions* list is defined as in Filtering-based routing), while an example is depicted in Fig. 10.3(d), where we assume SN/EN functions that assign subscription $x < 1$ and event $x = 0$ to node n_1.

Rendezvous-based routing has been introduced in [851], and recently many systems appeared following such a scheme (Scribe [146], Bayeux [911], Hermes [673], Meghdoot [345] and [56]). This approach is motivated by the fact that a controlled subscription distribution allows to better load balance subscription storage and management: all subscriptions matching the same events will be hosted by the same node, avoiding a redundant matching to be performed in several different nodes. Also delivery of events is simplified, consisting in the creation of single-rooted diffusion trees starting from target brokers and spanning all subscribers.

However, it is clear that defining the couple of $EN(e)$ and $SN(\sigma)$ functions so that they satisfy the mapping intersection rule is a non-trivial task. This implies defining a clustering of the subscription space, such that each cluster is assigned to a node that becomes the rendezvous for the subscriptions and events that fall into that cluster.

A rendezvous-based algorithm over a broker-based architecture does not handle well dynamicity: when a new node n joins the system, the whole partitioning criterion has to be rearranged among nodes. Moreover, subscriptions that map to n's partition have to be moved to n from the node that was previously in charge. Similarly, when a node leaves or crashes, the subscriptions that it stores should be relocated to another node. Unstructured networks are even less suitable in this sense because the system is highly dynamic and its size not known. On the contrary, the powerful abstraction realized by structured overlay networks greatly helps in the definition of the mapping functions, thanks to the fact that the fixed-size address space can be used as a target of the functions rather than the set of nodes. This allows the mapping to be independent from the actual system composition and not be influenced by changes in it.

Maybe the biggest drawback of rendezvous-based solutions is the restrictions it may impose to the subscription language. In general, mapping a multi-dimensional, multi-typed content-based subscription to the uni-dimensional or bi-dimensional numerical-only address space of structured overlays is not straightforward. While numerical range constraints can be intuitively handled, constraints over string attributes, like substrings, prefixes or suffixes, that are an important part of a content-based language, can be hardly reduced to numerical ranges, then they may be excluded from the subscription language.

As for performance, memory overhead depends on the mapping function used. In general, the mapping function should map a subscription to the lower number of nodes possible in order to satisfy the mapping intersection rule. It is natural though that "larger" subscriptions (i.e. matching more events) will be mapped to more nodes with respect to "smallest" ones. This allows also to share the load due to matching. Moreover, routing information should be preserved at a node to reach the rendezvous nodes.

Filter Merging

Filter merging, or aggregation, has been proposed as an optimization strategy for distributed pub/sub systems. Filter merging techniques combine filters to reduce the number of propagated filters and thus the size of distributed state. Merging and covering are needed to reduce processing power and memory requirements both on client devices and on event routers. These techniques are typically general and may be applied to subscriptions, advertisements, and other information represented using filters.

A filter-merging-based routing mechanism was presented in the Rebeca distributed event system [594]. The mechanism merges conjunctive filters using perfect merging rules that are predicate-specific. Routing with merging was evaluated mainly using the routing table size and forwarding overhead as the key metrics in a distributed environment. Merging was used only for simple predicates in the context of a stock application [594]. The integration of the merging mechanism with a routing data structure was not elaborated and we are not aware of any results on this topic. Recently, a framework for dynamic filter merging has been proposed that integrates with content-based routing tables, such as a poset or a forest organized based on the covering relation [797].

The optimal merging of filters and queries with constraints has been shown to be NP-complete [199]. Subscription partitioning and routing in content-based systems have been investigated in [852] using Bloom filters and R-trees for efficiently summarizing subscriptions.

Bloom filters are an efficient mechanism for probabilistic representation of sets, and support membership queries, but lack the precision of more complex methods of representing subscriptions. To take an example, Bloom filters and additional predicate indices were used in a mechanism to summarize subscriptions [812]. An Arithmetic Attribute Constraint Summary (AACS) and a String Attribute Constraint Summary (SACS) structures were used to summarize constraints, because Bloom filters cannot capture the meaning of other operators than equality. The subscription summarization is similar to filter merging, but it is not transparent, because routers need to be aware of the summarization mechanism. Filter merging, on the other hand, does not necessarily require changes to other routers. In addition, the set of attributes needs to be known a priori by all brokers and new operators require new summarization indices. The benefit of the summarization mechanism is improved efficiency, since a custom-matching algorithm is used that is based on Bloom filters and the additional indices.

On the Guarantee of Event Delivery

Let us point out that selective-based solutions are deterministic approaches to event routing, in the sense that they build event routing data structure to deterministically route event to its intended destinations. Nodes cooperate for letting these data structures do their best to timely track subscription changes. It is important to remark that deterministic event routing does not imply any deterministic guarantee on event delivery. There is indeed a non-zero delay between a change and the time in which the event routing data structure captures this change. During this delay deterministic approaches to event routing might become inefficient, in the sense that they can lead, on one hand, to event loss due to the fact that an event is routed to part of the overlay where there are no longer interested recipients (e.g., due to recent unsubscription) and, on the other hand, to not routing an event to an interested destination that just did the subscription [55]. Therefore, deterministic approaches to event routing are clearly best-effort in terms of delivery of events due to topology rearrangements.

Moreover, the effect of churn makes much more pronounced the discrepancy between the event routing data structures at a given time and the ones that would allow ideal deterministic event routing, amplifying, thus, the inefficiency of the event routing with respect to the delivery of events. Deterministic event routing approaches work therefore better over overlay infrastructures where the churn is mastered by some external entity. For example, in managed environments such as a broker overlay, the churn is very low and strictly under control of humans. In a peer-to-peer structured overlay, the churn effect is handled by the overlay infrastructure layer and then masked to the event routing level.

As the size and the dynamic of the system grow, the effect of churn can be disruptive in terms of delivery of events in deterministic event routing even in structured peer-to-peer networks [517]. This is why gossip-based (or epidemic) protocols have emerged as an important probabilistic event routing approach to cope with these large scale and dynamic settings [266].

10.5.3 Gossip-Based

In basic gossip-based protocols, each node contacts one or a few nodes in each round (usually chosen at random), and exchanges information with these nodes. The dynamics of information spread resembles the spread of an epidemic [222] and lead to high robustness, reliability and self-stabilization [88]. Being randomized, rather than deterministic, these protocols are simple and do not require to maintain any event routing data structure at each node trying to timely track churn and subscriptions changes. The drawback is a moderate redundancy in message overhead compared to deterministic solutions. Gossiping is therefore a probabilistic and fully distributed approach to event routing and the basic algorithm achieves high stability under high network dynamics, and scales gracefully to a huge number of nodes.[5] Specific gossip algorithms for pub/sub systems have been proposed in [261, 188, 39, 669, 845, 176, 57, 339]. For a deeper discussion of gossip-based dissemination algorithms, see Chap. 8.

In gossip protocols, the random choice of the nodes to contact can be sometimes driven by local information, acquired by a node during its execution, describing the state either of the network or of the subscription distribution or both.[6] In this case, we are in the presence of an *informed gossip protocol*. The algorithm presented by Eugster and Guerraoui in [261] (*pmcast*) is an example of informed gossip specifically targeted to pub/sub system. It follows a principle similar to that of filter-based routing: avoiding to gossip a message to not-interested subscribers. *pmcast* organizes processes in a hierarchy of groups. Groups are built and organized in hierarchies according to the physical proximity of nodes. Each process maintains in its view the identities and the subscriptions of its neighbors in a group. Special members in a group, namely delegates, maintain an aggregation of the subscriptions within a group and have access to the delegates view of nodes at adjacent levels of the tree. Events are gossiped throughout the tree. The membership information allows to exclude from gossiping the nodes that are not interested in an event.

---

[5] Gossiping has been also used to improve delivery guarantee of a filtering-based event routing protocol in [188].

[6] To help this process of acquiring information at each node some limited horizon advertising mechanism can be employed.

Costa and Picco [669] proposed a hybrid approach that mixes deterministic and probabilistic event routing. Subscriptions are propagated only in the immediate vicinity of a subscriber. Deterministic event routing leverages this subscription information, whenever available, by deterministically routing an event along the link a matching subscription was received from. If no subscription information exists at a given node, events are forwarded along a randomly chosen subset of the available links over the peer-to-peer overlay.

Voulgaris et al. [845] exploit a more complex architecture where content-based event diffusion is realized traversing multiple layers, each characterized by a different overlay technology. The lower layer exploits a randomized overlay to gossip continuously information about subscriptions among participants; a middle layer is used to maintain semantic links among participants sharing similar interest; finally, an upper layer connects in a ring-like structure all participants that should be notified about a specific event. Event diffusion is realized through gossiping till the point where semantic links can be used to directly address events.

Chockler et al. [176] use a different approach. The system they propose is based on a single overlay network where each participant manages both a set of randomized links and a set of semantic links through which it is connected to other participants sharing similar interests. Event diffusion is again realized leveraging semantic links as long as they are available, or resorting to gossiping when they are not.

Finally, Baldoni et al. [57] propose a two layer infrastructure for topic-based event diffusion based on a uniform peer sampling service. This service is leveraged to uniformly distribute information about participant interests. Event diffusion is realized in two steps: first a random walk traverses a low-level general overlay connecting all participants looking for someone subscribed to the target topic; once such a subscriber is found event diffusion continues at an upper layer leveraging an overlay connecting all participants subscribed to the same topic. This second step can be realized through various techniques like message flooding or gossiping.

10.5.4 Event Routing for MANETs

Wireless MANETs can support both deterministic and probabilistic event routing protocols that we presented till now. However as remarked in previous sections, while in wired networks all event routing algorithms are built on the top of a transport protocol using MAC broadcast only for local performance improvements, in a wireless network this sharp layering is no more a dogma due to battery drain and to the fact that unicast is expensive while multicast can be cheap. This is why event routing algorithms can also either rely on the MAC layer [386, 38, 54] or integrate with the classical MANET routing protocols such as MAODV [885, 589].

Huang and Garcia-Molina [387], Anceaume et al. [19] and Cugola et al. [671] present three algorithms for building and maintaining a tree event routing structure in a mobile ad hoc network on the top of a transport protocol. In particular, Cugola et al. describe an algorithm for restoring the event routing tables after a disconnection in a generic acyclic graph topology within a mobile ad hoc network. The paper advocates a separation of concerns between the connectivity layer and the event dispatching layer. The algorithm for repairing the event routing data structure works on the assumption that the tree is kept connected by some loop-free algorithm at the routing level.

The above-described event routing protocols, as well as the ones integrating with a MANET routing protocol, are subject to the problem described in Sect. 10.5.2 since they aim at building deterministic event routing structure over a MANET. This problem is amplified in this context by the frequent overlay topology changes due to mobility, besides the ones due to churn and subscription changes. This is why recent approaches to event routing in MANET rely directly on the MAC layer exploiting the broadcast nature of the medium at the same time [386, 38, 54]. For example, [38] and [54] are structureless in the sense that they do not maintain any deterministic data structure on the topology at a peer. Therefore, event routing in such cases can only be based on either gossip or on flooding. In particular Baldoni et al. [54] employ a form of informed flooding event routing based on the euclidean distance between two nodes to direct the event to the destination. This distance is estimated by counting the number of beacon messages missed from a given source. Each peer p periodically broadcast a message summarizing its local subscriptions. This allows mobile peers in the proximity of p to know p's subscription and to construct its own subscription table. When an event e arrives to a peer p it checks if there is a matching in its own subscription table and in the affirmative it broadcasts e with a delay proportional to the number of beacon messages missed by p from the peer matching e. If a peer in the proximity of p received e as well and heard the relay of e from p, it drops the planned e's forwarding. This creates a wave effect that brings most of the time the event to the intended destination very quickly.

10.5.5 Event Routing for Sensor Networks

Two contributions were proposed ([189, 355]) adapting to the context of wireless sensor networks event routing solutions introduced for wired networks. Costa et al. [189] propose a sensor-based implementation of their semi-probabilistic algorithm of [669], that exploits only broadcast for communication and introduces specific solutions for reducing the impact of packet collisions. Hall et al. [355] adapt the content-based networking protocol of [144] to sensor networks, in the form of a routing protocol which extends the acyclic overlay used in [144] with backup routes for handling permanent and transient failures. A further optimization is made, by assuming that the set of possible receivers (i.e., the base stations) is small and known by all nodes: this allows to evaluate the actual receivers directly on the publisher side.

10.6 Security

Security is an important requirement for pub/sub systems. An overview of pub/sub security topics was given in [848]. They propose several techniques for ensuring the availability of the information dissemination network. Prevention of denial-of-service attacks is essential and *customized publication control* is proposed to mitigate large-scale attacks. In this technique, subscribers can specify which publishers are allowed to send them information. A challenge-response mechanism is proposed, in which the subscriber issues a challenge function, and the publisher has to respond to the challenge. The use of the mechanism in a distributed environment was not elaborated.

The EventGuard system comprises of a set of *security guards* to secure pub/sub operations, and a resilient pub/sub network [770]. The basic security building blocks are tokens, keys, and signatures. Tokens are used within the pub/sub network to route messages, which is not directly applicable for content-based routing. Bogus messages are prevented by authenticating subscribers and publishers using *ElGamal* signatures.

Pub/sub broker networks are vulnerable to message dropping attacks. For example, overlays such as Hermes and Maia [665] may suffer from bogus nodes. The prevention of message dropping attacks has a high cost and only a few systems address them. The EventGuard uses an *r-resilient* network of brokers [770].

Secure event types and type-checking was proposed in [665]. Secure event type definitions contain issuer's public key, version information, attributes, delegation certificates, and a digital signature. Scope-based security was discussed in [279], in which trust networks are created in the broker network using PKI techniques.

10.7 Mobility Support

Most research on event systems has focused on event dissemination in the fixed network, where clients are stationary and have reliable, low-latency, and high bandwidth communication links. Recently, mobility support and wireless communication have become active research topics in many research projects working with event systems, such as Siena [135] and Rebeca [278]. A mobility-aware event system needs to be able to cope with a number of sporadic and unpredictable end systems, to provide fast access to information irrespective of access location, medium and time. Problems such as delayed events, events generated for offline systems and the delay posed by the transmission of events create synchronization and event delivery problems, that need to be solved.

User mobility occurs when a user becomes disconnected or changes the terminal device. *Terminal mobility* occurs when a terminal moves to a new location and connects to a new access point. Mobility transparency is a key requirement for the event system: it should be able to hide the complexity of subscription management caused by mobility. Mobile components typically require that the pub/sub topology

is updated and thus it is necessary to prove for a mobility protocol that the safety
properties are not violated, which is referred to as *mobility-safety* in [798]. Typically, a *stateful* mobility protocol is used that buffers messages for a disconnected
client.

In the simplest mobility support setting, a special broker is used to buffer messages for offline pub/sub clients. This was used in the Elvin event system that supports disconnected operation using a centralized proxy, but does not support mobility between proxies [742]. In a wide-area system, mobility support between proxies
is needed. A handover or handoff protocol allows clients to move between proxies.
Such a handover protocol can also be used for load balancing purposes.

The JEDI event system was one of the first pub/sub systems to support mobile
components in a hierarchical topology of event brokers [204]. JEDI maintains causal
ordering of events and is based on a tree-topology, which has a potential performance bottleneck at the root of the tree with subscription semantics. Handover is
easier to implement for the hierarchical topology than for a tree or graph topology,
because the root of the hierarchy can be used as an anchor point for mobility related
signalling.

Mobility support in a tree or graph based distributed event infrastructure, such as
Siena and Rebeca, is challenging because of the high cost of the flooding, and issues with mobile publishers. Content-based flooding is needed when filter covering
or merging optimizations are being used to compact routing tables. These optimizations lose information regarding the source of a subscription or advertisement, and
thus lead to more complicated handover protocols. Mobile publishers, on the other
hand, require that events published during mobility or after mobility are delivered
to subscribers who were active at the time of the handover and are still active after
the handover.

A generalized subscription state transfer protocol consists of four phases:

1. Subscriptions are moved from broker A to broker B.
2. B subscribes to the events.
3. A sends buffered notifications to B.
4. A unsubscribes if necessary.

The problem with this protocol is that B may not know when the subscriptions
have taken effect—especially if the routing topology is large and arbitrary. This is
solved by synchronizing A and B using events, which potentially involves flooding
the content-based network.

To solve this synchronization problem, the Siena event system was extended with
generic mobility support, which uses existing pub/sub primitives: publish and subscribe [135]. The mobility-safety of the protocol was formally verified. The benefits
of a generic protocol are that it may work on top of various pub/sub systems and
requires no changes to the system API. On the other hand, the performance of the
mobility support decreases, because mobility-specific optimizations are difficult to
realize when the underlying topology is hidden by the API.

In addition to Siena, several other event systems have been extended with a mobility support protocol. In the rest of this section, we will briefly cover mobility friendly systems and recent insights into this problem domain.

Rebeca supports both logical and physical mobility. The basic system is an acyclic routed event network using advertisement semantics. The mobility protocol uses an intermediate node between the source and target of mobility, called Junction, for synchronizing the servers. If the brokers keep track of every subscription the Junction is the first node with a subscription that matches the relocated subscription propagated from the target broker. If covering relations or merging is used this information is lost, and the Junction needs to use content-based flooding to locate the source broker [278].

JECho is a mobility-aware event system that uses opportunistic event channels in order to support mobile clients [168]. The central problem is to support a dynamic event delivery topology, which adapts to mobile clients and different mobility patterns. The requirements are addressed primarily using two mechanisms: proactively locating more suitable brokers and using a mobility protocol between brokers, and using a load-balancing system based on a central load-balancing component that monitors brokers in a domain.

Recent findings on the cost of mobility in hierarchical routed event infrastructures that use unicast include that network capacity must be doubled to manage with the extra load of 10% of mobile clients [124]. Recent findings also present optimizations for client mobility: *prefetching*, *logging*, *home-broker*, and *subscriptions-on-device*. Prefetching takes future mobility patterns into account by transferring the state while the user is mobile. With logging, the brokers maintain a log of recent events and only those events not found in the log need to be transferred from the old location. The home-broker approach involves a designated home broker that buffers events on behalf of the client. This approach has extra messaging costs when retrieving buffered events. Subscriptions-on-device stores the subscription status on the client so it is not necessary to contact the old broker. In this study the cost of reconfiguration was dominated by the cost of forwarding stored events (through the event routing network).

The cost of publisher mobility has also been recently addressed by Muthusamy et al. in [606]. They start with a basic model for publisher mobility that simply tears down the old advertisement and establishes it at the new location after mobility. Thus a specific handover protocol is not needed. They confirm the high cost of publisher mobility and present three optimization techniques, namely *prefetching*, *proxy*, and *delayed*. The first exploits information about future mobility patterns. The second uses special proxy nodes that advertise on behalf of the publisher and maintain the multicast trees. The third delays the unadvertisement at the source to exploit the overlap of advertisements, but does not synchronize the source and target brokers. The publisher mobility support mechanisms used in the study are not necessarily mobility-safe.

A formal discrete model for both publisher and subscriber mobility was presented in [798, 797]. In this work, two new properties are defined for the pub/sub topology, namely mobility-safety and completeness. A handover protocol is mobility-safe if it prevents false negatives. A topology or a part of a topology is complete if subscriptions and advertisements are fully established (propagated) throughout it.

Mobility-safety of a generic stateful handover was shown for acyclic pub/sub networks. The completeness of the topology is used to characterize pub/sub handover protocols and optimize them. One of the results of this work is that rendezvous-points are good for pub/sub mobility, because they can be used to limit signalling and flooding of updates.

10.8 Summary and Outlook

Publish/subscribe is now largely acknowledged as one of the most interesting paradigm for distributed interaction. However, the positive characteristics of a pub/sub system (such as scalability) are not directly inherited from the paradigm but has to be enforced by specific architectural and algorithmic solutions. Following this observation, the architectural model that we proposed in this chapter intends to critically classify and analyze the large amount of research works proposed for distributed event routing since then, pointing out the critical aspects of the different solutions, specifically in terms of interaction and dependencies among the choices made at each architectural layer. We believe that our classification can be of valuable importance to define new solutions for event routing algorithms and their mapping to the overlay infrastructure.

Acknowledgements

This work has been partially supported by the EU STREP projects SM4ALL and CoMiFin.

The authors wish to thank Emmanuelle Anceaume, Roberto Beraldi, Mariangela Contenti, Gianpaolo Cugola, Maria Gradinariu, Carlo Marchetti, Sara Tucci Piergiovanni and Roman Vitenberg with whom we worked around the theme of publish/subscribe in the last few years. The authors would like also to thank the attendees at the MINEMA Summer School on Middleware for Network Eccentric and Mobile Applications that was held in Klagenfurt in 2005. Discussions during those days were instrumental to form structure and content of this chapter.

Chapter 11
Tuple Space Middleware for Wireless Networks

Paolo Costa, *Vrije Universiteit, Amsterdam, The Netherlands*
Luca Mottola, *Politecnico di Milano, Italy*
Amy L. Murphy, *FBK-IRST, Italy*
Gian Pietro Picco, *University of Trento, Italy*

11.1 Introduction

Wireless networks define a very challenging scenario for the application programmer. Indeed, the fluidity inherent in the wireless media cannot be entirely masked at the communication layer: issues such as disconnection and a continuously changing execution context most often must be dealt with according to the application logic. Appropriate abstractions, usually provided as part of a *middleware*, are therefore required to support and simplify the programming task.

Coordination [549] is a programming paradigm whose goal is to separate the definition of the individual behavior of application components from the mechanics of their interaction. This goal is usually achieved by using either *message passing* or *data sharing* as a model for interaction. Publish/subscribe, described in Chap. 10 is an example of the former, where coordination occurs only through the exchange of messages (*events*) among publishers and subscribers. While message passing, in its pure form, is inherently *stateless*, data sharing enables coordination among components by manipulating the (distributed) *state* of the system. The tuple space abstraction, the subject of this chapter, is a typical example of a data sharing approach. The two models, publish/subscribe and tuple spaces, have sometimes crossed paths in the scientific literature: their expressive power has been compared on formal grounds in [127]; the limits of an implementation of a stateful tuple space on top of a stateless publish/subscribe layer has been investigated in [150]; some extensions of publish/subscribe with stateful features exist (e.g., the ability to query over past events as in [511]). A thorough discussion of the relationship between the two is outside the scope of this chapter, and hereafter we focus solely on tuple spaces.

Linda [310] is generally credited with bringing the tuple space abstraction to the attention of the programming community. In Linda, components communicate through a shared *tuple space*, a globally accessible, persistent, content-addressable data structure containing elementary data structures called *tuples*. Each tuple is a sequence of typed fields, as in ⟨"foo", 9, 27.5⟩, containing the information being communicated. A tuple t is inserted in a tuple space through an **out**(t) operation, and can be withdrawn using **in**(p). Tuples are anonymous, their selection taking place through pattern matching on the tuple content. The argument p is often called a *template* or *pattern*, and its fields contain either *actuals* or *formals*. Actuals are values; the fields of the previous tuple are all actuals, while the last two fields of ⟨"foo", ?integer, ?float⟩ are formals. Formals act like "wild cards", and are matched against actuals when selecting a tuple from the tuple space. For instance, the template above matches the tuple defined earlier. If multiple tuples match a template, the one returned by **in** is selected non-deterministically. Tuples can also be read from the tuple space using the non-destructive **rd**(p) operation. Both **in** and **rd** are blocking, i.e., if no matching tuple is available in the tuple space the process performing the operation is suspended until a matching tuple becomes available. The asynchronous alternatives **inp** and **rdp**, called *probes*, have been later introduced to allow the control flow to return immediately to the caller with an empty result when a matching tuple is not found. Moreover, some Linda variants (e.g., [717]) also provide *bulk operations*, **ing** and **rdg**, used to retrieve all matching tuples in one step.

The fact that only a small set of operations is necessary to manipulate the tuple space, and therefore to enable distributed component interaction, is per se a nice characteristic of the model. However, other features are particularly useful in a wireless environment. In particular, coordination among processes in Linda is decoupled in time and space, i.e., tuples can be exchanged among producers and consumers without being simultaneously available, and without mutual knowledge of their identity or location. This decoupling is fundamental in the presence of wireless connectivity, as the parties involved in communication change frequently due to migration or fluctuating connectivity patterns.[1] Moreover, tuple spaces can be straightforwardly used to represent the context perceived by the coordinating components. On the other hand, this beneficial decoupling is achieved thanks to properties of the Linda tuple space—its global accessibility to all components and its persistence—difficult to maintain in a dynamic environment with only wireless links.

In the last decade, a number of approaches were proposed that leverage the beneficial decoupling provided by tuple spaces in a wireless setting, while addressing effectively the limitations of the original Linda model. Our group was among the first to recognize and seize the potential of tuple spaces in this respect, through the LIME model and middleware [670]. This chapter looks back at almost a decade of efforts in the research community, by concisely describing some of the most representative systems and analyzing them along some fundamental dimensions of comparison. In doing so, it considers two main classes of applications that rely on wireless communication. First, Sect. 11.2 considers *mobile networks*, where the network topology is continuously redefined by the movement of mobile hosts. Then, Sect. 11.3 considers the more recent scenario defined by *wireless sensor networks* (WSNs), networks of tiny, resource-scarce wireless devices equipped with sensors and/or actuators, enabling untethered monitoring and control.

---

[1] A rather abstract treatment of coordination and mobility can be found in [714].

The structure of these two sections is identical. Each first provides a brief survey of representative systems in the corresponding class of wireless applications. Then, it elicits some recurring themes and dimensions of comparison, which are used to classify and compare the systems. Finally, a small case study is presented to show how tuple spaces can be used in the context of a realistic application for the wireless domain at hand. The case studies are borrowed from the work of the authors, and are based respectively on LIME [601] and its adaptation to WSNs, TeenyLIME [190].

Finally, Sect. 11.4 offers some concluding remarks.

11.2 Mobile Networks

This section discusses the applicability of tuple spaces to environments with moving hosts. It considers two primary mobility models, namely nomadic and mobile ad hoc networks (MANETs). In nomadic mobility, a wired infrastructure supports the connection of mobile devices to the wired network through base stations. Instead, MANET removes the infrastructure and nodes communicate directly only when they are within range.

11.2.1 Representative Systems

The following outlines the main tuple space approaches developed for mobile settings. Due to space constraints, this should not be considered an exhaustive list, but rather an outline of representative systems. The sequence of presentation roughly corresponds to the chronological order of appearance of each system.

TSpaces. After the initial enthusiasm in the 1980s with tuple spaces, the 1990s saw a resurgence of the model for distributed computing. IBM's TSpaces [819] was one of the first such systems, providing a client/server interface to a centralized tuple space, merging the simplicity and flexibility of the tuple space front-end with an efficient database back-end. TSpaces targets both fixed, distributed systems, as well as provides support for nomadic computing with devices such as hand held PDAs. In both cases, all data is centrally managed and clients remotely issue operations.

L^2imbo. While TSpaces relies on a centralized server, the L^2imbo platform [215] proposes a decentralized implementation in which each host holds a replica of the tuple space. The underlying mobility model is still nomadic, forcing mobile hosts to connect through a base station to gain full access to the data. Nevertheless, because the tuple space is replicated locally, hosts can perform limited operations even during disconnections.

The implementation leverages off IP Multicast to disseminate updates and ensure consistency among replicas. Tuple spaces are uniquely mapped to multicast groups and all interested hosts must be members. When a host joins a group, a local replica of the tuple space is created and all subsequent updates are received as multicast messages. To avoid conflicts, the authors introduce the notion of *tuple ownership*. While reading operations are always permitted, writing operations can be performed only after the ownership for the specific tuple has been acquired.

In L^2imbo all tuples are associated with a type, which is used, along with traditional field-based matching, by the **in** and **rd** requests to retrieve the desired tuples. To enable fine-grained searches, types are arranged in a hierarchy and a match occurs if a tuple of the type or any subtype is found.

LIME. The LIME model [601] defines a coordination layer that adapts and extends the Linda model towards applications that exhibit physical mobility of hosts in a MANET and/or logical mobility of software agents [297]. Given the topic of this chapter, hereafter we bias our presentation towards the former case. LIME, a mobile host has access to a so-called interface tuple space (ITS), permanently and exclusively attached to the host itself. The ITS, accessed using Linda primitives, contains tuples that are physically co-located with the host and defines the only data available to a lone host. Nevertheless, this tuple space is also transiently shared with the ITSs belonging to the mobile hosts currently within communication range. When a new host arrives, the tuples in its ITS are conceptually merged with those already shared, belonging to the other mobile hosts, and the result is made accessible through the ITS of each of the hosts. This provides a mobile host with the illusion of a local tuple space containing tuples coming from all the hosts currently accessible, without any need to know them explicitly.

LIME also augments the Linda model with the notion of *reaction*. A reaction $\mathcal{R}(s, p)$ is defined by a code fragment s specifying the actions to be performed locally when a tuple matching the pattern p is found in the shared tuple space. This effectively combines the proactive style, typical of tuple space interaction, with the reactive paradigm useful in the dynamic, mobile environment. Recent papers further enhance the LIME model by adding support for security [358], replication [599] and code deployment [668].

EgoSpaces. EgoSpaces [435] is a tuple space middleware similar to LIME exploiting a fully distributed architecture. In its model, the network is perceived as an underlying database of tuples. Each host defines its own tuple space by creating a *view*, i.e., a subset of the tuples available at other hosts, selected according to the specified constraints (e.g., host IDs, number of hops, tuple patterns).

Hosts interact with these views through the basic Linda operations and special constructs for event-driven communication. EgoSpaces also provides transactional support to ensure that a sequence of operations (e.g., a **rd** followed by an **in**) is executed atomically.

TOTA. In contrast to the previously described systems, TOTA [550] exploits a tuple-centric approach to support both pervasive computing and MANET mobility. In TOTA, tuples spread hop-by-hop among nodes according to the rules specified in the tuple itself. These rules may include, for example, the scope of the tuple (i.e., how many hops the tuple should travel) or the conditions for the propagation to occur (e.g, if a tuple denotes a fire alarm, it will be replicated only if the room

Table 11.1 Features of representative tuple space systems for mobile computing

	TSpaces [819]	L²imbo [215]	LIME [601]	EgoSpaces [435]	TOTA [550]
Architecture	Centralized	Decentralized	Decentralized	Decentralized	Decentralized
Mobility scenario	Nomadic	Nomadic	MANET	MANET	MANET
Context-awareness	N/A	QoS attributes	Context as data	Context as data	Context as data
Disconnected operation	None	Read-only	Yes	Yes	None
Atomicity	Yes (strong)	No	Yes (strong)	Yes (best-effort)	No
Reactions (to)	Yes (operations)	No	Yes (state)	Yes (state)	Yes (operations)
Scope	Whole TS	Whole TS	Federated TS or single host/agent	Programmer-defined views	Local or one hop
Other extensions	Probes, bulk, time-outs	Time-outs	Probes, bulk	Probes, bulk	Bulk

temperature is above a threshold). Tuples can also be modified along the way to take into account changing conditions (e.g., a tuple containing the temperature can average its value over all the readings found). Hosts can access the local or neighboring tuple spaces to retrieve, add, or remove tuples. Event-driven constructs are also present to react when a particular tuple is inserted in the tuple space.

11.2.2 Discussion

We now focus on the features that distinguish the above representative systems. Table 11.1 provides a concise summary while details are provided below.

Architecture and Mobility Scenario. The placement of the tuple space data plays a key role in the applicability of the model, especially in relation to the target mobility scenario. For example, the centralized, client-server model of TSpaces supports nomadic mobility for resource poor devices such as PDAs. It also allows the server to both take on the majority of the computation burden as well as provide optimizations to data access, e.g., with a database back-end, which is not possible on small, mobile devices.

Instead, to support MANETs, a single server will not always be reachable and each node essentially becomes a mini-server with part of the global data. This is the decentralized model adopted by TOTA as well as the transiently shared tuple spaces of LIME and EgoSpaces. While this fully distributed model requires direct interaction among nodes to share locally hosted data, at the same time it eliminates any central point of coordination, an essential requirement in the MANET environment.

L²imbo strikes a balance between the two approaches, supporting nomadic mobility with decentralized tuple spaces. The primary advantage is support for disconnected operation, as discussed later.

Context-Awareness. As outlined in [598], tuple spaces naturally support the requirements of context-aware applications. Indeed, context data can be stored in the tuple space just as any other data, as proposed by LIME, EgoSpaces and TOTA, leveraging the decoupled nature of Linda to separate the context data producers from the consumers.

L^2imbo takes this one step further by exposing context information critical to supporting quality of service in the mobile environment. This is accomplished through a set of monitors running on each host. *Connectivity* monitors check the connectivity between a pair of hosts and report quality (e.g., throughput). *Power* monitors, instead, observe the power level on the local host as well as on nearby hosts such that applications can employ an appropriate saving scheme. Finally, *cost* monitors track the communication costs between two hosts. Based on this information, applications can monitor local and remote resources and, if necessary, activate specific energy-saving policies.

Despite the suitability of the tuple space abstraction, the standard Linda matching based on types and exact values is often insufficient for context-aware applications as such queries frequently require range rather than exact value matching. For example, a query may look for any host within a 50 m radius of its current location. This has been addressed in LIGHTS [58], the tuple space underlying LIME. To support the needs of context-awareness, LIGHTS extends Linda matching semantics to use range matching, fuzzy logic comparison operators, and other extensions enhancing the expressiveness of tuple queries.

Disconnected Operation. Given the dynamicity of the mobile scenario, connectivity cannot be guaranteed at all times. In some systems such as TSpaces, disconnections are considered a fault, and disconnected clients cannot access the tuple space. Instead other systems adopt the view that disconnection is an expected event in mobile computing, and some amount of functionality must be provided even when a node is isolated from all others.

L^2imbo tolerates disconnected operations by keeping a replica of the entire tuple space on each host. Eventual consistency among different copies is ensured by a variant of IP multicast [285] in which each tuple space is bound to a multicast group to which every node accessing the tuple space belongs. While disconnected, a host can optimistically read tuples from its local cache, but it cannot remove any tuples. When a host reconnects to the network, it communicates to the multicast group the tuples created during the disconnection and receives notification of all tuples added/removed by other nodes.

In the MANET environment, LIME allows a node to operate freely on any tuples in its federated tuple space. When isolated, a node has access only to the tuples it hosts, but, still, any operations are possible. An extension [599] addresses replication at the tuple level based on patterns. This gives the individual agent control over the amount of data replicated and hence transmitted across the wireless links. The update policy is also flexible, allowing updates to propagate only from the master, from any newer version, or never.

Atomicity. A key feature of any modern distributed platform is the ability to perform multiple operations on a diverse set of hosts in a single atomic step. This is particularly true for the stateful model of tuple spaces because the lack of atomicity may lead to an inconsistent system state, e.g., when two hosts succeed in removing the same tuple.

To this end, mainstream systems such as TSpaces offer native transactional support by using a database server as a backend. Unfortunately, this approach is overly complex for systems of mobile devices with fluctuating connections. To overcome this, LIME exploits transactions only to manage group membership by implicitly electing a single leader for each group, responsible for ensuring the atomicity of tuple space *engagement* and *disengagement*, respectively the process of merging or breaking down the tuple space based on a change in connectivity. Conversely, atomicity for distributed operations (e.g., **in**) are guaranteed through a lightweight approach, relying on a combination of reactions and non-blocking probes. When an **in** operation is issued, if a matching tuple is found in the local tuple space, the operation immediately returns. Otherwise, the run-time installs a reaction for the same pattern. When this reactions fires, a non-blocking **inp** is issued to remove the tuple. If, in the meanwhile, the tuple has been withdrawn by another host, the reaction remains in place and the process continues to wait.

One drawback of the LIME approach, however, is that atomicity can be guaranteed only for system-supported operations. To address this concern, EgoSpaces proposes a traditional, yet costly, transaction-based mechanism. In addition, it provides a best-effort solution based on *scattered probes*. These operations are implemented such that all the hosts participating in the view are contacted one at a time and an empty set is returned if no matching tuple has been found. Scattered probes provide a weaker consistency because they are allowed to miss a matching tuple in the view. On the other hand, the implementation is both more efficient because transactions are not employed and more flexible because hosts are not suspended.

Reactions. Mobility challenges many of the assumptions made in traditional distributed computing. Of primary concern is that data is only transiently accessible due to changing connectivity among participants. This leads naturally to the introduction of some notion of reactive operation where, similar to event-based programming, a component can be notified when something interesting happens. In TSpaces and TOTA, nodes can be notified when certain *operations* (e.g., the insertion or removal of a tuple) are issued on the tuple space. However, the notification occurs only if the recipient is connected to the tuple space where the operation is issued. In LIME and EgoSpaces, instead, hosts are notified when data matching a given pattern appears in the host's federated tuple space. These *state-based* reactions are important in the MANET environment because two agents may not be connected when a tuple is inserted: the state-based semantics ensure that the reaction fires whenever a "new" relevant tuple is detected (e.g., when two nodes meet), and not just upon an insertion operation. Thus, being notified when data is *accessible* as opposed to when it is *inserted* provides a natural and very powerful programming primitive, useful in many mobile applications.

Scope. In large-scale mobile networks it is not practical to share the entire tuple space across all nodes. The overhead to route requests and replies would quickly drain all host resources. Therefore, most implementations limit the distribution to a single (physical) hop. TOTA introduces explicitly scoped operations, such as **read-OneHop**, while LIME allows the programmer to set the scope of an operation to a single, specific host.

EgoSpaces [435], instead, enables the programmer to flexibly define the *scope* of the shared tuple space, called a *view*. This is expressed through a *declarative specification* providing constraints over the properties of the underlying network (e.g., only nodes within 5 hops or 500 meters), on the host (e.g., only PDAs), and on the data (e.g., only location information).

Other Extensions. In addition to reactive operations and scoping, the transient accessibility of data changes the programming paradigm. In standard Linda programs for parallel computation, blocking operations are natural because, unless a process has data to work with, it should be suspended. In the mobile environment, instead, processes are often interactive or at least more flexible to the current context. Thus, if some needed data is not available, it may be better to switch to another task rather than block. Moreover, knowledge about the lack of data, which may also imply the unavailability of a host known to carry such data, may be important per se.

This kind of interaction is not supported by traditional blocking operations, thus most of the systems discussed in this section provide the non-blocking probes mentioned in Sect. 11.1, which query a tuple space and return either a matching tuple or a keyword indicating that no match exists. Such immediate return after the operation is issued gives the programmer a high degree of flexibility. An intermediate approach is taken in TSpaces and L^2imbo, where a timeout can be specified for (blocking) **in** and **rd** operations to avoid locking the process indefinitely.

Additionally, some applications logically create multiple versions of the same piece of information with each successive version invalidating the previous. For example, a location tuple constantly changes as a host moves through space. Each new location represents an update, replacing the now-irrelevant previous location. In most tuple space systems, it is only possible to remove the old data and insert new data, losing any logical connection between the two. Instead, it is meaningful to allow the data to be *changed*, associating it with the old data and at the same time identifying that it has been updated. Such a mechanism is provided by a variant of LIME [599] in which the user specifies a template for the old data together with the actual new data. The new data is distinguished from the old with a version number. This mechanism also serves as a building block upon which consistency between master and replicas is managed.

11.2.3 Tuple Spaces in Action

To illustrate concretely the benefits of the tuple space abstraction in the mobile environment, this section illustrates TULING [598], a sample application built on top

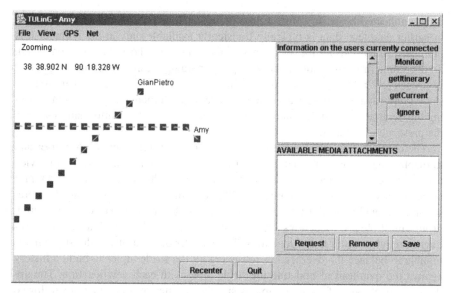

Fig. 11.1 Screenshot of a TULING user, Amy. While near a second user, GianPietro, his history and movement are visible, but once out of range, updates are no longer propagated and only the locally visible movements of Amy are displayed

of LIME for collaborative exploration of a space, emphasizing the exchange of context information. While offering a simple example, TULING demonstrates how the operations available in LIME are both natural and sufficient to provide the range of interaction necessary for exploiting context. Moreover, the interaction patterns of TULING are applicable to real-world collaborative applications, such as team coordination in a disaster recovery scenario.

Scenario and Requirements. TULING is intended to be used by multiple individuals moving through a common environment, each equipped with a GPS- and wireless-enabled PDA. Users see a representation of their current position as well as a trail representation of previous movements, as shown in Fig. 11.1. When a new user comes within range, her name is displayed and one of several monitoring modes can be selected, e.g., retrieving only the current location, tracking the location as it changes, or retrieving the entire movement history of the user. The choice is typically based on the tolerance for the overhead associated to each type of monitoring.

TULING also allows users to add annotations, such as a textual note or a digital photograph, to their own current location. These annotations are indicated on the display with a special icon: by clicking on the icon, the annotation can be viewed as long as the requesting user is connected to the user who made the annotation.

Design and Implementation. The design and implementation of TULING focuses on making various aspects of the application, e.g., location, other users, and annotations, available as state inside the tuple space. This choice makes all data accessible to all connected hosts. In contrast, in a model such as publish/subscribe, explored in earlier chapters, data is typically transiently available only at the moment it is published, thus limiting the awareness of the data to the hosts that are connected. To overcome this restriction, a query-response paradigm can be exploited in parallel to proactively retrieve all missed data. In contrast, the LIME model of transiently shared tuple spaces provides a single interface for both modes of interaction, unifying the treatment of new and stored data as well as system-level information such as the presence of other hosts.

The combined requirements to both monitor the current location of a user and to display the history information about the itinerary require that TULING provide access to the current and previous locations of a user. The *current location* is represented by a single tuple containing the GPS coordinates and a timestamp. To represent movement history, TULING uses a separate tuple template. The chosen solution groups multiple prior locations together into a single *stride* tuple that contains a sequence number and a list of locations. The number of locations in the stride list is tunable to balance the overhead of retrieving all the stride tuples to build a history against the overhead of updating the stride tuple with each new location. The updating of each type of location tuple is simply a matter of issuing an **out** followed by an **in**. This sequence ensures that at all times a location tuple is present in the tuple space. It is also worth noting that these operations are entirely local to the host where they are issued, therefore the overhead is minimal.

To access the location context of other hosts, TULING uses a combination of reactions and probe operations. Specifically, it uses a feature of LIME called the LimeSystem tuple space, a system-maintained tuple space that contains information about which hosts are currently connected. To display the name of a user within range, it employs a reaction on the LimeSystem tuple space. To give a feel for how simple it is to accomplish this operation, Fig. 11.1 shows the required code to register a reaction for the arrival of a new host. A similar process is required for reacting to the departure of a host.

Once the name of a user is displayed, the monitoring mode must be selected. As a result of LIME's transient sharing of the tuple spaces among hosts within range, the locations of all connected users are available to one another. To get the current position, a non-blocking read operation **rdp**, restricted in scope to the selected host, is issued for the location tuple. A probing read is used to prevent the system from blocking in the case where the host disconnects before the read operation completes. Similarly, retrieving the itinerary requires a bulk read operation, **rdg**, on the stride tuples for the selected host. Instead, monitoring a user is accomplished by installing a reaction on the location tuple. Each time a new location is inserted, the reaction fires remotely, a copy of the tuple is sent to the registered user, and their local display is updated. When two hosts disconnect causing their tuple spaces disengage and become no longer shared, the reaction ceases to fire. Nevertheless, when they come back within range, the reaction is automatically reinstalled by the system and updates propagate once again. It is worth noting that a similar reaction for location tuples scoped on the entire federated tuple space can be installed to display the locations of *all* users who come within range. In this case, the individual per-user

---

Algorithm 11.1: Code to react to the arrival of a new user using the `LimeSys-temTupleSpace`.

```
public class  NewHostListener implements ReactionListener{
   NewHostListener() {
      LimeSystemTupleSpace lsts = new LimeSystemTupleSpace();
      // HOST arrival pattern
      Tuple newHostPattern =
            new Tuple().addActual("_host")
                       .addFormal(LimeServerID.class);
      lsts.addListener(
            {new LimeSystemReaction (
                       newHostPattern,
                       this,
                       Reaction.ONCEPERTUPLE)});
   }
   void reactsTo(ReactionEvent re) {
      // display name of the new host arriving
      // the second field of re.getEventTuple() contains the host name
   }
}
```

---

reactions described above are unnecessary. The trade-off, however, is the inability to control the overhead as all users are unconditionally monitored.

The annotation feature is similarly supported with a combination of a reaction scoped over the entire federated tuple space to track which annotations are accessible, and a **rdp** to retrieve the contents of a requested annotation. Importantly, the implementation separates the knowledge of the existence of an annotation from the annotation itself, using two different tuple formats. This was motivated by the observation that the annotation contents themselves may be large, e.g., a digital photograph. Because a LIME reaction retrieves a copy of the matching tuple, reacting to the annotation contents would involve transferring the entire annotation whether or not it will be used, unnecessarily using the wireless communication media. Instead, by reacting to a small tuple which is essentially a reference to the actual annotation, the reaction processing remains efficient. The result, however, is the restriction that annotations can only be viewed while users are connected; a reasonable compromise for effectively managing overhead.

Experience with TULING clearly demonstrates the effectiveness of LIME to support context aware interactions in the MANET environment. The simple combination of key LIME operations, such as the probing **inp** and reactions, give the programmer rich mechanisms to interact with both the application and context data.

11.3 Wireless Sensor Networks

Wireless sensor networks (WSNs) pose peculiar challenges, only partially overlapping with those of mobile networks. Although communication occurs wirelessly,

nodes tend to be static. Moreover, WSN devices typically offer much fewer resources than those employed in mobile networks. As a result, a good fraction of
the programming effort often focuses on low-level concerns such as resource management. Abstractions such as tuple spaces can help programmers to address the
requirements of WSN applications by raising the level of abstraction and hiding
distribution.

Differently from the approaches described in the previous section, in WSNs few
works provide a genuine tuple space abstraction to application programmers. Nevertheless, a relevant fraction of existing approaches leverages off the same first principles. The systems surveyed in this section indeed provide *data-centric* programming
abstractions where the location of data, as well as the identity of the individual devices, plays only a secondary role. On top of this feature, they enable various forms
of *data sharing* among different devices. Consequently, distributed interactions occur implicitly in accessing some piece of data that programmers cannot a priori
locate on some specific device. Blending these features in a single programming
framework finds fertile ground in WSNs, where data is of paramount importance to
application developers.

The structure of presentation is similar to the one we followed for mobile computing in the previous section. Section 11.3.1 describes exemplary approaches from
the current state of the art. Again, the choice of systems to be discussed does not
pretend to be exhaustive: the goal is to give the reader the insights necessary to
appreciate how the driving concepts of tuple spaces have been applied in WSNs.
Next, Sect. 11.3.2 illustrates the key features of the approaches described, comparing them against each other. Finally, Sect. 11.3.3 discusses a small case study to
provide a concrete example of how tuple spaces are applicable to the WSN domain.

11.3.1 Representative Systems

This section surveys systems that either provide a tuple space abstraction to application developers, or more generally take inspiration from the key features of tuple
spaces. In doing so, the discussion is limited to approaches geared towards programming individual WSN devices. Alternative paradigms have been explored in WSNs,
whose goal is to give programmers a way to program the network as a whole. These
approaches, commonly termed "macroprogramming" [342], radically depart from
traditional programming. Therefore, they are not directly comparable with the ones
discussed next.

Abstract Regions. Welsh et al. [861] propose a set of general-purpose communication primitives providing addressing, data sharing, and aggregation among a given
subset of nodes. A *region* defines a neighborhood relationship between a specific
node and other nodes in the system. For instance, a region can be defined to include
all nodes within a given number of hops or within physical distance d. Data sharing
is accomplished using a tuple space-like paradigm by giving developers language
constructs to read/write $\langle key, value \rangle$ pairs at remote nodes. In a sense, this resembles

the **rd** and **out** operations in traditional tuple space middleware, although the data format and matching is clearly much less expressive. Dedicated constructs are also provided to aggregate information stored at different nodes in a region. Moreover, a lightweight thread-like concurrency model, called Fibers, is provided for blocking operations. By their nature, Abstract Regions target applications exhibiting some form of spatial locality, e.g., tracking moving objects, or identifying the contours of physical regions.

Agilla. The work in [287] presents a middleware system for WSNs that adopts a mobile agent paradigm [297]. Programs are composed of one or more software agents able to migrate across nodes. In a sense, an Agilla agent is similar to a virtual machine with its own instruction set and dedicated data/instruction memory. Coordination among agents is accomplished using a tuple space. Agents insert data in a local data pool to be read by different agents at later times. The data of interest is identified using a pattern matching mechanism, in a way similar to what is described in Sect. 11.1. In Agilla, the use of tuple spaces allows one to decouple the application logic residing in the agents from their coordination and communication. At the same time, tuple spaces also provide a way for agents to discover the surrounding context, e.g., by reading tuples describing the most recent sensed values. Reactive applications requiring on-the-fly reprogramming of sensor nodes (e.g., fire monitoring) are thus an ideal target scenario for Agilla.

ATaG. The Abstract Task Graph (ATaG) [45] is a programming framework providing a mixed declarative-imperative approach. The notions of *abstract task* and *abstract data item* are at the core of the ATaG programming model. A task is a logical entity encapsulating the processing of one or more data items, which represent the information. The flow of information between tasks is defined in terms of *abstract channels* used to connect each data item to the tasks that produce or consume it. To exchange data among tasks, programmers are provided with the abstraction of a shared data pool where tasks can output data or be asynchronously notified when some data of interest is available. This style of interaction is similar to tuple spaces for mobile networks when using reactive operations, as described in Sect. 11.2.2, although in ATaG this is limited to triggering notifications when other processes perform a write, and the data of interest is determined solely based on its type, rather than an arbitrary pattern. The ability to isolate different processing steps in different tasks makes ATaG suited to control applications requiring multi-stage processing, e.g., road traffic control [590].

FACTS. Terfloth et al. [800] propose a middleware platform inspired by logical reasoning in expert systems. In FACTS, modular pieces of processing instructions called *rules* describe how to handle information. These are specified using a dedicated language called *RDL* (ruleset definition language), whereas data is specified in a special format called a *fact*. The appearance of new facts trigger the executions of one or more rules, that may generate new facts or remove existing ones. Facts can be shared among different nodes. The basic communication primitives in FACTS provide one-hop data sharing. Facts are reminiscent of tuples as a way of structuring the application data, while the triggering of rules in response to new facts is similar to the execution of reactions in mobile tuple space middleware. Reactive applications such as fence monitoring [868] are easily implemented using FACTS, thanks to the condition-action rules programmers can specify.

Hood. The programming primitives provided by Hood [864] revolve around the notion of neighborhood. Constructs are provided to identify a subset of a node's physical neighbors based on application criteria, and to share data with them. A node exports information in the form of *attributes*. Membership in a programmer-specified neighborhood is determined using *filters*. These are boolean functions that examine a node's attributes and determine, based on their values, whether the remote node is to become part of the considered subset. If so, a mirror for that particular neighbor is created on the local node. The mirror contains both *reflections*, i.e., local copies of the neighbor's attributes that can be used to access the shared data, and *scribbles*, which are local annotations about a neighbor. Hood can be seen as a tuple space system where **out** operations are used to replicate local information on neighboring nodes, and filters take care of the matching functionality. Thanks to the support of multiple independent neighborhoods, Hood is applicable in diverse settings ranging from target tracking applications to localization mechanisms and MAC protocols [864].

TeenyLIME. Inspired by LIME, TeenyLIME [191] offers a Linda-like interface to programmers. To make the tuple space paradigm blend better with the asynchronous programming model of most WSN operating systems (e.g., TinyOS [370]), operations are non-blocking and return their results through a callback. In TeenyLIME, tuples are shared among neighboring nodes. *Reactions* are provided to allow for asynchronous notifications in case some specific piece of data appears in the shared tuple space. In addition, several WSN-specific features are made available to better address the requirements of sensor network applications. For instance, a notion of *capability tuple* is provided to enable on-demand sensing. This can save the energy required to keep sensed information up to date in the shared tuple space in the absence of any data consumer. Similarly to Hood, TeenyLIME reaches into the entire stack, providing constructs to develop full-fledged applications as well as system-level mechanisms, e.g., routing protocols. However, TeenyLIME specifically targets sense-and-react applications, (e.g., HVAC in buildings [223]), where its reactive and WSN-specific features provide a significant asset.

11.3.2 Discussion

The approaches outlined above are possibly more heterogeneous than those we discussed in Sect. 11.2. Table 11.2 summarizes the most important similarities and differences.

Reactive vs. Proactive Operations. The ability to react to external stimuli is of paramount importance in WSNs. Likewise, being able to proactively influence the data shared with other processes is fundamental to achieve coordination through the

Table 11.2 Features of tuple space systems for wireless sensor networks

	Abstract Regions [861]	Agilla [287]	ATaG [45]	FACTS [800]	Hood [864]	TeenyLIME [191]
Reactive vs. Proactive	Proactive	Both	Reactive	Reactive	Proactive	Both
Push vs. Pull	Push	Both	Push	Push	Push	Both
Data Filtering	Shared variables	Pattern matching	Named channels	Fact constraints	Application-level filtering	Pattern matching w/value constraints
Data Processing	Reduce operator	N/A	N/A	N/A	N/A	N/A
Scope	Programmer-defined	Local	Programmer-defined	One-hop	One-hop	One-hop

tuple space. Agilla and TeenyLIME provide both modes of operations, i.e., the traditional tuple space operations to express proactive interactions along with a notion of *reaction* inspired by similar functionality in the mobile setting, as described in Sect. 11.2.2. The semantics provided, however, are generally weaker. For instance, no guarantees are provided on whether a reaction can fire multiple times for the same tuple. Similarly, the RDL language in FACTS allows for the specification of programmer-provided conditions for a rule to fire. Nevertheless, in these cases, implementing reactive operations raises issues with their semantics that are difficult to face on resource constrained devices. For instance, in the presence of multiple reactions (rules) being triggered simultaneously, the aforementioned systems provide no guarantees w.r.t. the order they are scheduled. This issue is partially solved in ATaG by forcing channels to behave in a FIFO manner, and imposing a round-robin schedule across different channels. Differently, both Abstract Regions and Hood provide only proactive operations. In Abstract Regions, the only way of observing a change in the shared data is to proactively read the value of a variable. Similarly, in Hood the local reflections must be manually inspected to recognize some change in the shared data. Both systems are therefore significantly less expressive, and lead to more cumbersome programming whenever some form of reactive behavior is required.

Proactive Push vs. Proactive Pull. Although the approaches described above do provide some notion of data sharing, none of them completely abstracts away the location of data. As a result, proactive operations can occur with different modes of operations. Existing systems operate either in a push manner—where the data producers move the data towards the data consumers—or in a pull manner, i.e., where data stays with the data producer until a consumer explicitly retrieves them. ATaG and FACTS provide only push primitives. In the latter, for instance, facts can be written at remote nodes but cannot be retrieved from them. The rest of the systems described, instead, provide both modes of operation. Agilla and TeenyLIME provide simple variations of the traditional tuple space API, offering pull operations such as **rd** as well as push ones, e.g., **out**. In Abstract Regions, remote variables can be read and written, which provides a way to describe both push and pull interactions.

Likewise, although the normal operational mode in Hood is to periodically push a node's attributes towards its neighbors using a form of beaconing, a node can also request an on-demand update of a neighbor's data, a functionality analogous to a pull operation.

Data Filtering. The pattern matching mechanism in traditional tuple spaces provide an expressive way to identify the data of interest. Agilla and TeenyLIME retain the same by-field, type-based matching semantics, but also provide the ability to specify further conditions on the value of the data themselves other than simple equality. For instance, in TeenyLIME it is straightforward to specify a pattern identifying a tuple whose first field is a 16 bit integer with a value above a given threshold. A similar feature is also present in FACTS, though the processing to enforce the condition on the fact repository is partially left to the application programmer. The other systems considered, instead, provide simpler—but less expressive—mechanisms for identifying the data of interest. In Abstract Regions, variable names are, in a sense, the only data filtering mechanism available; similar considerations hold for ATaG and Hood. Furthermore, in all these latter approaches the type of data to be shared is essentially decided at compile time. Therefore, the conditions to filter data cannot be changed by the application while the system is running.

Data Processing. Embodying constructs dedicated to elaborate data in the programming model can help WSN programmers to deal with commonly seen processing patterns, e.g., averaging the reading of a number of sensor nodes. The only work explicitly addressing this concern is Abstract Regions. The presence of the `reduce()` operator in the API allows programmers to apply a given operator to all shared variables of a given type, and possibly to assign the result to a further shared variable. The set of operators available, however, strictly depends on the specific implementation used in support of the region at hand. Some operators, (e.g., MAX) can be straightforwardly implemented for various kinds of regions. Instead, others need customized, per-region implementations: for instance, computing the correct outcome of AVG requires keeping track of duplicates.

Scope. In Sect. 11.2.2, it was observed that providing a shared tuple space spanning all the nodes in the system is often prohibitive in mobile networks. This issue is brought to an extreme in WSNs, due to the limited communication abilities of typical WSN devices. Consequently, in the current state of the art, the span of the shared memory space turns out to be quite limited. Most approaches provide one-hop data sharing, or even just local sharing as in the case of Agilla. Notable exceptions are Abstract Regions and ATaG. In both cases, however, dedicated routing support must be provided to enable data sharing beyond the physical neighborhood of a node. In Abstract Regions, each different region requires a dedicated implementation. For instance, an n-hop region can be implemented using limited flooding, whereas a region defined in terms of geographic boundaries can be implemented using GPSR [444]. This same protocol is used in ATaG when the span of the data pool is defined based on the locations of nodes assigned a given task. Alternatively, the data pool in ATaG can be shaped according to logical attributes of the nodes (e.g., their type) using the routing provided by Logical Neighborhoods [590].

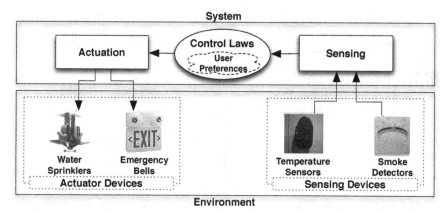

Fig. 11.2 Emergency control in buildings

11.3.3 Tuple Spaces in Action

This section illustrates how the driving concepts of tuples spaces can be applied to a paradigmatic sensor network scenario. We discuss the implementation of a sense-and-react application using TeenyLIME. In similar scenarios, nodes hosting actuators are deployed alongside sensing devices. The system is designed to react to stimuli gathered by sensors and affect the environment by means of actuators.

Scenario and Requirements. Consider an application for *emergency control in buildings* whose main functionality is to provide guidance and first response under exceptional circumstances, e.g., in case of fire. The application logic features four main components, illustrated in Fig. 11.2. The *user preferences* represent the high-level system goals, in this case, the need to limit fire spreading. *Sensing devices* gather data from the environment and monitor relevant variables. In this case, smoke and temperature detectors recognize the presence of a fire. *Actuator devices* perform actions affecting the environment under control. In the scenario at hand, water sprinklers and emergency bells are triggered in case of fire. *Control laws* map the data sensed to the actions performed, to meet the user preferences. In this case, a (simplified) control loop may activate emergency bells when the temperature increases above a safety threshold, but operate water sprinklers only if smoke detectors actually report the presence of fire.

The characteristics of the scenario make programming similar systems a challenging task:

- *Localized computations* [256] must be privileged to keep processing close to where sensing or actuation occurs. It is indeed unreasonable to funnel all the sensed data to a single base-station, as this may negatively affect latency and reliability without any significant advantage [11].

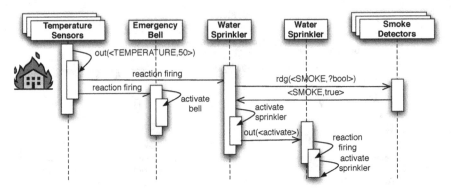

Fig. 11.3 Sequence of operations to handle a fire. Once notified about increased temperature, a node controlling water sprinklers queries the smoke detectors to verify the presence of fire. If necessary, it sends a command activating nearby sprinklers

- *Reactive interactions*, i.e., actions that automatically fire based on external conditions, play a pivotal role. In this case, a temperature reading above a safety threshold must trigger an action on the environment.
- *Proactive interactions*, however, are still needed to gather information and fine tune the actuation about to occur. For instance, the sprinklers in the building ask for smoke readings before taking any action.

The stateful nature of the tuple space abstraction naturally lends itself to addressing the above requirements, as the actions taken on the environment must be decided based on the current sensed state. In addition, tuple spaces make it easier to express the coordination required in these scenarios, e.g., since they provide both proactive and reactive operation. Alternative paradigms, such as publish/subscribe, only partially meet the above challenges, being essentially biased towards either reactive or proactive interactions.

Design and Implementation. In our design, sensed data and actuating commands take the form of tuples. These are shared across nodes to enable coordination of activities and data communication. Figure 11.3 illustrates how proactive and reactive interactions in TeenyLIME are used to deal with the possibility of a fire. Both emergency bells and water sprinklers have a reaction registered on their neighbors, watching for temperature tuples over a given threshold. This is accomplished using TeenyLIME's value matching functionality, described in Sect. 11.3.2. Temperature sensors periodically take a sample and pack it in a tuple, which is then stored in the local tuple space. This operation, by virtue of TeenyLIME's one-hop sharing, automatically triggers all the aforementioned reactions in case of a positive match.

However, different types of actuator nodes behave differently when high temperatures are detected. The node hosting the emergency bell immediately activates its attached device. Instead, the water sprinkler node proceeds to verify the presence of fire, as shown in Fig. 11.3. The latter behavior, specified as part of the reaction code, consists of proactively gathering the readings from nearby smoke detectors,

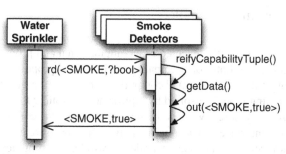

Fig. 11.4 Processing of capability tuples

using a **rdg** over the entire shared tuple space. If fire is reported, the water sprinkler node requests activation of nearby sprinklers through a two-step process that relies on reactions as well. The node requesting actuation inserts a tuple representing the command on the nodes where the activation must occur. The presence of this tuple triggers a locally-installed reaction delivering the activation tuple to the application, which reads the tuple fields and operates the actuator device accordingly.

Based on the above processing, smoke detectors need not be monitored continuously: their data is accessed only when actuation is about to occur. However, when a sensed value is requested (e.g., by issuing a **rd**) fresh-enough data must be present in the tuple space. If these data are only seldom utilized, the energy required to keep tuples fresh is mostly wasted. An alternative is to require that the programmer encodes requests to perform sensing on-demand and return the result. To avoid this extra programming effort, TeenyLIME's capability tuples can be used as described in Sect. 11.3.1. Here, a capability tuple is used as a placeholder to represent a node's ability to produce data of a given type, without keeping the actual data in the tuple space. When a query is remotely issued with a pattern matching a capability tuple, a dedicated event is signaled to the application. In response to this, the application takes a fresh reading and outputs the actual data to the tuple space. The sequence of operations is depicted in Fig. 11.4. Note how, from the perspective of the data consumer, nothing changes. Instead, at the data producer, capability tuples enable considerable energy savings as the readings are taken only on-demand.

The simple yet expressive programming primitives provided by TeenyLIME allow programmers to express complex interactions in a few lines of code. [191] compares TeenyLIME-based implementations of various applications and system-level mechanisms, e.g., routing protocols, against functionally equivalent nesC implementations. Results indicate that TeenyLIME yields cleaner and simpler implementations. For instance, the number of lines of code written by programmers using TeenyLIME is usually half of the corresponding counterpart implemented in nesC.

11.4 Summary and Outlook

Tuple spaces were originally invented as a computational and programming model for parallel computing. Later on, companies such as IBM proposed tuple spaces as a programming model for distributed computing. The next natural step is, therefore, for the programming models inspired by tuple spaces to follow the evolution of networking towards untethered, wireless communication. In this context, simplicity of the programming interface and decoupling in time and space are the most significant advantages. Nevertheless, the characteristics of wireless communication demand a re-thinking and extension of the base model made popular by Linda.

This chapter concisely presented the state-of-the-art concerning middleware platforms based on the tuple space abstraction and expressly designed for wireless scenarios. By analyzing and comparing representative systems it provided the reader with a wide perspective on the efforts in this specific field. Moreover, by exemplifying the use of tuple spaces in realistic application case studies, it conveyed concretely the power of this abstraction.

Although the presentation is structured along the two application scenarios defined by mobile computing and wireless sensor networks, this does not necessarily imply that these two realms must be treated separately. Indeed, in our own work we explored a two-tier approach where sensor data is available to mobile data sinks, from which it is available to other sinks in range. In this model and system, called TinyLIME [207], the transiently shared tuple space originally introduced by LIME is the unifying abstraction enabling seamless access to information spanning different kinds of networks. Indeed, bridging the physical world where users live and move with the virtual world enabled by sensing and actuating wireless devices is a new frontier of computing [408], and one where the simple and effective data sharing paradigm fostered by tuple spaces can give a fundamental contribution.

Chapter 12
Security Middleware for Mobile Applications

Bart De Win, *DistriNet, Katholieke Universiteit Leuven, Belgium*
Tom Goovaerts, *DistriNet, Katholieke Universiteit Leuven, Belgium*
Wouter Joosen, *DistriNet, Katholieke Universiteit Leuven, Belgium*
Pieter Philippaerts, *DistriNet, Katholieke Universiteit Leuven, Belgium*
Frank Piessens, *DistriNet, Katholieke Universiteit Leuven, Belgium*
Yves Younan, *DistriNet, Katholieke Universiteit Leuven, Belgium*

12.1 Introduction

Over the last decade the popularity of mobile devices has increased enormously. Initially, personal managers and mobile phones were designed as closed, dedicated devices. More and more, these devices have evolved into general purpose instruments that can be extended at user's will (a.o. via proper software development kits). This has lead to the current generation of smartphones and full-blown personal information management systems. At the same time, the information managed by the devices has evolved from limited and personal to general purpose and business-centric and, consequently, they constitute a core component of daily life.

These evolutions have had a significant impact on the security and privacy features of these devices. While rather simple, low-protected security models were provided initially, the current devices have evolved into natural extensions of personal computing platforms offering advanced, fine-grained data and software protection. Compared to personal desktops, however, the big challenge is due to the limited hardware protection models (such as data protection in memory) and dito computational resources that are available for the software security measures to build on. Consequently, the protection models of these devices have always been more restricted and targeted towards a specific setting. In this context, security middleware is to be interpreted as a broad category of security enhancements for applications (and their data) on mobile devices with limited capabilities.

In this chapter, the state-of-the-art in software protection for mobile devices is discussed. First, the security characteristics of mobile devices (as opposed to regular desktop systems) are elaborated upon by eliciting and illustrating specific types of threats. This should help in understanding and appreciating the particular difficulties of these platforms and allow one to draw connections between related problems on specific devices. Second, in the wide range of protective measures for mobile devices, we focus on two recent security enhancements that address some of these threats and, hence, improve the protection of applications on these devices: execution memory protection and security by contract. These techniques are comple-

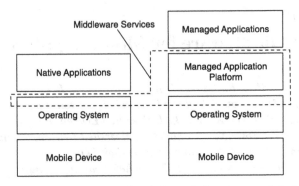

Fig. 12.1 Mobile application platforms: native platforms (left) and managed platforms (right)

mentary in the sense that the first technique focuses on applications running on native platforms, while the latter improves security properties for managed platforms. Besides motivating and discussing the general approach of these techniques, their usefulness for mobile devices in particular is highlighted as well.

The structure of the rest of this chapter is as follows. In Sect. 12.2, the security landscape for mobile devices is discussed with a strong focus on general security threats. Additionally, the characteristics of the security architectures of these platforms is briefly discussed, but an exhaustive comparison is out of the scope of this chapter. Sections 12.3 and 12.4 elaborate on two specific techniques, execution memory protection and security by contract respectively. Section 12.5 concludes and provides some ideas for future work.

12.2 The Security Landscape for Mobile Devices

12.2.1 Overview of Mobile Device Platforms

Since mobile devices are becoming small but powerful general-purpose computers, mobile platforms are getting more advanced and they are adopting many of the features found on typical desktop platforms. Mobile software platforms can be divided in two important classes: native platforms and managed platforms (see Fig. 12.1).

In native platforms, applications are compiled into machine code and are executed directly on the processor of the device. Native applications make efficient use of the limited resources that are available, and therefore this has been the traditional approach of developing mobile applications. The most important native platforms in the market today are Symbian (version 9), Windows Mobile (version 5 and 6), Garnet OS (previously known as Palm OS version 5), various flavors of linux and, recently, the version of Mac OS X found on Apple's iPhone.

In managed platforms, applications are compiled into bytecode and executed on a virtual machine instead of on the real device. As a consequence, applications are portable to any device that has a native implementation of the virtual machine. Because of the variety of mobile devices, managed platforms have become very important in the mobile development space. The two best known managed platforms for mobile devices are the .NET Compact Framework [866] and the Mobile Information Device Profile (MIDP) [785] of Java Micro Edition (Java ME). Java ME consists of the Connected Device Configuration (CDC) [784] for more powerful devices and the Connected Limited Device Configuration (CLDC) [783] for limited devices. MIDP is the profile that specifies Java ME runtime within CLDC for the mobile devices that fall within the scope of this chapter.

As indicated in Fig. 12.1, both native and managed platforms include middleware services for developing mobile applications. The virtual machine of a managed platform is just one example of such a middleware service. Others are offered to developers either explicitly through APIs or transparently in the form of compiler extensions.

A managed platform always needs to run on some native underlying platform. This means that it is possible that native applications and managed applications coexist on the same device. Usually, managed platforms solely define the middleware layer between the applications and the native platform. Java ME and the .NET Compact Framework are examples of this approach. Other platforms define both the middleware layer and the native platform that is used. In these platforms, it is usually the case that only managed applications are supported. An example of such a combined approach is Google's Android Platform, which defines both a Java-based middleware layer and an underlying linux-based operating system.

12.2.2 Protection of Mobile Platforms

Just as any other platform that executes general-purpose applications, mobile platforms need security mechanisms to protect the resources on the devices. Two important techniques can be distinguished: memory management and software protection.

Native platforms need to ensure that applications cannot access memory that belongs to other applications or to the platform itself. This separation is guaranteed by *memory management*. Some of the older mobile device platforms such as Palm OS have no form of memory protection at all, which makes these platforms inherently insecure. However, most modern native platforms (Windows Mobile, Symbian) do provide basic memory protection. One of the important strengths of managed platforms (.NET Compact Framework, Java ME) is that they have a strong memory protection model because applications have no direct access to the memory and because these platforms guarantee type safety.

Other resources on the device such as personal data, wireless networking, or SMS messages are accessed via APIs and are protected by the *security architecture* of the platform. Over the years, security concerns on mobile devices have gotten more attention and the security architectures in the latest versions of current mainstream platforms such as Symbian, Java ME and the .NET Compact Framework are be-

Table 12.1 A summary of the security issues

Issue	Description
I1	Important security mechanisms found on full-scaled platforms are omitted
I2	The developer has very little options in customizing security mechanisms
I3	APIs can be protected but often in an all-or-nothing way
I4	It is hard to securely store sensitive data
I5	A mobile device is subject to different kinds of threats

coming more powerful. These security architectures mainly focus on reducing the privileges under which applications execute. Although some security architectures are more advanced than others, the underlying security model they implement is similar. The platform APIs are divided in two or more trust levels and each application is assigned a certain level of trust based on credentials that are shipped with the application (typically a signature from the developer). The security architecture enforces that applications cannot invoke APIs of a higher trust level than their own. Untrusted applications are prohibited or are executed in a sandbox to ensure that they can only use a minimal and tightly-controlled set of resources.

Sometimes, for instance in the case of MIDP and Symbian, the platform realizes the aforementioned model by an underlying permission-based access control mechanism. Instead of grouping API operations in trust domains, operations have one or more required permissions. Domains are then defined indirectly in terms of permissions and applications that belong to a certain domain get the permissions of the domain. This approach supports more fine-grained specification of granted behavior (by giving an application an additional permission). Symbian and MIDP also support the dynamic assignment of permissions based on a prompt with the user. These dynamic permissions are only valid for a single execution of an operation or until the application quits, based on a policy that is included with the platform.

12.2.3 Security Issues

It is understandable that mobile platforms have simpler security models due to the limited resources that are available to them. However, users and developers must realize that on these platforms there are a number of important security issues. In this section, a number of general security issues are discussed that are found in today's mainstream mobile device platforms. Table 12.1 gives an overview of these issues. In the rest of this section, each of these issues is discussed in more detail.

I1: Important Security Mechanisms are Omitted. Many mobile platforms are effectively stripped-down versions of full-scale platforms that run on normal desktop computers, for instance: mobile linux-based devices, the mobile vs. the regular version of Mac OS X, Java ME vs. Java SE and the .NET Compact Framework vs. the full .NET Framework. The full versions of these platforms have extensive security architectures and offer many security mechanisms and countermeasures.

When comparing the scaled-down versions of these platforms with their full-sized counterparts, it is notable that many of these security mechanisms are omitted or replaced by much less powerful variants because of resource constraints. However, since mobile devices have become general-purpose computing devices with a high degree of connectivity, they are susceptible to many of the same threats as found on full-scaled platforms. Two examples are given below.

Many security problems on full native platforms are caused by poorly programmed software that leads to buffer overflows. By feeding a vulnerable application a carefully crafted input, an attacker can overwrite certain memory locations in the system. This enables the attacker to execute arbitrary code, potentially resulting in system compromise. Since these kinds of attacks are among the oldest and best known, most modern operating systems such as Windows Vista, Mac OS X and linux contain at least some countermeasures to prevent buffer overflows from occurring, or making them harder to exploit. Mobile device platforms on the other hand, have no support for any of these countermeasures. The recent hacking of Apple's iPhone for unlocking purposes illustrates that buffer overflow-based attacks are not unrealistic on mobile devices.

One of the important security mechanisms that are found on full-scaled versions of managed platforms is access control based on stack inspection [288], in which the effective permissions of a subject that invokes an operation is determined by inspecting the execution context. By doing so, an untrusted application cannot gain more privileges by accessing a resource indirectly via a trusted application or module. Although mobile devices run untrusted applications and dynamically downloaded code, the MIDP and the .NET Compact Framework do no support stack inspection.

I2: Flexibility and Extensibility are Limited. The security architectures on mobile platforms are deliberately kept simple. One of the simplifications with respect to full-scaled platforms is that the flexibility and extensibility of the security architectures are limited. On full-scale platforms, the user or the administrator has the flexibility of changing the security policies that are enforced to reflect evolving or unanticipated security requirements. Mobile platforms typically use hard-coded policies or policies that are specified once by the manufacturer of the device. Moreover, full-scale platforms allow the extension of the security architecture to support the enforcement of different or more advanced security policies, whereas the security architectures on mobile platforms are not extensible at all. If developers are faced with the limits of an inflexible security architecture, they need to fall back to implementing the necessary security mechanisms in their applications, which is clearly not desirable.

In Symbian, for instance, applications used to be able to write and read all data on the device, including data from other applications or from the operating system. Since version 9, Symbian includes access control for data (this is called *data caging*) that limits the directories in which an application can read and write. The policy that states which directories are accessible is hardcoded and the only way that applications can be given access to more data is by giving them access to all data on the system.

A good example of limited extensibility can be found in Java ME's MIDP profile. The full-scaled version of Java can be extended with customized SecurityManagers or LoginManagers in order to change the way in which security decisions are made or in which users are authenticated. None of these features can be found in the security architecture of MIDP. Moreover, the store of trusted certificates on which the access control model is based, cannot be changed. Finally, it is impossible for developers to extend permissions in MIDP.

I3: Access Control Towards APIs is Coarse-Grained. Apart from some exceptions (Palm OS), most mobile platforms are able to limit access to individual API operations (see Sect. 12.2.2). Applications are placed in a trust domain based on a digital signature. Often, users can decide to trust an unsigned application anyway. Applications in a low trust domain are executed in a sandbox that restricts access to sensitive API operations (e.g., only applications in the mostly trusted domain can send SMS messages). An example of this approach is Windows Mobile, which defines two security domains: *trusted* applications have full system privileges and *normal* applications have no access to sensitive operations on the API. The operating system maintains a privileged and an unprivileged keystore. A security policy configures the mapping of privileged, unprivileged and unsigned applications to the trusted or normal domains. This mapping can be influenced by a user prompt.

For the user that wants to run a downloaded application, this model is too coarse-grained. Either the application is trusted and can do anything (which requires blind trust in the developer of the application), or the application is not trusted and it cannot invoke any sensitive operations. Moreover, the decision on whether to trust an application or not solely depends on the signature of the application.

Some platforms (MIDP and Symbian) have a more fine-grained security model. The assignment of applications to security domains is still based on code signing, but these platforms are able to identify, assign and revoke individual permissions. Because of this reason, these platforms can offer support for dynamically raising the permissions of an application at runtime based on prompting the user. For instance, in MIDP, the user can be prompted when an untrusted applications wants to make a network connection. The user can then grant the application this permission temporarily (once or until the application quits).

I4: Privacy Is Poorly Protected. By now, many mobile platforms have included support for access control to APIs. However, the protection of sensitive data on the device is often very poor.

Palm OS and Windows Mobile have no way of letting a user limit application access to his/her data. MIDP is unable to protect data on the device except for data that is accessed via the PIM API: it is possible to deny read and write access to contacts, meetings and todo's. Symbian supports a simple form of access control towards data (this is called data caging). More specifically, applications have their own /private directory that is readable and writable, system files /sys cannot be read or written and there is a publicly readable directory /resources that can only be written to by the system. However, all other directories on the file system are fully accessible by all applications.

I5: Mobile Devices Have Different Threats. In many respects, mobile devices are playing catch-up with desktop computers: they have gotten powerful processors, more memory and they can be always-online. This evolution has undoubtedly made mobile devices vulnerable to the same kinds of threats than desktop systems. However, it is important to realize that due to their specific nature, there are important differences in the kinds of threats to these devices.

First of all, many known threats to normal computers (such as self-propagating worms) target server processes that offer a certain network service. On mobile devices, it is much less likely that there are any servers that can be targeted. The only kind of server processes that are commonly found in this setting are the short-distance wireless communication (Bluetooth and Infrared) processes of the operating system. Therefore, the focus of the threats on mobile devices shifts completely from servers to applications.

Secondly, the high degree of connectivity (WiFi, wireless broadband, cellular networks, Bluetooth, Infrared) in combination with the mobility of the device increase the attack surface in comparison with desktop computers and opens a whole new family of threats. For instance, there have been exploits for Symbian that use SMS or MMS messages as a propagation vector [269, 270], or that allowed an attacker to take over complete control of the device via Bluetooth [268], allowing him to initiate calls and send SMS messages.

A third category of threats on mobile devices are denial-of-service threats. Because of resource limitations, mobile device platforms often very poor process management models. Therefore, simply feeding a badly written application with corrupt data that causes it to crash or hang, can often bring the whole system to a halt, forcing the user to reset its device. The most extreme example is Palm OS, which has no support for multitasking and cannot stop unresponsive applications at all.

12.3 Protection for Native Platforms: Memory Protection

As discussed in Sect. 12.2.3, issue I1, mobile devices will often run stripped down versions of desktop operating systems, making them vulnerable to the same type of threats. Applications that run natively on a mobile device, are often written in C or C++ and as a result are vulnerable to buffer overflows and similar vulnerabilities which could allow an attacker to execute arbitrary code on the device.

A buffer overflow occurs when a program writes past the bounds of an object in memory and starts to overwrite adjacent objects. By overwriting memory addresses (e.g., stored code pointers), an attacker may be able to control the execution flow of a program. This could allow the attacker to execute arbitrary code with the privileges of the application that is being executed. A typical type of buffer overflow is the stack-based buffer overflow, where the program will overflow past the bounds of a stack-allocated array of characters. The stack is used to facilitate the execution of functions and recursion: it contains the local variables of each function, together with the return address (the address of the instruction at which execution must resume once the function has terminated) and the value of several registers that must preserve their values after the function has finished executing. If a buffer overflow occurs in a local variable, the attacker can overwrite the return address. When the function terminates, it will transfer control to the instruction at the return address. If attackers make the return address point to their injected code (provided as data input to the program), they can force the program to execute arbitrary code.

In August 2006, a number of vulnerabilities [646] were discovered in LibTIFF.[1] LibTIFF is used in a number of desktop operating systems, like Linux and Mac OS X. However, it is also used on the Apple iPhone, where this vulnerability was widely exploited by users of the iPhone [588] to perform a "Jailbreak".[2] This vulnerability can be triggered in both MobileMail (the iPhone mail client) and MobileSafari (the iPhone web browser) and as a result is remotely exploitable by letting the user browse to a site containing a specific TIFF file or by emailing a TIFF file to the user.

This vulnerability was also present on another mobile device: the Sony PlayStation Portable, where it was also exploited to allow behavior not condoned by the manufacturer. In this case, the vulnerability was used to gain more permissions to allow users to run homebrew[3] games.

These two examples show that software originally designed for desktop environments is being ported widely to these new devices, resulting in the same types of vulnerabilities being present in these devices. As more and more of these devices enter into the market, more of these types of vulnerabilities will be discovered and exploited.

12.3.1 Existing Countermeasures

Many countermeasures exist on desktop operating systems that can prevent these kind of attacks. As such, they can help in providing protection against issues I1 and I5 as discussed in Sect. 12.2.3. An extensive survey can be found in [891]. Very few countermeasures have been ported to mobile devices however. In this section the few countermeasures that do exist are described and their shortcomings are discussed. In the next section, other existing countermeasures and issues that may exist when porting them to mobile devices are discussed.

Non-executable Memory. These countermeasures will make data memory non-executable. Most operating systems divide process memory into at least a code (also called the text) and data segment and will mark the code segment as read-only, preventing a program from modifying code that has been loaded from disk into this

---

[1] LibTIFF is a library for reading and writing TIFF files, a popular image format.

[2] By default, it is not possible for users to install additional native applications on the iPhone. The term Jailbreaking refers to the escaping of these limitations, allowing users to gain full control of the device.

[3] Homebrew games are games which are typically produced by consumers and are generally not authorized (or digitally signed) by the manufacturer of the product, resulting in the need to circumvent security restrictions on the device before they can be played.

segment (unless the program explicitly requests write permissions for the memory region). As such attackers have to inject their code into the data segment of the application. As most applications do not require executable data segments as all their code will be in the code segment, some countermeasures mark this memory as non-executable, which will make it harder for an attacker to inject code into a running application. A major disadvantage of this approach is that an attacker could use a code injection attack to execute existing code. One type of such an attack is called a return-into-libc attack [869]. Instead of injecting code on the stack and then pointing the return address to this code, the desired parameters are placed on the stack and the return address is pointed to existing code (a simple example is to call the libc wrapper for the *system()* system call and to pass it an executable that will execute the attacker's code as an argument). This is also the attack used in the original iPhone exploit: the stack is marked as non-executable so a return-into-libc attack was performed which would execute a number of library functions to gain the desired results. A later exploit uses the stack-based buffer overflow to perform a return-into-libc attack, which copies the injected code onto the heap and then transfers control flow there.

Stack Cookies. The observation that attackers usually tried to overwrite the return address when exploiting a buffer overflow led to a string of countermeasures that were designed to protect the return address. One of the earliest examples of this type of protection is the canary-based countermeasure [195]. These countermeasures protect the return address by placing a value before it on the stack that must remain unchanged during program execution. Upon entering a function the canary is placed on the stack below the return address. When the function is done with executing, the canary stored on the stack will be compared to the original canary. If the stack-stored canary has changed an overflow has occurred and the program can be terminated. A canary can be a random number, or a string that is hard to replicate when exploiting a buffer overflow (e.g., a NULL byte). StackGuard [196, 195] was the first countermeasure to use canaries to offer protection against stack-based buffer overflows, however attackers soon discovered a way of bypassing it using indirect pointer overwriting. Attackers would overwrite a local pointer in a function and make it point to a target location, when the local pointer is dereferenced for writing, the target location is overwritten without modifying the canary. Propolice [259] is an extension of StackGuard, it fixes these type of attacks by reordering the stack frame so that buffers can no longer overwrite pointers in a function. These two countermeasures have been extremely popular: Propolice has been integrated into the GNU C Compiler and a similar countermeasure has made it's way into Visual Studio's compiler [115, 332]. The countermeasure which was integrated into Visual Studio is also used on mobile devices running Windows CE 6.

12.3.2 Applicability of Desktop Countermeasures

Many of the countermeasures that are currently in use on desktop systems can be ported to embedded devices. However, due to limited memory and processing power, efficiency becomes a more important concern on these devices. Many countermeasures are also designed with a specific architecture in mind, making it hard to apply them in a mobile setting. Several other limitations in the architectures on mobile devices may also be important issues when trying to port countermeasures to these architectures. For example, many architectures on mobile devices have no support for paging, making it hard to implement a countermeasure like address space layout randomization (ASLR)on these devices. In this section different countermeasures that exist for desktop operating systems are discussed and the advantages, disadvantages and possible problems if they were to be implemented on mobile devices are examined. While the general techniques of these countermeasures are discussed, porting a specific implementation of a technique discussed here may require significant additional effort however.

Safe Languages. Safe languages exist in all types and forms on both desktop operating systems, many of these languages will prevent memory safety problems by removing control from the programmer (e.g., Java removes pointers from the programmer's control), by inserting extra checks or a combination of both. Many such languages exist, however since this section focusses on memory errors, only languages that stay as close as possible to C and C++ are examined. These safe languages are referred to as safe dialects of C. Some dialects [615] will only need minimal programmer intervention to compile programs, while others [424, 572] require substantial modification. Others [474] severely restrict the C language to a subset to make it safer or will prevent behavior that the C standard marks as undefined [637].

Some of these languages have a relatively high overhead, however most have acceptable overhead. The main disadvantage of using these languages on mobile devices is the same as for desktop devices: programmers must learn and port their existing code to this new language.

Boundschecking. Bounds checkers provide extensive protection against exploitation of buffer overflows: they check array indexation and pointer arithmetic to ensure that they do not attempt to write to or read from a location outside of the space allocated for them. Two important techniques are used to perform traditional bounds checking: adding bounds information to all pointers and adding bounds information for objects. In the first technique, pointers contain extra information about the object they are referring to. In the second technique the objects themselves will contain the extra information. Boundscheckers that use either technique will generally suffer from a high computational and memory overhead, making them less suited for mobile devices.

Obfuscation of Memory Addresses. Memory-obfuscation countermeasures use an approach that is closely related to stack cookies: their approach is also based on random numbers. These random numbers are used to 'encrypt' specific data in memory and to decrypt it before using it in an execution. These approaches are currently used for obfuscating pointers (XOR with a secret random value) while in memory [198]. When the pointer is later used in an instruction it is first decrypted in a register (the decrypted value is never stored in memory). If an attacker attempts to overwrite the pointer with a new value, it will have the wrong value when decrypted. This will most likely cause the program to crash. A problem with this approach is that XOR encryption is bytewise encryption. If an attacker only needs to overwrite 1 or 2 bytes instead of the entire pointer, then the chances of guessing the pointer correctly vastly improve (from 1 in 4 billion to 1 in 65000) [15]. If the attacker is able to control a relatively large amount of memory (e.g., with a buffer overflow), then the chances of a successful attack increase even more. While it is possible to use better encryption, it would likely be prohibitively expensive since every pointer needs to be encrypted and decrypted this way. While the prototype implementation in [198] is fairly efficient because of the increased use of registers by the modified gcc compiler, this type of protection could turn out to be expensive on other architectures.

Address Space Layout Randomization. ASLR is another approach that makes executing injected code harder. Most exploits expect the memory segments to always start at a specific known address. They will attempt to overwrite the return address of a function, or some other interesting address with an address that points into their own code. However for attackers to be able to point to their own code, they must know where in memory their code resides. If the base address is generated randomly when the program is executed, it is harder for the exploit to direct the execution-flow to its injected code because it does not know the address at which the injected code is loaded. Shacham et al. [745] examine limitations to the amount of randomness that such an approach can use.[4] Their paper also describes a guessing attack that can be used against programs that use forking as the forked applications are usually not rerandomized, which could allow an attacker to keep guessing by causing forks and then trying until the address is found. It may not be possible to implement this type of countermeasure on architectures which do not have support for paging. Architectures that have limited address space (e.g. 16-bit architectures), may also not be able to benefit from this approach.

Instruction Set Randomization. ISR [63, 446] is another technique that can be used to prevent the injection of attacker-specified code. Instruction set randomization prevents an attacker from injecting any foreign code into the application by encrypting instructions on a per process basis while they are in memory and decrypting them when they are needed for execution. Attackers are unable to guess the decryption key of the current process, so their instructions, after they've been decrypted, cause the wrong instructions to be executed. This prevents attackers from having the process execute their payload and has a large chance crashing the process due to an invalid instruction being executed. However if attackers are able to print out

---

[4] This limitation is due to address space limitations in 32-bit architectures: often countermeasure will limit randomness to a maximum amount of bits, which will be less than 32 bit, making guessing attacks a possibility.

specific locations in memory, they can bypass the countermeasure since the encryption key can often be derived from encrypted data (since most countermeasures will use XOR). Other attacks are described in [859]. The current implementations are proof of concept implementations and suffer from high overheads. However, a CPU could be designed which supports this kind of countermeasure. Given the security problems with this approach, a CPU supporting a stronger encryption than XOR, as described in [382], may be more desirable, which is clearly difficult to accomplish on mobile devices.

Separation and Replication of Information. Countermeasures that rely on separation or replication of information will try to replicate valuable control-flow information or will separate this information from regular data [893, 894, 62]. This makes it harder for an attacker to overwrite this information using an overflow. Some countermeasures will simply copy the return address from the stack to a separate stack and will compare it to or replace the return addresses on the regular stack before returning from a function. These countermeasures are easily bypassed using indirect pointer overwriting where an attacker overwrites a different memory location instead of the return address by using a pointer on the stack. More advanced techniques try to separate all control-flow data (like return addresses and pointers) from regular data, making it harder for an attacker to use an overflow to overwrite this type of data. These countermeasures could easily be applied to mobile architectures given their low performance overhead. Care must be taken however in environments where memory is scarce as these countermeasures will tend to have some memory overhead when replicating or separating the information.

Paging-Based Countermeasures. Paging-based countermeasures make use of the Virtual Memory Manager, which is present in most modern architectures. Memory is grouped in contiguous regions of fixed sizes (e.g., 4 Kb on Intel IA32) called pages. Virtual memory is an abstraction above the physical memory pages that are present in a computer system. It allows a system to address memory pages as if they are contiguous, even if they are stored on physical memory pages that are not. An example of this is the fact that every process in Linux starts at the same address in the virtual address space, even though physically this is not the case. Pages in most architectures can have specific permissions assigned to them: Read, Write and Execute. Many of the countermeasures in this section will make use of paging permissions or the fact that multiple virtual pages can be mapped onto the same physical page on Intel IA32. These countermeasures could be applied to mobile devices that have support for paging. However, they will require architectures with support for paging or for applying permissions to those pages. These may not always be available in an architecture running on a mobile device.

Execution Monitors. Execution monitors monitor specific security relevant events (like system calls) and perform specific actions based on what is monitored. Some monitors try to limit the damage a successful attack on a vulnerability could do to the underlying system by limiting the actions a program can perform. These are called policy enforcers. In general, enforcement is done through a reference monitor where an application's access to a specific resource[5] is regulated. They can be very coarse or very granular. An example of a coarse grained policy enforcer is one that ensures that a program executes it's system calls in a given order (these can be learned from previous runs of the program). If the program tries to execute a different system call, it will be denied. Attackers can however perform a mimicry attack, where they mimic the behavior of the original program, while only providing different arguments for the system call. The more granular a policy enforcer is, the harder a mimicry attack becomes: a very fine grained enforcer could force the attacker to only execute the existing code [1]. The more fine grained policy enforcers could suffer from higher overhead, which may be a problem in a mobile setting. While the more coarse grained enforcers will usually not suffer from as high overheads, they can easily be bypassed by a determined attacker.

Fault isolators are a second type of execution monitor: they ensure that certain parts of software do not cause a complete system (a program, a collection of programs, the operating system, ...) to fail. The most common way of providing fault isolation is by using address space separation. However, this causes expensive context switches to occur that incur a significant overhead during execution. Because the modules are in different address spaces, communication between the two modules also incurs a higher overhead. For that reason, software fault isolation [846] exists, which allows for better performance. Software fault isolation can be an efficient way of preventing errors from causing an entire system to fail, however it may be incomplete when faced with a determined attacker. Ensuring proper isolation even from an attacker may add additional overhead.

Hardened Libraries. Hardened libraries replace library functions with versions that contain extra checks. An example of these are libraries that offer safer string operations: more checks will be performed to ensure that the copy is within bounds, that the destination string is properly NULL terminated (something *strncpy* does not do if the string is too large). Other libraries check whether format strings contain '%n' in writable memory [711] (and will fail if they do) or will check to ensure that the amount of format specifiers are the same as the amount of arguments passed to the function [197]. Since these libraries operate at a higher level than most other countermeasures described in this section, applying them when programming applications for a mobile device should not pose any significant problems. As with their desktop variants, the programmer has to use these libraries correctly and consistently, possibly modifying existing code to make use of these libraries.

Runtime Taint Trackers. Taint tracking is an important type of countermeasure for web-based vulnerabilities. It is ideally suited for detecting cross-site scripting, SQL injection, command injection and other similar vulnerabilities in web applications [672, 875]. These taint trackers instrument the program to mark input as tainted.[6] If such input is used in a place where untainted data is expected (like an SQL query), an error is reported. Taint tracking can also be used to detect memory errors. In this case, the taint tracker will generate an error when an trusted memory

---

[5] The term resource is used in the broadest sense: a system call, a file, a hardware device, ...

[6] Tainted data is data which is untrusted, usually derived from input.

location (like a return address) has been modified by tainted data. One important limitation with these taint trackers is that they suffer from false positives. Such a false positive can occur when a tainted data is used in a place where untainted data is expected but is not actually vulnerable to attack. For example, if a format string is derived from tainted data and used as format specifier to *printf*, however a check has occurred to ensure that this tainted data is in fact benign. A taint tracker may report this as a vulnerabilty, while it is in fact safe code. These countermeasures will generally also suffer from a significant overhead, both in terms of performance and memory usage, which may not be acceptable in a mobile device.

12.3.3 Model-Based Countermeasure Design

An important issue with porting these desktop applications to embedded devices is the fact that many desktop countermeasures were designed with a specific architecture or operating system in mind. This can make porting more difficult and prone to being bypassed. When a countermeasure is ported, the countermeasure developer must ensure that the countermeasure can not easily be bypassed on the new platform. An example of such porting going wrong occurred when Microsoft ported the StackGuard countermeasure [195] from GCC to Visual Studio [116]. The specifics of the Windows operating system were not taken into account, which resulted in attackers being able to bypass the countermeasure by ignoring the return address and continuing to write on the stack until they overwrote the function pointers used for exception handling. Subsequently, an exception would be generated by the attackers and their injected code would be executed [522]. A possible way to prevent these types of problems when porting countermeasures to a new platform is to use machine model aided countermeasures [892].

Model-based countermeasure design builds a model of the execution environment of the program based on the memory locations and abstractions that influence the execution flow. This abstract high-level model contains memory locations and abstractions that can be used by an attacker to directly or indirectly influence the control flow of a particular application, supplemented with the locations that could lead to indirect overwriting of these memory locations. Finally, these are supplemented with contextual information: what these specific memory locations are used for at different places of the execution flow and what operations are performed on them. This machine model[7] allows a designer of countermeasures to view a platform in a more abstract way and as a result more effort can go into designing countermeasures rather than understanding obscure, possibly insignificant, platform details. It also allows a designer to take into account what the effects of a particular countermeasure are on a platform before having to implement it.

---

[7] This is model is a high level representation of a platform and should not be confused with an executable model of a specific program (e.g. a bytecode representation of a Java program).

12.4 Protection for Managed Platforms: Security by Contract

As explained in Sect. 12.2, the typical way security is enforced on a modern mobile device is by checking that an application is certified by some trusted third party. If the application is signed by a trusted third party, it is allowed to run. If the signature cannot be verified, the user is asked whether the application should be run or not. This signature-based system doesn't work well in the case of mobile code. The first problem is that the decision of allowing an application to run or not, is too difficult for a user to make. He would like to run an application as long as the application doesn't misbehave or doesn't violate some kind of policy. But he is in no position to know what the downloaded application exactly does, so he cannot make an educated decision to allow or disallow the execution of a program. A second problem is that certifying an application by a trusted third party is rather expensive. Many mobile application developers are small companies that do not have the resources to certify their applications. A third, and perhaps most damning, problem is that these digital signatures do not have a precise meaning in the context of security. They confer some degree of trust about the origin of the software, but they say nothing about how trustworthy the application is. Cases are already known where malware was signed by a commercial trusted third party [696]. This malware would have no problems passing through the mobile security architecture, without a user noticing anything.

Recent research [739, 224] has addressed these issues by working out a security-by-contract (SxC) paradigm for the development, deployment and execution of mobile applications. The key idea behind SxC is to have a system that can automatically prove that a mobile application will not try to access resources to which it *doesn't* have access, or that it doesn't try to abuse resources to which it *does* have (limited) access. SxC supports a number of different mechanisms to prove this claim. The application is allowed to run if one of these mechanisms can certify that the application will not violate the policy. As such, the SxC paradigm is particularly suited to address issues I2 and I3 as discussed in Sect. 12.2.3.

One of the characteristics of a managed platform is that it offers an environment where so-called 'safe languages' can be used to guarantee type soundness and memory safety. These two features make it much easier to prove certain security-related properties. Because proving properties about unknown code is of central importance in this new SxC paradigm, it works best in managed environments.

12.4.1 Policies and Contracts

Loosely speaking, a system policy is a set of rules to which an application must comply. These rules usually limit the access of an application to a specific part of the system API. For instance, there could be a set of rules to prohibit applications from accessing the network or to limit the access to the file system. These accesses are the *security-related events*. One could think of a policy as an upper-bound description of what applications are allowed to do.

Contracts are very similar, but instead of defining the upper-bound of what an application *can* do, it describes the upper-bound of what the application *will* do. It's the 'worst case' scenario of security-related behavior of an application. The contract is typically designed by the application developer and is shipped together with the application as metadata.

An example of a policy could be *"An application can use the GPS device, cannot place a phone call, and can send a maximum of 1000 bytes over the network"*. If an application arrives that comes with a contract that says *"This application will use the GPS device, and will send at most 200 bytes over the network"*, then the application should be allowed to run because it complies with the policy.

Policies are in essence the system view of the device's security requirements. This is why they are often called *system policies*. Likewise, contracts are the application view of the security requirements. They are also called *application contracts*.

12.4.2 Policy Enforcement

The main focus of an SxC system is to make sure that a mobile application will never violate the system policy. A number of different *enforcement technologies* can be used to ensure this, each with different benefits and drawbacks. Some of the most common techniques are discussed here.

Matching The basic idea behind policy and contract matching, is to automatically check whether an application contract is subset of the system policy. If the policy is the upper-bound of what the application *can* do, and the contract is the upper-bound of what the application *will* do, then having an application contract that is a subset of the policy means that the application will never do anything that violates the policy.

Different matching algorithms exist, ranging from simple algorithms like identical matching or hash-based matching, to more advanced implementations like simulation matching or language inclusion matching. Identical matchers compare the application contract and system policy on a byte-per-byte basis. If they are exactly the same, the contract and the policy are said to match; otherwise, the algorithm cannot guarantee that they match. Hash-based matching is a bit smarter than identical matching, but not by much. A hash-based matcher will break the contract and policy in smaller pieces, and will check that the different pieces of the contract are all present in the policy. This is normally done by generating a hash of every piece, and comparing the hashes of the contract with the hashes from the policy. A disadvantage of these two types of algorithms is that they only work for the most trivial cases of policy-contract matching. For instance, they cannot successfully match the simple policy *"An application cannot send more than 1000 bytes"* with a contract *"This application will not send more than 200 bytes"*. Fortunately, the more advanced forms of matching, like simulation matching or language inclusion, are able to successfully handle these cases.

Static analysis and proof-carrying code   A second option is to automatically inspect the binary code of the application and to construct a proof that explains *why* the application will comply with the contract. This process is called 'static analysis'. However, static analysis is still a developing discipline and current implementations often need some kind of input from the application developer (e.g., code annotations). Furthermore, it is also a complex and slow process. Statically analyzing a non-trivial application requires a lot of computational power, which is not practical on a mobile device.

In proof-carrying code [614], the result of a static analysis (performed by an expert on a powerful computer) is stored and distributed along with the application. When the application arrives on the mobile device with this extra metadata, the problem of constructing a proof is reduced to verifying a proof. Since proof verification can be done more efficiently than proof generation, it is realistic to implement a proof-checker on a mobile device.

Inlining   Another enforcement technology is policy inlining [255]. During the inlining process, the system goes through the application code and looks for calls to security-related events (SRE). When such a call is found, the system inserts calls to a monitoring component before and after the SRE. This monitoring component is a programmatic representation of the policy. It keeps track of the policy state and intervenes when an application is about to break the policy. As such, the application is made to comply with a contract that is equivalent to the system policy.

The biggest advantage of this technique is that it can be used on applications that are deployed without a contract or any other additional metadata. It can be used as a fall-back mechanism for when the other approaches fail and it can also ensure backwards compatibility. A disadvantage is that the application is modified during the inlining process, which might lead to subtle bugs. Also, the monitoring of the SREs comes with a performance hit. The performance hit is strongly dependent on the complexity of the policy being enforced. However, the decrease in performance is often relatively small.

Digital signatures   Finally, it should be mentioned that the classical mobile security architecture, based on digital signatures, can also be modeled within an SxC system. A trusted third party could be used to certify that an application complies to a given policy. This trusted party would have a different public/private key pair for every different policy it certifies. An application developer would send his application to this trusted party for compliance certification with one of the trusted party's policies. When the application is found to comply with the policy, it gets signed with the key corresponding to that policy. The signature can then be stored in the application metadata.

Notice the subtle difference between the meaning of the digital signature in the SxC system and in the classical mobile security architecture. Both systems make use of the exact same mechanism, but on the SxC system, the signature tells more about the application than simply its origin. It certifies that the application will not violate a specific policy. It certifies that this application will not harm the

Fig. 12.2 The application development life cycle

mobile device, whereas a signature in the classical security architecture gives no precisely specified guarantees.

The upside of using this enforcement technology is that it is relatively simple to implement and use. However, third party certification can be costly, and requires trust in the certifying party.

Working prototypes of the SxC architecture are available for both the .NET Compact Framework and the Java Micro Edition [739, 224]. The .NET implementation implements the SxC system by modifying the Windows Mobile application loader. When an application is started, it is sent to a custom application loader instead of the regular loader. This custom loader tries to enforce the policy by using one of the above enforcement techniques. The Java implementation works in a very similar fashion, with only some low-level technical differences.

12.4.3 The SxC Process

To take full advantage of this new paradigm, applications have to be developed with SxC in mind. This means that some changes occur in the typical *Develop-Deploy-Run* application life cycle. Figure 12.2 shows an updated version of the application development life cycle.

The first step to develop an SxC compliant application, is to create a contract to which the application will adhere. Remember that the contract represents the security-related behavior of an application and specifies the upper-bound of calls made to SREs. Designing a contract requires intimate knowledge of the inner workings of the application, so it's typically done by a (lead-)developer or technical analyst. Once the initial version of the contract has been specified, the application development can begin. During the development, the contract can be revised and changed when needed.

After the application development, the contract must somehow be linked to the application code in a tamper-proof way. One straightforward method to do this, is by having a trusted third party inspect the application source code and the contract. If they can guarantee that the application will not violate the contract, they sign a combined hash of the application and the contract. Another way to link the contract and the code, is by generating a formal, verifiable proof that the application complies with the contract, and adding it to the application metadata container. When this step is completed, the application is ready to be deployed. The application is distributed together with its contract and optionally other metadata such as a digital signature from a third party or a proof.

When the program is deployed on a mobile device, the SxC framework checks whether the application contract is compatible with the device policy. All supported policy enforcement technologies are tried in order, until one of the techniques can certify that the contract indeed matches with the policy. If none of the formal enforcement technologies can certify that the application will comply with the policy, the application can be inlined.

When an application exits the SxC-cycle, the user can be assured that either some formal or trust-based proof is available that guarantees the application's compliance with the system policy, or that the application has been rewritten to ensure that it will never violate the policy.

Having an SxC platform on a mobile device alleviates most of the problems discussed in Sect. 12.2.3 for managed applications. The standard security mechanisms that are missing on mobile devices are replaced with an extensible new mechanism, offering fine-grained control over which API functions are allowed to be called. Experiments have shown that this new security infrastructure performs well enough to be more than workable on today's devices.

12.5 Summary and Outlook

Over the last and coming decades, the use of mobile devices and their supporting platforms have evolved considerably and, hence, the security middleware thereon has to adapt accordingly. This chapter has discussed the state-of-the-art in security middleware for mobile devices. In a first part of the chapter, the basic protection mechanisms present on these devices have been overviewed and their specific character has been discussed by highlighting a number of current limitations and problems. A second part has elaborated on two protection techniques that can resolve a number of these problems. Security by contract improves device and data protection by empowering end-users to strictly control the behavior of mobile applications either via a number of enforcement techniques, including static analysis or via runtime enforcement. Memory protection, on the other hand, guarantees that the run-time memory of an executing application cannot be accessed, or tampered with, by malign applications or users.

The challenges that lie ahead of us are manifold. Firstly, given the fact that these devices constitute a core part of our daily lives, and since they impact both business and personal life, data protection on these devices (both via API's as well as via direct access) is becoming a critical success factor. At the same time, the tighter coupling of mobile devices with other personal and company systems will have to be realized, both from a functionality perspective as well as from a security perspective. An example of the latter is the more straightforward deployment of company-wide security policies. Secondly, the ongoing proliferation of platforms has lead developers towards web applications as a new model of applications for mobile devices. From a security perspective, this means that new efforts will have to be spent in securing the web browsers on these devices, which again are typically more limited that full-blown variants. Also, the grafting of general purpose security mechanisms upon different platforms could help in supporting this situation. Finally, the close connection between mobile devices and their users warrants more research on improving the usability of the security middleware on these devices, and in particular the trade-off with functionality and performance.

Acknowledgements

This research is partially funded by the Interuniversity Attraction Poles Programme Belgian State, Belgian Science Policy, and by the Research Fund K.U. Leuven.

Chapter 13
Dynamic Adaptation

Paul Grace, *Computing Department, Lancaster University, UK*

13.1 Introduction

Mobile computing is characterized by *dynamic change to the operational environment*; examples of this include: *i*) changes in environmental context, such as geographic location or network type (e.g. moving from an infrastructure wireless network to an ad hoc wireless network); *ii*) resource fluctuation e.g. available battery power or network bandwidth; or *iii*) change in network connectivity. In the face of such change, mobile systems must dynamically adapt their behavior, often to continue operating or to maintain the required levels of service.

In this chapter we focus on how software adaptation changes the run-time behavior of a distributed system. Here, software modules are added, removed and replaced while a system operates continuously. A software module encapsulates implementation relating to a particular concern (e.g. a protocol or message marshalling); these can be in the form of software components, micro-blocks, or aspect-oriented compositions. Although the styles of modules differ, the algorithms to adapt them typically consider one or more of the following properties; these are general to all types of distributed system, however this chapter investigates the extent of how they are applied in mobile computing.

Safety   Before adaptation, the system is placed in a state (often termed *quiescence* [476]) such that the future adaptation will not cause erroneous behavior or lose critical information.

Consensus   In a multi-party system the adaptation decision must be agreed upon by the interested parties.

Consistency   Multiple adaptations (which may be concurrent) must not place the system into an inconsistent state.

Rollback   If the adaptation fails it must be possible to return the system to its original executing state.

The primary role of middleware is to simplify the development of distributed systems; however, middleware also has a pivotal role in supporting dynamic adaptation; it is ideally situated to monitor changes, and manage individual adaptations.

This allows the developer to focus on application logic, rather than complex run-time changes. The middleware itself may be self-adaptive i.e. it transparently adapts its own behavior to provide the best level of distribution services in the face of environmental change; or the middleware may respond to and manage application-specific adaptations defined in declarative policies. Three common approaches to perform adaptations of this type within middleware are:

Reflection   this is the capability of a system to reason about itself and act upon this information to adapt its own behavior dynamically.

Dynamic Aspect-Oriented Programming (AOP)   this allows concerns that typically crosscut implementation (such as security, persistence, logging, and monitoring) to be adapted at runtime.

Policy-based approaches   these provide declarative descriptions of reconfigurations to enact either parameterized (pre-defined adaptation points) or software module adaptation.

In this chapter, we investigate how each of these state-of-the-art solutions have been applied in the domain of mobile computing; these serve to illustrate to the reader how to perform real world software adaptations, and also help identify which technology is appropriate for particular application scenarios. To conclude the chapter, we identify that as adaptive systems become more complex, novel software engineering approaches are required to help developers solve these future challenges.

13.2 Dynamic Adaptation in Action

Mobile computing presents many challenges that can be addressed using dynamic adaptation. Here mobile users experience a range of heterogeneous devices communicating via ad hoc and fixed infrastructure wireless networks. In this section we examine two use-cases illustrating examples of dynamic adaptation in this environment; notably, these promote *self-managing* or *autonomic* behavior [450], which removes the need for human intervention to resolve faults or performance issues.

Case Study 1: Protocol Heterogeneity Encountered by the Mobile User. Figure 13.1 illustrates three locations (the user's home, the user's office, and a coffee bar close to the office) where the mobile user can communicate with application services using portable devices and wireless networks. Three applications reside across the three locations; these applications encompass a range of application styles that are typically utilized in mobile settings e.g. data information, ad hoc chat messaging, etc. However, the spontaneous nature of interaction means that the communication protocols involved cannot be determined until runtime. For example, the services are implemented using different interaction protocols (e.g. SOAP, CORBA, etc.) and are advertised using different discovery protocols (e.g. UPnP and SLP). Hence, adaptation local to the mobile device is required to adapt the middleware to ensure services can be found and interacted with irrespective of the protocols involved.

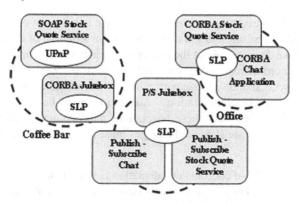

Fig. 13.1 Communication protocol heterogeneity

This should be performed safely and consistently to ensure the behavior of the device is not compromised; for example, by reconfiguring a protocol before all current requests have completed.

Case Study 2: P2P Mobile Multimedia. Many mobile applications involve collaboration between groups of mobile users. For example, a multimedia conferencing application involves devices connected via ad hoc wireless networks receiving a shared multicast video stream. The nature of this environment means that the application is subject to fluctuations in environmental context and network Quality of Service (QoS). When the level of available network bandwidth decreases the data must be encoded using a lower quality filter. For this, the filter encoding data on the source must be dynamically adapted, but also the decoding filters on the peer receivers must also be adapted in order for all members of the conference to be able to view the same video frames. This is an example of distributed adaptation across a system to improve operation performance. An important requirement is that the adaptation across multiple hosts must be carried out in a co-ordinated manner, ensuring that no member is left in an inconsistent software configuration.

Analysis. These case-studies illustrate two distinct classes of adaptation: *node-local adaptation* and *distributed adaptation*. Node-local adaptation describes software adaptation upon a single machine (or address space); the decision to adapt is made from information local to the individual node, and only modules on that host are changed. Case study 1 is an example of node-local adaptation, the client device detects the service types in the current environment and adapts its own software implementation accordingly. Alternatively, distributed adaptation (illustrated by case study 2) describes the co-ordinated adaptation of software modules that reside across network connected devices; using global context information, nodes must reach some consensus about the appropriate behavioral adaptation to take, and ensure that this is executed in a safe and valid manner. Research into practical solutions for co-ordinated adaptation have investigated how to provide the following properties:

Open access   Required to inspect the structure and behavior of co-ordinating distributed nodes. The information about the current structure of distributed nodes can help make decisions about the appropriate adaptation to employ.

Consensus   Coordinating middleware nodes require mechanisms to make decisions about what actions to take; particularly what reconfigurations are required.

Safe, valid, distributed reconfiguration   Adaptations must be made to distributed topologies of components. These adaptations must be made when the distributed system is in a safe state, and changes must be validated once the update is complete.

However, flexibility of coordinated adaptation is important in the face of diverse deployment domains. For example, some consensus and reconfiguration protocols may be too resource-intensive for limited resource, ad hoc environments (indeed the complexity of ad hoc networks has seen the prevalence of node-local adaptations in this domain). Similarly, placing the system in a safe state may range from making individual nodes safe, through to making an entire network simultaneously safe. In the subsequent sections, we examine a range of solutions in this spectrum.

13.3 Reflective Middleware

Reflection is a technique that first emerged in the language community to support the design of more open and extensible languages; it promotes the concept that a system or computer program should be able to inspect itself and act upon itself. This philosophy has been embraced within the field of system software development, where reflective operating systems [884] and reflective middleware [469] have emerged. The following quote from Paddy Maes' seminal paper [541] accurately describes the key elements of typical reflective systems in these domains:

> "A reflective system is a system which incorporates structures representing (aspects of) itself. We call the sum of these structures the self-representation of the system. This self representation makes it possible for the system to answer questions about itself and support actions on itself. Because the self-representation is causally-connected to the aspects of the system it represents, we can say that: i) the system always has an accurate representation of itself. ii) The status and computation of the system are always in compliance with this' representation. This means that a reflective system can actually bring modifications to itself by virtue of its own computation."

Therefore, the design of a reflective system is based upon a suitable causally connected self representation (CCSR) to support developers in performing adaptations; for example, the architectural topology of connected software components is a good representation for adapting components dynamically. Further, an interface describing operations to inspect and change the self-representation known as the Meta-Object Protocol (MOP) [453] is also provided at the meta-level. To maintain the CCSR, the process by which the base-level is made tangible at the meta-level is known as *reification*; also *absorption* is the technique that ensures changes to the CCSR are executed at the base-level.

Early work in the field of reflective middleware identified that well-established middleware such as CORBA, EJB and DCOM maintain a black-box philosophy, whereby a fixed service is available to users, and it is typically impossible to view or alter this implementation [469] if and when the environment conditions changes. Reflective middleware seeks to overcome this by creating *open* platforms that are: *configurable*, to meet the needs of a given application domain, *dynamically recon-figurable* to enable the platforms to respond to changes in their environment, and *evolvable* to meet the needs of changing platform design. This openness makes re-flective middleware especially amenable to the domain of mobile computing.

13.3.1 OpenCom and ReMMoC

To illustrate how reflection works in practice we first describe the OpenCom [194] model, which uses reflection to dynamically adapt software components. Subse-quently, we demonstrate how OpenCom is used to build the ReMMoC Reflective middleware [325], which operates on mobile devices to help applications overcome the problems of protocol heterogeneity in the mobile computing domain (as de-scribed in the first case study in Sect. 13.2). Notably, OpenCom supports node-local self-adaptation to allow ReMMoC to dynamically change between different middle-ware protocols at run-time.

Software components are defined as *"a unit of composition with contractually specified interfaces, which can be deployed independently and is subject to third party creation"* [791]; these provide the benefits of configurability, re-configurability and re-use. OpenCom components are language-independent encapsulated units of functionality and deployment that interact with other components exclusively through "interfaces" (provided behavior) and "receptacles" (interfaces that make explicit the dependencies of a component on other components). Bindings are con-nections between interfaces and receptacles. These components reside within the OpenCom kernel (illustrated in Fig. 13.2), which presents a runtime API for third-party creation, destruction, connection and disconnection of components at runtime.

As seen in Fig. 13.2, the OpenCom kernel can be extended with multiple types of reflective CCSRs each presenting a distinct MOP to the user for adapting different views of the component system. Three examples of these are:

Interface MOP   this supports inspection of a component's provided and required interfaces. Typically, you can examine the operations available on these inter-faces, and or dynamically invoke one of the operations.

Architecture MOP   this accesses the software architecture of the kernel repre-sented by a component graph (a set of connected components, where a connec-tion maps between a required and provided interface in the same address space). Hence, the architecture MOP can be used to both discover and make structural changes at run-time.

Interception MOP   this enables the dynamic insertion of interceptors, which sup-port the insertion of pre- and post- behavior on to interfaces. These interceptors

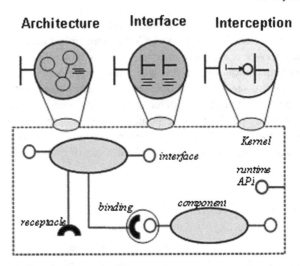

Fig. 13.2 The OpenCom component model

are executed before each operation invocation of an interface, and after the operation has completed

The final important element of OpenCom is the component framework (CF), which introduces the concept of a composite component that acts as the life support environment for a particular domain of middleware functionality i.e. the framework manages and adapts the local configuration of components within it. Frameworks improve the scalability of adaptations across a system; as the number of components in the kernel increases it becomes harder to perform safe adaptations, because a thread of control may involve many dependent components. Hence, CFs scope adaptation, and provide mechanisms to adapt safely within the framework's particular domain. The important properties of a framework include:

- A local architecture MOP to inspect and dynamically reconfigure the framework's subset of the complete architecture.
- A set of rules for valid dynamic reconfigurations; invalid attempts are prevented and the framework rolls back to its last safe state.
- A quiescence mechanism that places the CF in a state ready to be adapted. For this purpose, each component framework provides a readers/writers lock to access components within the framework. Every non reflective operation accesses the lock as a reader (there can be n readers using the lock at any time). Any reflective call to change the configuration accesses the lock as a writer (a single writer can access the lock when there are no readers). Hence, there is no executing thread within the framework for the duration of the adaptation.

The ReMMoC middleware is composed of two OpenCom frameworks (as illustrated in Fig. 13.3), which perform adaptations in two separate but related dimensions: *i*) reconfigurable resource discovery protocols, and *ii*) reconfigurable

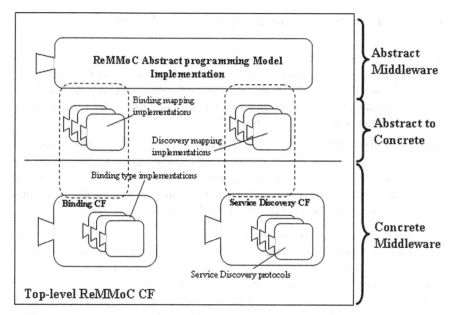

Fig. 13.3 The ReMMoC architecture

interaction protocols. The *discovery framework* monitors service discovery protocol messages that are multicast in the node's local networks. When a new protocol is detected the component implementation of that protocol is plugged into the discovery framework (this is performed using the create and connect component operations of the framework's architecture MOP); multiple protocols can reside side-by-side, hence if SLP (Service Location Protocol) and UPnP (Universal Plug and Play) are being used to advertise services then the framework reconfigures itself to include both protocols. The second feature of ReMMoC is the abstract programming model that hides the heterogeneity of protocol implementations from the application developer, e.g. when a service is advertised using the abstraction API this is mapped onto the underlying protocols; if SLP and UPnP are configured then the service will be advertised using both.

The second ReMMoC framework is the *binding framework*, which dynamically reconfigures itself to allow mobile clients to bind and interoperate with application services implemented upon heterogeneous interaction paradigms (i.e. Remote Method Invocation, Publish-Subscribe, Asynchronous Messaging). When a service is invoked, ReMMoC passes information from the service discovery framework to correctly reconfigure the correct binding protocol, e.g. if the found service is a SOAP service, the framework reconfigures to the SOAP RPC protocol. In all cases, ReMMoC uses reflection (the architecture MOP) to determine the current state of the binding framework, and then calculates the correct adaptation steps to achieve the required component configuration.

13.3.2 Further State of the Art Reflective Middleware

A number of important reflective middleware solutions have now emerged. Here we briefly examine two examples that have been utilized in the mobile computing domain. These illustrate different strategies for performing both node-local and distributed adaptations. Other influential systems include Open ORB [93], dynamic TAO [468], Rapidware [721], and DREAM [495].

K-Components [233]. are a distributed component model designed to support the development of loosely-coupled, self-adapting applications. A K-Component is a single component that resides on a single host; then multiple K-Components across hosts are connected (using remote procedure call bindings) to build distributed applications. Each component is made up of two key elements:

- The architecture meta model (AMM) which provides similar behavior to the component frameworks of OpenCom. The internal composition of the component (in terms of the interconnected sub components) is reified and this can be used to perform node-local adaptation using architectural reflection.
- Adaptation contracts are reflective autonomous programs related to individual K-Components. These consist of rules (typically in the form of event-condition-action policies) that state what self-adaptive behavior of the AMM to take when certain events or context changes occur.

K-Components support distributed adaptation using a decentralized co-ordination model. The components asynchronously exchange events; this can include contextual information, or changes within individual components. Based upon this information a local node can decide to adapt itself. For example, case study 2 in Sect. 13.2 would be implemented as follows: the K-Component changing the service filter would exchange events with the K-Components of the other devices in the conference application, stating that the media filter had changed; based upon this information the remaining nodes adapt themselves using architectural reflection. K-Components have been successfully applied in the mobile computing domain, notably autonomically adapting routing protocols in mobile ad hoc networks when conditions change [234].

ExORB [713]. is an adaptive middleware for cellular telephones. Such devices require a middleware infrastructure to make it possible for device carriers to configure new software, upgrade existing software and repair software bugs without manual intervention. For this, ExORB reifies the platform's state, logic and component architecture at run-time so that it is available remotely to be controlled by a third-party. The base-level consists of three elements:

Micro Building Blocks (MBB)   The smallest addressable functional unit in the system i.e. a single operation or method. Local state of the MBB is held separately in a system provided storage area. Therefore, when an MBB is replaced, the state need not be transferred to the new unit, rather it simply accesses the existing state.

Action specifies the order in which MBBs execute   Hence these define the system
logic. An action in DPRS is a deterministic directed graph, whose nodes are the
operational units and the edges are the execution transitions.

Domain aggregates collections of related MBBs   These store both the list of building blocks and the corresponding list of actions, plus the localized state of the
domain. Hence, collections of building blocks can be treated as single units (to
be suspended, resumed, inserted and removed).

The ExORB middleware can provide different flavors of Object Request Broker
architecture e.g. a CORBA broker, or an XML RPC broker. An example adaptation
of one of these systems is as follows: an MBB producing the IIOP header begins to
produce faulty messages and is replaced by a new version of the MBB. For this the
3rd party suspends the execution of the appropriate collection of building blocks and
replaces a single MBB before restarting the middleware. Hence, ExORB provides
a mechanism to perform 3rd party node-local adaptation remotely, rather than self-
adaptive behavior.

13.3.3 Analysis

Reflective middleware solutions now follow a well-established design approach that
typically combines software components and reflection; which in turn offers a prin-
cipled solution for dynamic adaptation within middleware implementations. This
has been shown to be successful for a number of ORB and message broker implementations, increasing flexibility, supporting self-healing and importantly providing
self-optimizing behavior. However, these solutions have generally focused on node-
local adaptation; i.e., where each node makes local decisions to adapt based upon
environmental context inputs and locally maintained policies. Such approaches do
not fully support next generation self-managing applications: e.g., autonomic computing, peer-to-peer computing, ubiquitous computing and ad hoc mobile computing, which all fundamentally require the co-ordinated adaptation of middleware behavior across distributed nodes. K-Components is an exception here, however its
loosely co-ordinated, decentralized approach is limited in that not all distributed
adaptations can be guaranteed to be carried out. Hence, alternative transaction-like,
consensus-based, co-ordinated adaptations are required to ensure that application
behavior is not disrupted when performing distributed adaptations (the development
of such protocols in ad hoc wireless networks remains an interesting challenge).

13.4 Policy-Based Middleware

Rather than perform adaptation of middleware behavior that is transparent to the
application (e.g. as performed by the majority of reflective middleware), policy approaches have identified that the application is often in a better position to determine

how to adapt to context changes. Hence, systems promoting application-aware adaptation have emerged [728]. These generally allow the application to state its rules for adaptation as a policy that can be interpreted by the underlying middleware. For each particular condition the matching rule is applied to change the middleware behavior. There is no common methodology to implement the adaptations; these are specific to individual systems e.g. they could be parameterized pre-determined reconfiguration sequences that can be invoked, or describe a more general script for changing software modules or services. A common pattern for describing policies is the event-condition-action (ECA) rule. The event part specifies the context change that triggers the invocation of the rule; the condition part tests if this context change is satisfied, which causes the description of the action (the adaptation) to be carried out.

13.4.1 Adaptation of Micro-protocol Frameworks

Ensemble [837], APPIA [581] and Cactus [167] are adaptive communication protocol frameworks. These support group and multimedia computing abstractions between a co-ordinating set of peers. Although originally designed for Internet scale computing, they can also be utilized in more dynamic, mobile environments. In these systems, a communication stack contains the implementation of a protocol composed of a set of micro-protocols; where an individual micro-protocol implements a small portion of communication e.g. de-fragmentation, reliability, ordering, etc. Figure 13.4 illustrates example blocks in a group protocol e.g. UDP transport, PING finds the initial membership of a multicast group, NAKACK is a negative acknowledgment protocol, UNICAST offers reliable unicast similar to TCP, and GMS manages group membership. These building blocks can be configured in many ways to create protocols with different properties; the application developer can then select the protocol by declaratively describing the stack structure.

Notably, the communication stacks can also be adapted at runtime. An early method for performing this was provided in the Ensemble system on a per-message basis. Here, individual nodes reconfigure themselves to match the protocol configuration of the node sending the message in the channel. For this purpose, a configurator micro-protocol sits at the top of the stack (shown in Fig. 13.4). Each incoming message is processed header by header (each micro-protocol corresponds to an individual header); if the protocol stack structure does not match the header structure at any point, then the message is passed directly to the configurator who calculates what the structure should be and then reorganizes the stack accordingly [837]. In Fig. 13.4 the incoming message from host 2 to host 1 contains a GMS header that is not in host 1's stack; hence the configurator will insert the GMS layer into the stack (based upon the header sequence) and then reprocess the message.

However, this approach limits the type of adaptations possible to evolutions or version upgrades of protocols. Hence, [715] presents a policy framework for defining more flexible context-aware adaptations of communication channels. Here a

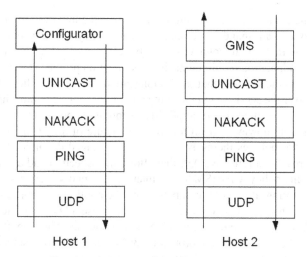

Fig. 13.4 Per message adaptation of a micro-protocol stack

centralized adaptation manager is made up of a context monitor, a reconfiguration manager and a set of policies describing adaptation actions to take. Individual nodes in the communication channel are composed with local context sensors (to detect local changes to be sent to the central manager) and local reconfiguration agents (to adapt the individual protocol stack). As with previous policy based adaptation systems, the adaptation manager reacts to changing context information from the channel according to ECA policies. The reconfiguration manager co-ordinates with the local reconfiguration agents on each node to ensure that the same micro-protocol reconfiguration (e.g. addition or removal of micro-protocols) is safely and consistently executed across the communication channel. However, the use of a centralized manager in ad hoc wireless networks is infeasible (as every node cannot be guaranteed to be able to communicate with the manager); hence, further research is required into decentralized mechanisms of co-ordination.

13.4.2 Further State of the Art Policy Middleware

CARISMA [137] is a reflective policy-based framework for adapting the behavior and operation of an underlying middleware platform. In this case, node-local adaptation of the XMIDDLE data-sharing middleware [556] for mobile devices in ad hoc wireless networks. The platform concentrates specifically on the important issue of how context information (e.g. device context such as power and memory, and external context such as network connection, bandwidth, and location) affects the performance of a mobile application and how middleware adaptations (specified declaratively in a policy) can be performed to maintain the best level of performance in the face of these changes.

In a specific context, one application may require the middleware to behave in a particular way e.g. an image processing application may ask to display pictures in black and white rather than color when the battery power is low. Each application describes their adaptation requirements in an application profile. This contains associations between the services that the middleware delivers, the policies that can be applied to deliver the services and the context configurations that must hold in order for a policy to be applied. Every application submits its policy to the middleware upon initialization, however, given the dynamic nature of mobile applications it is expected that the policies themselves need to be changed dynamically. Therefore CARISMA provides a reflective API that allows introspection and dynamic reconfiguration of this policy. CARISMA also manages the end-system resources of the mobile device being utilized by competing mobile applications. Different policies have different non-functional requirements e.g. the present Quality of Service is different, and they also utilize different amounts of resources. The resolution of these conflicts is resolved by an auction protocol [136]. Each application submits a bid for resource use citing non-functional concerns e.g. security, performance, availability etc. The resource goes to the highest bidder. In a similar fashion, reflection allows the application to dynamically change the non-functional properties of its bid if its requirements dynamically change.

The Lancaster Context Architecture [247] similarly resolves conflicting adaptation actions described in policies e.g. a request to reduce power consumption enforces applications using the network to postpone their activities, as a consequence the network bandwidth increases and this could trigger a request to applications to utilize the spare bandwidth (i.e. the two are in conflict). The architecture provides a common space for co-ordinated system wide interaction between adaptive applications and a complete set of context attributes. The system they present decouples adaptation policies and mechanisms. The context space acts as a repository for context information, storing information from the device monitors, applications and middleware for use by adaptation strategies. The adaptation control module is a key component of the architecture driven by a set of adaptation policies; it is responsible for co-ordinating adaptations and resolving potential conflicts. Furthermore, it is identified that the decoupling property of the architecture allows it to be integrated with a range of existing platforms such as event-based, tuple space and object-based middleware.

13.4.3 Analysis

Performing system level adaptations is a complex task e.g. using reflection to adapt middleware. Policy approaches raise the level of abstraction to enable declarative approaches for defining application-specific adaptation of middleware behavior. For example, reflective middleware can be described using the policies described in these state of art solutions. Hence, the premier solution for policies is to loosely couple them to both the middleware they are adapting and the dynamic adaptation framework; for example, Gridkit [326] is a reflective middleware solution based upon CARISMA's policy approach. In this way, application policies can be used to adapt different middleware, and similarly a well known policy language can be re-used across different adaptation technologies (e.g. reflection or micro-protocols). A limitation of policies is that they are often pre-defined, static, and/or slow to update at run-time; hence, they are not well suited to highly dynamic and unknown environments where some degree of learning is required to drive self-adaptive behavior.

13.5 Dynamic Aspect-Oriented Programming (AOP)

Aspect Oriented Programming (AOP) [454] is a software engineering approach designed to tackle the problems of tangled code i.e. the basic functional implementation of your component becomes tangled with additional code for features such as security, persistence, logging, and monitoring. Developers often implement these features in an ad hoc manner across the system, which leads to increased system development, debugging, and evaluation time because of the increased system complexity. Therefore, AOP supports the concept of separation of concerns to counter this problem; i.e. individual concerns such as security and monitoring code are not implemented within the base code, rather these are each implemented as software modules known as aspects that can then be woven into the base code at compile time. Aspects are composed of two important parts:

pointcut expression   this describes locations in the code or system where aspect behavior (implementation) should be composed. These individual positions are known as *join points*. The join point model differs depending upon the type of system; for example, when weaving aspects into an object-oriented language: method calls, field accesses and object instantiations are suitable join points.

advice behavior   this describes the implementation behavior (often a single method) to execute at every join point. Each advice can be executed before, after, or around (before and after) the actual method invocation at the join point. For around behavior, a proceed keyword identifies the point in the advice code to switch control to the actual method.

Original aspect technologies such as AspectJ [455] weave aspects statically at compile-time; however, this approach is unsuitable for dynamic adaptation. Hence, dynamic AOP solutions have emerged; these offer identical software composition methodologies but the weaving of aspects is performed dynamically, either when the system is loaded into memory [105] (advices are added to the executable code before loading) or at run-time (advices are dynamically reconfigured).

13.5.1 Invasive and Non-invasive Weaving

The instrumentation of dynamic adaptation of aspect modules can be performed using either of two approaches: *invasive* and *non-invasive* weaving. This refers to whether the advice behavior is added within a software module. For example, invasive dynamic AOP in component architectures requires weaving code within the base component implementation, i.e. behind the interface contracts. Typically code re-writing techniques, such as bye-code rewriting as supported by tools such as Javassist [172] are used to instrument this behavior.

Prose (PROgrammable extenSions of sErvices) [628] is one such tool for performing invasive, dynamic weaving of aspect behavior at runtime. The solution is specific to the Java programming language; it is common for invasive approaches to be tied to particular languages because they require code modification. In this case, a Prose aspect is implemented as a Java class which contains both a description of the pointcut (where the behavior is to be applied) and the advices containing the actual cross-cutting implementation. These advices are woven to join points relevant to Java code i.e.: method calls, field accesses, exceptions and object instantiations. Hence, the byte code of the corresponding classes is rewritten dynamically to ensure that when the flow of program control accesses one of these join points then the advice behavior is correctly executed.

Non-invasive approaches utilize the component interfaces as join points, and hence these aspects are implemented as interceptors, or compositional filters on the interfaces. Lasagne [814] is an AOP framework that introduces aspects dynamically at system run-time in a non-invasive manner. The aspect-oriented approach of Lasagne is based upon extensions, where an extension encapsulates a slice of behavior that updates multiple components at the same time. For example, an authentication extension may crosscut a number of components involved in a client-server request (potentially distributed). Only when the extension is applied across the complete system has the new non-functional service been dynamically added to the system. The extension is implemented by a set of wrappers, where a wrapper is the per-instance implementation of the aspect that is to be applied at each component. The dynamic insertion of wrappers is non-invasive; Lasagne dynamically alters the message flow of the system to be directed to the wrappers before the component interface (in a similar technique to message interception).

13.5.2 State of the Art Dynamic AOP Middleware

Aspect-Oriented middleware solutions fall into two distinct categories. The first uses aspects to modularize crosscutting middleware functionality, so that evolution and customization of the middleware implementation is straightforward. For example, [898] modularizes the crosscutting concerns of a CORBA ORB. The second provides a distributed computing abstraction for deploying aspects e.g. a security aspect deployed across a set of co-ordinating peers, or an authentication aspect for a client-server session. Dymac [487], and JAC [654] are examples of middleware providing distributed aspects. In both categories, dynamic adaptation is important for performing node-local and distributed adaptation e.g. adapting the behavior of the distributed authentication module. We now examine three systems in detail as examples of both local and distributed adaptation.

MIDAS [679]. is a middleware layer developed at ETH Zurich for providing run-time extensions to mobile computing applications. The authors identify that mobile applications must adapt and extend themselves to their current environment conditions. For example, a PDA may interoperate with application services from different mobile locations. However, in these different locations, different functionality may be required to interact with the local services e.g. encryption layers must be added to allow interaction to happen at one location, whereas billing modules must be included to pay for services at another location. The application cannot carry every piece of possible code around with them, nor can the developer plan for every interoperation eventuality. Therefore, it is the role of MIDAS to add the functional extensions to the developer's basic code implementation (in this case service interoperation) at run-time. When a function extension is required it is downloaded to the MIDAS middleware, which then dynamically weaves the code into the base application at run-time using the PROSE tool. Hence, MIDAS performs node-local dynamic adaptation at individual mobile devices using an invasive weaving mechanism.

AspectOpenCOM [327]. is an extension of the OpenCom approach from Lancaster that supports non-invasive dynamic weaving of aspects. In addition to component compositions, aspects can be composed at the base-level. Pointcuts identify interface join points only, i.e. aspect behavior can only be attached around operation calls made at either receptacles or interfaces, not within the component implementation. This advice behavior is implemented within special component types known as aspect components, and the interception MOP of OpenCom is then used to attach advice behavior non-invasively; an interceptor redirects the control flow to the appropriate aspect advice when the join point is fired.

AspectOpenCOM supports fine-grained adaptation of the deployed aspects in the address space. For example, when a new aspect is woven into a system containing existing aspect composition, a reordering of advices at a shared join point may need to occur. Consider a client-server system with authentication, caching, logging, and encryption aspects. Initially no aspects are woven into the system. However, when the mean execution time of client requests deteriorates beyond some predetermined threshold due to network latency, a cache aspect is woven into the system. This aspect intercepts client requests and checks a local cache to see if the same request has already been issued. Later, when the system must operate in a secure mode, an authentication aspect is dynamically woven into the system; this consists of an advice that denies the client access to the server until they provide correct identification credentials. When aspects execute at the same join point, the order in which their respective advices are executed may be critical for the correct operation of the system. If the cache advice is executed before the authentication advice, clients

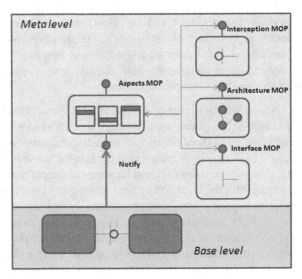

Fig. 13.5 The AspectOpenCOM architecture

are able to get access to resources without authenticating themselves. As such, the only correct way is that the authentication advice executes before the cache advice. Hence, these must be resolved through dynamic adaptations.

To support these types of adaptation, AspectOpenCOM presents an Aspect MOP (seen in Fig. 13.5); this self-representation reifies how the aspects in the system are composed with components, and then allows third party configurators to inspect and adapt this aspect behavior. The MOP provides a set of operations including: adapting advice behavior (e.g. ordering), adding and removing aspects, and adapting the pointcut expression to alter the join points where aspects are deployed. Hence, AspectOpenCOM supports non-invasive, node-local adaptation.

DyRES [815]. is a framework for performing safe, consistent adaptations of distributed aspects. The DyRES developers have identified that aspect frameworks have largely ignored the safety aspects of dynamic adaptation, and as such it is likely that inconsistent states will emerge. For example, a mobile application consisting of multi-parties who communicate via SMS messages may experience message loss when the network performance degrades. When this is detected a reliable messaging protocol is added dynamically across all the nodes as a new crosscutting aspect. To ensure that this is performed consistently across the nodes, DyRES provides three key properties: i) the co-ordinating nodes are placed in a safe quiescent state, ii) aspects are added and removed dynamically, and iii) the adaptation actions are co-ordinated across the hosts.

The framework is independent of the type of aspect compositions; hence, it can work with both invasive and non-invasive approaches (this is because it relies only on being able to communicate with the system-specific adaptation protocols);

to illustrate this the framework has been implemented atop both JBoss Dynamic AOP [414] and Spring [790].

13.5.3 Analysis

Dynamic AOP is essentially an extension to reflective programming, which promotes a higher level composition abstraction that hides the complexities of reflection. However, it can be seen from the work in AspectOpenCOM that the two technologies can be effectively combined in dynamic AOP solutions i.e. interception MOPs can be used to add aspects to a system in a non-invasive fashion; and these aspects can themselves be reified and adapted using reflection.

Dynamic AOP is a less mature domain than reflection, and the majority of solutions have generally focused on adaptation on local nodes, and not concentrated on techniques for safe adaptation; this is especially important for invasive approaches where a change to a class may produce version problems at run-time. Importantly, DyRES presents initial work in this area identifying how safe adaptations of aspects can be made across distributed hosts. Further, Dynamic AOP has had limited impact on the field of mobile computing (solutions have generally addressed the problems of traditional client-server systems). However, AOP has a significant role to play in this field, as the mobile domain must address cross-cutting concerns such as security; this is well demonstrated by the MIDAS middleware.

13.6 Summary and Outlook

This chapter has demonstrated that dynamic adaptation is an important solution to address many problems emerging in the next-generation of distributed applications that involve highly heterogeneous environments and dynamic users. A number of adaptation solutions are now available; these range from reflective management of software modules such as components and aspects, through to policy described adaptation of micro-building blocks. These complimentary approaches illustrate how adaptation can be instrumented at run-time; it is important to identify that adaptation is independent of the software composition mechanism; i.e. similar patterns are employed to detect when and how to adapt, and then perform this in a safe consistent manner, irrespective if the adaptation involves software components, aspects or micro-blocks. Adaptation is also general to all classes of distributed systems; however, when considered in the field of mobile computing the mechanisms for adaptation must consider the characteristics of the environment e.g. frequent disconnection and failure potential. As such, future research into ensuring safety, consensus, consistency and integrity of adaptations in these environments is essential.

A further issue with emerging dynamic middleware is that multiple domains of middleware behavior (discovery, security, mobility, etc.), each with its own increasing number of adaptation rules, make it difficult for a developer to first design and manage the potential dynamic adaptations. These may be in conflict or interfere with one another e.g. an adaptation in one domain may cause other domains to adapt and vice versa - potentially creating a feedback loop. Hence, we believe that advanced software engineering approaches are required to better design and maintain dynamic software. Once such approach is the use of model-driven engineering (MDE) techniques; here a higher-level abstraction (the model) is provided to describe adaptations [77]. Tools can then be used to investigate potential conflicts introduced by a new adaptation policy, and also verify that an adaptation will not disrupt the operation of the system before it is deployed.

To summarize, the two key areas of future research into dynamic adaptive systems are:

1. Providing flexible and configurable adaptation technologies for ad hoc wireless environments.
2. Promoting advanced software engineering solutions to enable developers to handle the complexity of concurrent adaptations.

Acknowledgements

The author would like to acknowledge his colleagues: Gordon Blair, Geoff Coulson, Nelly Bencomo and Francois Taiani at Lancaster; and Wouter Joosen, Bert Lagaisse, and Eddy Truyen at K. U. Leuven who helped formulate many of the ideas in this chapter. The author also acknowledges the ESF and the Minema scientific network for the opportunities to disseminate and enhance early versions of this work.

Part IV
Applicative Issues

The last part of the book focuses on aspects that are specific to some concrete application areas. In particular, we gather here chapters that discuss issues that are relevant to build context-aware applications, the problem of self-configuration, cyber foraging, and specific issues in the application area of vehicular networking.

- The part begins with Chap. 14, where P. Eugster, B. Garbinato, and A. Holzer address the issue of building context-aware applications, i.e., applications that can explicitly learn about a part of their deployment and execution context (individual or social), and act accordingly. The authors focus in particular on the problem of making context information available to the programmer and propose a categorization of middleware solutions to this problem.
- Entering an ad hoc network requires some negotiation, for example to ensure that each device will own a unique address. Ideally, the negotiation should be performed without user intervention, in what is usually called self-configuration. The problems raised when trying to achieve self-configuration are addressed by J. Manner in Chap. 15.
- Chapter 16 addresses cyber foraging, the process of opportunistically using resources available in the surroundings. J. Porras, O. Riva, and M. Kristensen give an overview of the cyber foraging process, with a special focus on the challenges that arise in every step of the process, and offer an overview of some existing prototype systems that support cyber foraging.
- Finally, Chap. 17, by A. Senart, M. Bouroche, V. Cahill and S. Weber, discusses recent results on vehicular networks and applications. These networks pose a number of specific challenges, due to the high speed and predefined movement pattern, as well as to the real-time communication requirements that are essential to safety applications.

Chapter 14
Middleware Support
for Context-Aware Applications

Patrick Th. Eugster, *Purdue University, USA*
Benoît Garbinato, *University of Lausanne, Switzerland*
Adrian Holzer, *University of Lausanne, Switzerland*

14.1 Introduction

With computing devices becoming more mobile and pervasive, a stronger interaction between an application and its changing environment opens new horizons in terms of application functionalities. Location-based applications, such as GPS navigation systems, are good examples of how information provided to an application on its surroundings offers new kinds of functionalities. Location is one of many environmental variables that might influence the behavior of an application. The notion of *context* encompasses these variables in the broad sense.

14.1.1 Definition of Context

There exist many definitions of the notion of *context* in the literature, some more general than others. In [732] for instance, context variables are categorized along three dimensions: *where you are, who you are with and what resources are nearby.* Similarly, the authors of [225] suggest that *any information about the user and the environment that can be used to enhance the user's experiences* is part of the *user's context.* Later, in [3], the same authors propose a definition of context as *any information that can be used to characterize the situation of an entity,* an entity being about anything. In [121], the notion of context is generalized as *a combination of elements of the environment that the user's computer knows about.* These different definitions all aim at specifying context in the very large sense.

In this chapter we rely on a more restricted definition of context, specifically tailored to mobile or ad hoc settings. That is, we consider distributed objects that typically run on small independent communication-enabled mobile devices. Such objects are also sometimes simply referred to as *peers.* The *context* of a peer is then characterized by a set of parameters, external to the object implementation, that might influence its behavior. We divide context information into two main categories, based on their scope.

- **Individual context.** Roughly speaking, the *individual context* of a peer represents its egocentric view of the world. That is, it gathers all parameters representing information about its environment directly accessible to the object, *without any interaction with other peers*. Such context parameters typically capture the state of the device on which the object is executing (e.g., its storage capacity, available memory) or some physical value measured by the device in its surrounding environment, e.g., temperature, location, etc.

- **Social context.** Intuitively, the *social context* of a peer represents its awareness of the existence of other peers. That is, a social context gathers parameters representing context information provided by other peers. In some sense, this context adds a communication dimension to the individual context, allowing for instance a peer to detect the presence of other peers, estimate their proximity, exchange information about individual contexts or services they can offer, etc.

14.1.2 Context-Aware Applications & Middleware

In traditional, context-agnostic applications, the context information introduced above is not accessible to the application code. In contrast, a context-aware application is an application that can explicitly (1) learn about a part of its context (individual or social), and (2) act accordingly. Examples of such applications are location-based games [92, 799, 843], proximity meeting/dating applications [753] where users can be notified when someone matching a desired profile stands nearby, or smart information applications [302] where relevant information is displayed according to user location, etc.

These applications are believed to be very promising [557] and their proliferation is assured to increase if adequate programming support is provided to leverage the burden of implementing them. Along that line, various middleware solutions have been proposed to address the issues raised by the programming of context-aware applications, some of them surveyed in [464, 52]. Even though they all target applications of similar type, these solutions stem from quite different approaches to addressing the main issues.

14.1.3 Middleware Classification

In this chapter, we classify such middleware solutions from the viewpoint of a developer of context-aware applications. In doing so, we examine how each solution answers the three following questions:

- *What context information is supported?* This question helps characterize how each middleware solution models individual or social contexts. When it comes to mobile ad hoc settings, the level of available *resources* is a good example of individual context. The *proximity* of other peers is then a typical example of social context.

- *What level of programming support is provided?* This question helps capture what each middleware solution provides to developers of context-aware applications, in terms of *Application Programming Interfaces* (APIs). A key question here is how well such specialized APIs are integrated with other more classical APIs, e.g., basic communication facilities.

- *What middleware architecture is adopted?* This question helps characterize and compare the architecture qualities of each solution. Key questions here are for example the level of decentralization or the level of portability offered by each middleware, giving us an indication of the technological settings that middleware is aimed at.

14.2 Context Information

This section presents the studied middleware solutions, with respect to the type of context they support. As mentioned previously, we focus on a special subset of context parameters, i.e., those that are particularly relevant to mobile or ad hoc settings. By particularly relevant contexts, we mean those that are made dynamic by the very nature of the setting. In our case, applications typically run on a (1) mobile device with (2) limited resources and wireless communication capabilities, and (3) interact in a peer to peer fashion with other devices in the mobile ad hoc network (MANET). From this description, we can point out contexts that will specifically be altered by mobile or ad hoc settings. First off, device mobility implies change of device location, which advocates for *location* monitoring. Then, the wireless capability of devices implies that power resources eventually dry up and that communication links are not fixed, which is a motivation to monitor local *resources*. Finally, communication in an ever changing network suggests that it is useful to monitor the *presence* of peers in the network and their *proximity*. Figure 14.1 depicts the four context types we focus on, namely: two types of individual context information (location and resource) and two types of social context information (peer presence and peer proximity).

14.2.1 Individual Context

Platforms supporting individual context are concerned with information coming from the peer itself, i.e., information that requires no interactions with other peers.

Individual Context	Social Context
⊕ Location	👁 Peer Presence
🔋 Resource	💻💻 Peer Proximity

Fig. 14.1 Context classification

They provide answers to the peer's egocentric questions: *where do I stand?* and *what is my profile?* The former refers to the geographical location of the peer and the latter to the computing resources locally available to the peer. Hereafter, we overview platforms specially geared at providing answers to these two questions.

Where do I stand? (location). Obtaining location information is a central issue in context-aware applications, in mobile or ad hoc settings. It is therefore not surprising that most platforms supporting these applications take location into account to a considerable extent. Applications sensitive to location typically react to the position of the device, with respect to some predefined sets of locations sometimes called *landmarks*. A good example of absolute location usage can be found in some automated museum guide, which displays information about an artwork as soon as the user stands in front of it. When manipulating location information, the first step is to gather some form of geographical coordinates, via one or more sensors (GPS, Bluetooth or Wifi beacons, etc.), in order to answer to the question: *where do I stand?*

In order to make this step transparent, *MiddleWhere* [686] offers a service interface that encapsulates the location detection technologies and that separates them from the application. Similarly, the Java Micro Edition (ME) [781] platform provides a Location API [543] that is independent of the actual global positioning technology. In particular, application developers using the Java ME location API can define the desired granularity and quality of location information and are notified via a callback when the locations changes. So, any middleware that relies on this API, such as *PERVAHO* [267], implicitly offers location support, even though it is not the primary goal of such middleware.[1] *CASS* [272], and *SOCAM* [337] also offer individual context aggregation facilities, but these are not restricted to location. Both aggregate context information from various sources in a central server to form high-level context useful to applications. The weather is an example of such high-level context based on an aggregation of low-level sensor information, e.g., temperature, humidity and light intensity. *SOCAM* encapsulates context in a *service locating service*. To retrieve user location, the application developer only needs to query this service, which will return the reference to a *context provider* able to provide this information.

Cooltown [67, 462], *CAMUS* [376], and *CARMEN* [76] use location for specific types of applications or network settings. *Cooltown* focuses on location-dependent web page browsing applications and enables peers to gather location information

---

[1] *PERVAHO* is discussed in details in Sect. 14.2.2, as its primary focus is on supporting mobile applications with an awareness of their social context.

from different types of sources, such as infrared, barcodes, optical recognition, or electronic tags. The location information contains an URL, which is used to access the location specific web page. Similarly, *CARMEN* is targeted at wireless Internet networks where peers get disconnected due to mobility. Once reconnected, a peer gets different information depending on its location and depending on the device on which it is used. Finally, *CAMUS* is designed for so-called Ubiquitous Robotic Companion (URC) networks. Such networks rely on a mixture of mobile robots, sensors and high-performance fixed and wired computers, made accessible through a web gateway. The location of a peer is used to filter the information it can access.

***What is my profile?* (resources).** Several middleware solutions are specialized in the manipulation of local resources. These solutions typically rely on some reflective or meta-level architecture. *MobiPADS* [157] is such an example: it provides an API that allows to retrieve resource-related context, such as CPU, memory, storage, network, and battery power via a *contextListener* interface. Project *AURA* [307] also provides support for resources and location. It uses a *wireless bandwidth advisor* to collect information about the network connectivity, and a *people locator* service to determine user location. *AURA* also tries to capture user intent and act upon it. An example of application scenario using *AURA* is the following. Jane is at Gate 23 of some airport, waiting for her flight, and she wants to send an e-mail with a large attachment. *AURA* notices that the meager available bandwidth will not allow her to send her message before her plane leaves. But scanning the airport's wireless facilities, Aura discovers that around Gate 15 a much greater bandwidth is available and hence advises Jane to go there, if she want to send her message in time. *INFOWARE* [240] and *CARISMA* [137] both give access to local and also remote resources, such as battery power, bandwidth and accessibility to available services. The solutions are detailed in the following section.

14.2.2 Social Context

Platforms supporting social context are concerned with information coming from other peers in the network. They provide answers to the questions: *who is around?* and *how close are you?* The first question refers to the presence of peers in the network and the second to the proximity of those peers. Hereafter, we present platforms specially geared at providing answers to these two social questions. Note that knowledge about peer proximity implies that peer presence can be sensed, but not the opposite.

Peer Presence: *Who is around?* When supporting peer presence, a middleware answers the question: *who is around?* This question is usually expressed in terms of some predefined physical location, such as a room, a building, or a landmark. Some middleware solutions use the fixed location with a stationary computer as starting point and discover users reaching or leaving the location. So they somehow narrow the question down to: *who is here?* When users are at that location, they can typically share information. The *Context Toolkit* described in [214] provides facilities to discover peer presence and peer activity around a widget placed in a predefined location. *CoBra* [166] is another example of such a middleware. Central to its conception is the notion of *smart spaces* and their attached *context brokers*, which gather contextual information about the space and the nodes located in them. *Mobile Gaia* [170] also allows devices to join a *personal space*. Once such a personal space is set up, a discovery service invites nearby devices to join the space and share resources and services. A location service is then responsible for location-awareness in the space. *Hydrogen* [373] provides individual context aggregation, including information on the device and the user besides information on location, time, and network. It also provides means to gather and aggregate context of surrounding devices.

Other middleware solutions are less restrictive and allow any peer to start sensing the network and define the location independently from the sensing peer itself. Those solutions rephrase the question to: *who is there?* For example, *SpatialViews* [627] allows peers to access all other peers located in a geographically determined *space*, such as a room in a given building. *LimeLite* [286] provides a similar feature in the sense that it enables automatic agent discovery and sorting, based on application rules and location. Similarly, *CoWSAMI* [33] allows to gather relevant data from nodes in the network based on their location, for example the traffic status of cars on the route from New York downtown to the JFK airport. The *Context-aware publish/subscribe* [293] service adds a contextual dimension to classic publish/subscribe schemes, in that publications and subscriptions can be restricted to a certain determined geographical location. Only subscribers located in that location can receive the published message.

Peer Proximity: *How close are you?*   When sensing peer presence at a defined location, *here* or *there*, the important aspect is *absolute* location of peers. These aspects are particularly interesting in office settings, with buildings and rooms. But in other settings, such as a sports event or a music festival,[2] it can be useful to use *relative* location, typically expressed in terms of the *proximity* between peers. This boils down to answering the question: *how close are you?* This question is of interest for applications sensitive to information coming from peers located nearby. In some systems, the notion of proximity can be quantified by the application developer, while in others proximity is limited to the peer's communication range.

Both *INFOWARE* and *CARISMA* provide support for the second type of proximity, i.e., they enable peers to sense resources available on their device and on other devices located within their communication range. They focus on resources such as battery power, network bandwidth and available services. *INFOWARE* aims at achieving resource availability prediction. This middleware keeps track of resources available in its neighborhood. When the application needs to access a resource to perform a task, *INFOWARE* will direct it to the neighbor who is the least likely to leave the neighborhood before the task is finished. The idea is based on the observation that mobile devices co-located for a certain period of time usually belong

---

[2] Or any other public event taking place in the open.

to a same person or at least to people working together in a team. Therefore, such devices are likely to stay together for a longer period of time. Along the same line, *CARISMA* provides application developers with primitives to describe rules on how applications should behave depending on the current resource status. Examples of such rules can be: if the *bandwidth* < 40% then use plain messages to communicate, if *bandwidth* > 40% then use encrypted messages, if *battery* < 20% then use plain messages. Such rules can create conflicts; in the previous example there is a conflict when the *battery* level is under 20% and the *bandwidth* is over 40%. *CARISMA* focuses on such runtime conflicts and proposes a micro-economic approach to resolve them.

CORTEX [765], *STEAM* [560, 562], *PERVAHO* [267], and *EgoSpaces* [434], provide support for the first notion of proximity, i.e., they allow application developers to customize the notion of proximity beyond their communication range. Examples of application scenarios using such context information can be a shop sending discounts to anyone within a few hundred yards to encourage people to come for shopping, or an ambulance sending emergency messages to cars located a few hundred yards in front of it. To facilitate the development of such applications, *CORTEX*, *STEAM* and *PERVAHO* all offer geographically scoped communication primitives based on publish/subscribe schemes. In these platforms, publications can be restricted to a range around the publisher. *PERVAHO* allows in addition subscriptions to be restricted around their subscriber. Unlike the *Context-based Publish/Subscribe* service presented previously, where events are bound to a certain absolute location, *STEAM* and *PERVAHO* restrict events to a relative location—in the case of *STEAM* around the publisher, and in the case of *PERVAHO* both around the publisher and the subscriber. *CORTEX* relies on the notion of Sentient Objects (SO), which can (1) sense the environment via sensors or other SOs, (2) aggregate context information in order to derive higher level contexts, (3) perform context-based reasoning and (4) provide some output via a publish/subscribe scheme. More specifically, CORTEX provides several Component Frameworks (CF) based sentient objects, such as a context CF or a publish/subscribe CF. The latter is used as a communication layer to share context with peers. The publish/subscribe scheme is similar to *STEAM*, as events can also be geographically scoped around the publisher. *EgoSpaces* is yet another example of middleware encapsulating peer proximity. This solution allows a node to access not only data available locally, but also data available to other nodes located in the network. The aggregated data is called a *view*. Such a view has two specifications, its scope and its content. The content specifies the kind of data that is aggregated, and the scope sets a geographical constraint around the querying node: only data available to nodes located within the defined geographical range is aggregated. *EgoSpaces* as well as the previously introduced *LimeLite* inherit from the *Linda* tuple space [310], where producers and consumers can interact by inserting tuples into, and extracting tuples out of a shared repository called a *tuple space*. Unlike *EgoSpaces* and *LimeLite*, the original *Linda* tuple space does not support context-awareness.

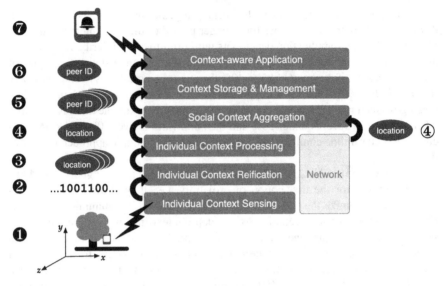

Fig. 14.2 A layered view of a context-aware application

14.3 Programming Support

Developers of context-aware applications must face different recurrent challenges to go from the context sensing to the context usage in their application code. So, in order to discuss how these challenges are addressed by existing middleware solutions, we adopt an approach similar to the one proposed by the OSI [912] model. That is, we decompose these solutions as a stack, where each layer processes information provided by the layer underneath and delivers the processed information to the layer above it, through well-defined Application Programming Interfaces (APIs). Figure 14.2 presents this layered decomposition and illustrates it with an example of proximity-based messaging application.[3] This application displays all peers located within a determined range and allows the user to send messages to any of them.

At the bottom of the stack is the *individual context sensing* layer, which takes input from the physical environment (❶) and transmits raw data (❷) to the *individual context reification* layer. This second layer is in charge of transforming the raw data stream into context objects (❸), easily manipulable by the developer. In our example, these objects are simply locations. The individual context reification layer then forwards context objects to the *individual context processing* layer, which acts like a context filter or selector. That is, this layer refines local contexts into relevant information (❹), e.g., by choosing information coming from the most reliable sensor, in case several are available, or by only relaying certain information. In our example, we are only interested in information about location changes. So, if the peer stays

---

[3] Our layered approach is somewhat similar to the one proposed in [9], although the latter proposes no layer concerned with social context processing issues.

still, the individual context processing layer works as a buffer and simply discards contextual information until the peer moves and a new location is received.

The *social context aggregation* layer is in charge of collecting relevant local context information, as well context information coming from other peers via the network (④). In our example, since we are interested in peer proximity, this layer compares its locations with the locations received from other peers, in order to gather the identifiers of all peers located within the determined range. These peer identifiers (❺) are then sent to the *context storage and management* layer, where they are persisted. The context-aware application can then query the context storage and management layer and obtain the list of peers in the neighborhood, in order to display them and access a specific peer (❻) to whom the user wants to send a notification message (❼).

Various Levels of Programming Support. In the following, we categorize middleware solutions into three different levels of programming support, depending on which layers they encapsulate, as shown in Fig. 14.3. A middleware providing *low-level programming support* offers only the individual context processing layer and the individual context reification layer. Then, a middleware providing *mid-level programming support* also integrates a social context processing layer. Finally, a middleware offering *high-level programming support* adds a context storage and management layer to the previous layers.

14.3.1 Low-Level Programming Support

Middleware solutions offering low-level programming support merely provide APIs that allow the application developer to access local context information in a (1) transparent manner based on (2) some selection criteria. *MiddleWhere* is a good example of such a middleware. Its aim is to separate applications from location detection technologies in order to make the different technologies transparent to the application developer via a location service interface, similarly to the Java ME location API. Raw location information is extracted from different sensors and sent to a spatial database. Then, a reasoning engine computes this information to relay adequate information to the subscribers of the location service. To determine the quality of location information, *MiddleWhere* uses three characteristics: (1) the resolution of the sensor (10 meters for GPS *vs.* 200 meters for the triangulation method), (2) the probability of correctness of the result based on the number of sensors able to detect the device in the area of interest and (3) the result freshness which is based on the elapsed time since the last sensor reading. Similarly, *Cooltown* offers support for obtaining location information via different sensors, such as Bluetooth beacons, bar codes, electronic codes or optical recognition. *CARMEN* aggregates local context information, e.g., about available resources or the current location; this information is provided by the *Event Manager*, which performs low-level context composition. The *Context Manager* then provides a list of services accessible to the user based on location, device capabilities, and possibly other application policies.

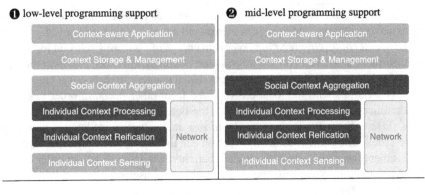

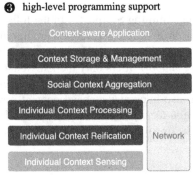

Fig. 14.3 Highlighted layers encapsulated by (❶) low-level programming support, (❷) mid-level programming support, and (❸) high-level programming support

14.3.2 Mid-Level Programming Support

Middleware solutions offering mid-level programming support provide APIs that allow the application developer to obtain processed social context information. Such middleware solutions typically add distributed communication to low-level programming support and thus enable to gather surrounding contextual information. There exists quite a few platforms integrating a social context aggregation layer, but not all of them provide the same set of APIs. Some are oriented towards peer discovery, others towards data sharing, or simply communication.

Peer Discovery. Several middleware provide means to discover other peers in the network, based on some criteria. *LimeLite* for example enables automatic discovery of peers fitting a certain profile, offering certain services, or standing at a specific location. *LimeLite* considers each peer as a distinct individual and provides advanced matching patterns that allow agents to target their operations to a sub-group of nodes satisfying some criteria. Similarly, *SpatialViews* is a high-level language that makes it possible to define spaces based on available services and geographical zones. An iterator can then be used to find peers located in the space and providing the defined services. Another example is the *Context Toolkit*, which provides a way to hide the complexity of the sensors being used thanks to abstractions called *context widgets*. Such widgets notify the application, via callbacks, as soon as relevant information is updated. *Context Toolkit* encompasses a *IdentityPresence* widget used to report the arrival and departure of people at a determined location.

Data Sharing. Middleware solutions providing data sharing facilities have a transparent data aggregation scheme, where peers can access remote data as it was local. The *EgoSpaces* platform for example has such a feature: it aggregates local and remote data in what is called a *view*. Such a view has two specifications—its scope and its content. The content specifies the kind of data that is aggregated, and the scope sets a geographical constraint around the querying node, so only data available to nodes located within a defined geographical range is aggregated. *Mobile Gaia* is user centric and allows a user to discover personal devices located in the neighborhood, via a so-called *personal space*. Mobile Gaia supports large collections of heterogeneous personal devices: as users meet, their personal spaces merge to share resources. The *INFOWARE* middleware follows a different approach: it aims at achieving resource availability prediction. That is, it keeps track of available remote and local resources and uses the neighborhood history to predict future neighborhood states.

Communication. Several middleware solutions propose context-aware communication tools to propagate and gather information in the network. Such tools add context rules to a communication protocol. In these middleware solutions, communication can only occur between peers located in a certain geographic setting. *STEAM* proposes a publish/subscribe service where publications are restricted to a certain geographical range around the publisher. Beyond this range, messages are not delivered to subscribers anymore. *PERVAHO* offers a location-based publish/-subscribe service. This service extends the traditional notion of content-based publish/subscribe services by (1) adding a location filter and (2) introducing the notion of persistent publications. To filter location, both publications and subscriptions are associated with geographical ranges around their publishers, respectively their subscribers. Such publication spaces and subscription spaces determine the geographical validity of an event. That is, in order to deliver a publication to a subscriber, both publisher and subscriber must be physically located in the intersection of the publication space and of the subscription space. Compared with *STEAM*, *PERVAHO* adds persistent publications and the notion of subscription space.

The *Context-aware publish/subscribe* service is also based on a classic content-based publish/subscribe service and exhibits the following context-based semantics. Alike in the *PERVAHO* platform, published events can be persistent or non persistent, and they are associated with a *context of relevance*, representing the space in which, according to the publisher, the events will be relevant; similarly, subscriptions are associated with a *context of interest*. A match occurs when both these contexts have a non empty intersection. This feature implies that if a publication has a worldwide context of relevance, all subscribers with a content matching subscriptions will deliver the publication no matter how small their context of interest

is. In order to avoid spamming, a subscription domain can be added to a publication, in order to restrict it to subscribers matching a certain context and similarly, a
publication domain can be added to subscriptions.

14.3.3 High-Level Programming Support

Middleware solutions offering high-level programming support provide APIs for
data aggregation & storage or for high-level context management. Hereafter we first
present middleware solutions focusing on data aggregation & storage, then solutions
focusing on *high-level context management*.

Data Aggregation & Storage. Middleware solutions offering a storage API usually aggregate different types of processed context information in some knowledge
database, which can be accessed by the application either synchronously through
a query language or asynchronously through a callback scheme. *CoWSAMI* offers
a good example of such a solution, with its *Context Manager* that stores different
types of context information and that can be accessed via SQL queries. The Context Manager delegates the different context discovery task to a lower layer called
the *Context Collector*, which uses logical context rules to gather context information. Context rules specify what kind of context needs to be aggregated under what
conditions. For example, when trying to gather information about the traffic on a
predefined route, these rules can be used to aggregate *velocity* information from
cars driving *on that route*.

CASS, CAMUS, and *CoBra* propose a similar integration based on a more centralized approach. Context information is gathered using a rule engine and stored
in a centralized database or via broker, from which it can be lated retrieved and
manipulated using SQL. On the contrary, Hydrogen offers a decentralized *Context
Server*, which aggregates information about the local and surrounding contexts of a
device and which can be accessed by any service running on the device. Hydrogen
offers both a synchronous and asynchronous way for services to retrieve data from
the server.

High-Level Context Management. High-level management APIs usually build
upon some storage facility and provide some extra features to the application developer. For example, middleware solutions such as *SOCAM, CORTEX,* and *Mobi-
PADS* all make it possible to manipulate context in a hierarchical way, where context information processed by the middleware can either be used by the application
through a storage API, or can be reused as an input to compute a higher level context
information. In *SOCAM, context providers* relay raw context information to a *context interpreter*, which in turn provides refined context information to the service or
to another context interpreter. To refine the raw context, the context interpreter uses
a rule-based reasoner associated with a Knowledge Base (KB). Information can be
retrieved using a set of API's provided by the KB. *CORTEX* offers a similar feature
with its Sentient Object (SO) model where SOs can sense the environment via sen-

sors or other SOs to refine already processed context. Similarly, *MobiPADS* gathers contextual events in a *Environment Monitor*, such events being filtered through low-level event conditions. The *Environment Monitor* then relays its output either (1) to another *Environment Monitor*, in order to provide input for higher level context information, or (2) to the MobiPADS system itself, in order to influence application behaviour.

CARISMA provides another kind of high-level context management, specifically aimed at resolving conflicting context-based rules. That is, it provides an *abstract profile syntax* that allows to define an application's policy or behavior, depending on contextual events. If more than one contextual event occur simultaneously at runtime, the triggered actions may lead to conflicts. Conflicts can either be intra-peer, e.g., Alice has a *bandwidth* > 40% and *battery* < 20%, both leading to contradictory actions, the former triggering the use of encrypted message to communicate, the latter the use of plain messages, or inter-peer, e.g., Alice's *bandwidth* > 40% and Bob's *battery* < 20%. *CARISMA* proposes a micro-economic approach to tackle such conflicts and to reach a consensus with a mutual perception of benefit. The scheme proposed is based on an *auction protocol*.

The *AURA* system supplies yet another approach and focuses on user intent. *AURA* aims at reducing user distractions by creating an adaptive environment. Its architecture is particular in the sense that it provides a layer on top of the application to predict user intent. This layer, called *Prism*, is used for task management. It takes in information from a context observer and takes control over the current running tasks and reconfigures them if needed. For example, the system can find out about the plan of a user to move to another location or to start a certain activity by examining the electronic agenda. Then, the system can proactively configure the user's computing environment to best suit its future needs.

14.4 Middleware Architecture

We use the architecture dimension to measure the ability of existing middleware solutions to support heterogeneous computing environments. In doing so, one classifies these solutions along three architectural dimensions, namely (1) decentralization, (2) portability and (3) interoperability. Decentralization measures a platform's dependence on specific components. Intuitively, a centralized middleware, i.e., one that relies on some central entity, is more vulnerable and less flexible than a decentralized one, which does not depend on any fixed component. Portability classifies platforms in two groups: portable platforms, can run on many different operating systems, and operating system dependant platforms, which can only run on few operating systems (usually one). Interoperability then measures the ease with which a platform can communicate with heterogeneous software components. Ideal interoperable platforms can communicate with many different applications, regardless of the operating system they are built on or of the programming language they are written in.

14.4.1 Decentralization

The decentralization of a middleware solution indicates its resilience to network topology changes. The more a middleware is decentralized, the more it is resilient to such changes. Centralized solutions tend to make strong assumptions about the network topology, as they often rely on some fixed network infrastructure or at least on the central role played by some predefined components. As a consequence, these solutions may stop working properly as soon as the connection to that central component is lost. Decentralized platforms, on the other hand, weaken these assumptions, by considering a network of peers that can equally contribute to the accomplishment of the tasks. Decentralized architectures are usually adopted when designing a middleware to be deployed in a ad hoc environment, where communication occurs spontaneously and in a peer-to-peer fashion.

Decentralized Middleware. Not surprisingly, the surveyed platforms providing proximity-awareness and geared towards MANETs tend to be decentralized. *CORTEX, STEAM, EgoSpaces*, and *PERVAHO* all support proximity-awareness and exhibit a decentralized architecture.[4] Other solutions also geared towards MANETs and exhibiting the same decentralized architecture include: *Context-aware publish/subscribe, LimeLite, SpatialViews, INFOWARE* and *CARISMA. Limelite* specifies the two assumptions it makes about the network: (1) the rate of configuration changes must be small compared to the network latencies to avoid that the majority of message gets lost, and (2) the broadcast range of all devices should be the same. While the second condition will not cause *LimeLite* to crash, it may lead to inconsistencies. *Context-aware publish/subscribe* and *INFOWARE* are both based on an ad hoc routing protocol. The former is based on the geocast routing protocol [544], whereas the latter relies on the *Ad hoc On demand Distance Vector* routing protocol (AODV [664]). *CoWSAMI* is an example of a middleware solution that proposes a mixed approach with a decentralized architecture that can optionally rely on a fixed device located on the Internet.

Centralized Middleware. Generally, middleware solutions offering context storage facilities use a centralized data persistence component. This central server is used to gather contextual information from mobile peers and to make it accessible to others. Such solutions include: *CAMUS, CASS, CoBra, MobiPADS, Hydrogen*, and *SOCAM*. Other centralized solutions are specially geared at wireless Internet scenarios with special interest in single-user context-aware applications. *CARMEN, Cooltown*, and *AURA* are examples of such solutions, which depend on a gateway to the Internet. Another example of a somewhat centralized platform is *Mobile Gaia* with its *personal spaces*: such a space represents a collection of co-located peers which can share data. Personal spaces are set up by a *coordinator* that manages the space and plays the role of a server for that particular space.

---

[4] The *PERVAHO* middleware actually exists in two flavors, one relying on a server-based implementation and the other on a fully decentralized architecture.

14.4.2 Portability

Platforms implemented in a portable programming language such as Java [782] can more easily be deployed on heterogenous operating systems. Many platforms such as *CAMUS, CARISMA, CARMEN, CoWSAMI, EgoSpaces, Hydrogen, LimeLite, MobiPADS, PERVAHO, SOCAM,* and *SpatialViews* are based on Java for that reason. To cope with resource constraints associated with mobile devices, such middleware solutions usually rely on a subset of the Java Standard Edition (SE) specifically designed for mobile applications and dubbed Java Micro Edition (ME) [781]. *PERVAHO* extends the Java ME platform with a ready-to-use Location-based publish/subscribe service. *SpatialViews* provides a Java language extension to help specifying geographical locations called *views* and to allow exploring peers located in those views through a specialized iterator. To achieve this, *SpatialViews* introduces a homemade compiler extending the native Java compiler. *CAMUS* and *SOCAM* both use the web ontology language OWL [71] and the Jena [139] semantic web framework for Java to represent and manipulate context information. *SOCAM* is built on top of the open service gateway initiative (OSGi [647]), which is a Java embedded server providing remote management of context-aware services. Finally, *CORTEX* is an example of middleware that targets a specific operating system and that is thus not portable. More precisely, *CORTEX* is specifically implemented for Windows CE [571]—Microsoft's operating system for small devices.

14.4.3 Interoperability

Interoperability indicates the capability of a middleware to interact with services implemented in different languages on different operating systems. Platforms using standardized interoperable data formats, such as the extensible markup language (XML [117]), to communicate with their surroundings offer an interesting interoperability capability. Several platforms, such as *CAMUS, CARISMA, Context Toolkit, Cooltown, CoWSAMI, Hydrogen, MobiPADS,* and *SOCAM,* precisely use XML to support interoperability. Communication between different network components is then achieved through a web protocol, so information formatted in XML can be read by any type of program offering a web network connection.

14.5 Summary and Outlook

This chapter surveyed various middleware platforms specifically aimed at mobile or ad hoc settings, with respect to three questions: (1) *what context is supported?* (2) *what level of programming support is provided?* and (3) *what architecture is implemented?* Several answers were proposed to these three questions, which are summarized in Fig. 14.4.

What context	What programming support	What architecture
🔋 Resources	▬▬▬ Low-level Integration	☕ Portable
◉ Location	▬▬▬ Mid-level Integration	🔖 Interoperable
👁 Peer Presence	▬▬▬ High-level Integration	⊡┄⊡ Decentralized
💻💻 Peer Proximity		

Fig. 14.4 Middleware classification dimensions

Figure 14.5 summarizes the result of this survey by indicating for each middleware what are its answers to the above three questions. The figure clearly shows that none of the surveyed platforms addresses all needs. Rather, they specialize in certain types of applications, and provide services tailored to those families of applications. Some of these answers seem to be however correlated, making trends visible through two major clusters of middleware solutions: those targeting individual context-aware applications and those targeting social context-aware applications.

14.5.1 Individual Application Support

A first cluster groups middleware platforms targeting exclusively individual context-aware applications. The profile of platforms in this cluster is: centralized architecture (11 platforms) and context support limited to individual context (9 platforms out of 11). Platforms grouped around this cluster are: *AURA, CAMUS, CARMEN, CASS, CoBra, Context Toolkit, Cooltown, Middleware, Mobile Gaia, MobiPADS,* and *SOCAM*. Most of them target applications that can rely on a gateway to the Internet, which explains the centralized nature of their architecture. This centralized architecture enables context storage with little effort—hence the incentive for a high-level of integration (6/11). Interestingly, most of these platforms are only interested in individual context. However, there are some centralized platforms providing some social support. *CoBrA, Context Toolkit,* and *Mobile Gaia* belong to this category along with the centralized version of *PERVAHO* and with *CoWSAMI*, when it can access a web server to support its activity.

14.5.2 Social Application Support

A second cluster groups middleware platforms built on a decentralized architecture (11 platforms), offering support for social context (11 platforms out of 11) and providing generally mid-level programming support (7 platforms out of 11). It is interesting to note that all platforms providing support for peer proximity

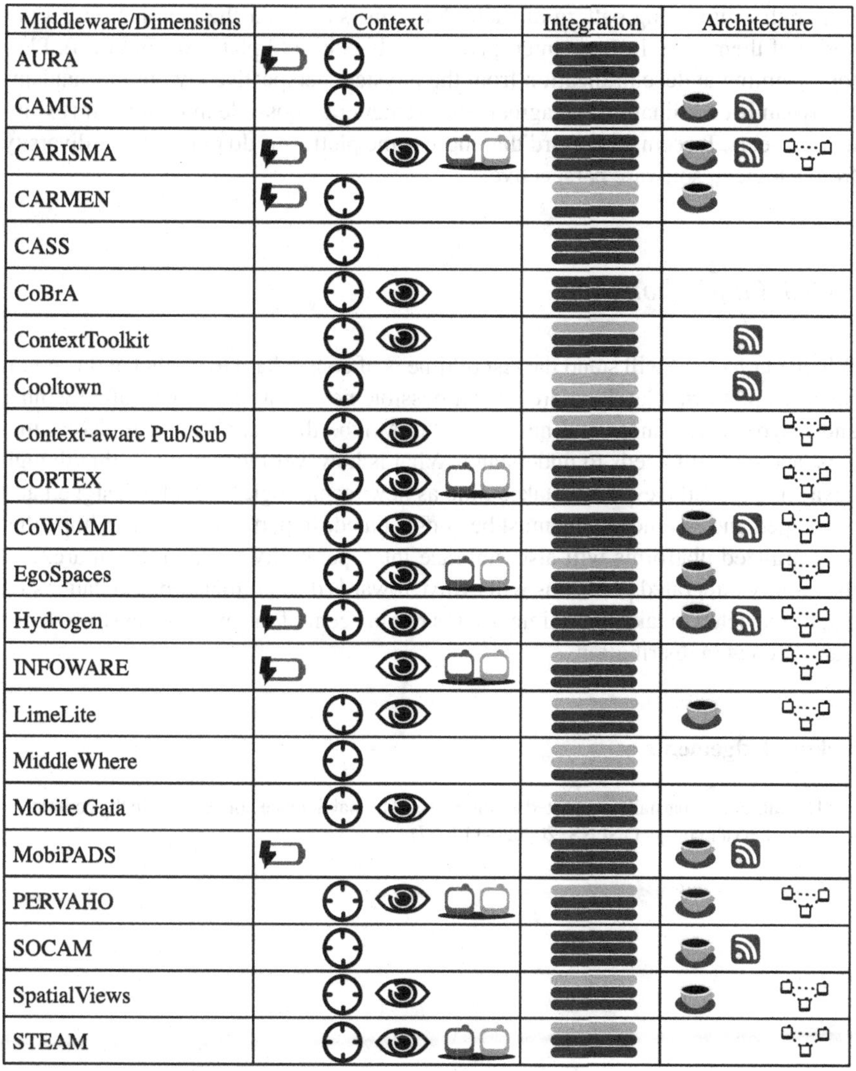

Middleware/Dimensions	Context	Integration	Architecture
AURA			
CAMUS			
CARISMA			
CARMEN			
CASS			
CoBrA			
ContextToolkit			
Cooltown			
Context-aware Pub/Sub			
CORTEX			
CoWSAMI			
EgoSpaces			
Hydrogen			
INFOWARE			
LimeLite			
MiddleWhere			
Mobile Gaia			
MobiPADS			
PERVAHO			
SOCAM			
SpatialViews			
STEAM			

Fig. 14.5 Middleware classification grid

are also decentralized (6 platforms out of 6). Platforms falling in this cluster are: *CARISMA*, *CORTEX*, *Context-aware publish/subscribe*, *CoWSAMI*, *EgoSpaces*, *Hydrogen*,
INFOWARE, *LimeLite*, *PERVAHO*, *SpatialViews*, and *STEAM*. All these platforms target applications running in a MANET setting. This explains why those platforms are built to be resilient to changes in network topology (decentralized) and tend to both leverage and facilitate peer interaction (social context). Indeed, very little can be achieved individually in MANETs.

A rather surprising pattern of such platforms is however the fact that only a minority of them is built on an interoperable architecture (3 platforms out of 8). Platforms aiming at decentralization from the physical perspective lay out mechanisms for dynamic coordination and agreement to counteract possible individualism exhibited by peers. It seems awkward that these same platforms do not cater for diversity from a logical, software perspective.

14.5.3 Final Thoughts

Which platform(s) will stand the test of time is of course hard to predict at this point. Since no platform clearly addresses all possible needs, there is no absolute winner and maybe there cannot and never will be. Maybe there are natural tradeoffs that exist and one still needs to understand. What is however more likely is that though advancing steadily, evolution takes time as new technologies must be designed and developed and advancements must be consolidated. In particular, one can hope that decentralized platforms will also embrace interoperability more in the future, and that highly integrated platforms will evolve towards decentralization. The latter step can be expected to take more time, as it is reminiscent of the everlasting struggle for transparency in distribution.

Acknowledgements

The authors are partially supported by the Swiss National Science Foundation, in the context of the Pervaho project (SFNS 200020-112057).

Chapter 15
Autoconfiguration and Service Discovery

Jukka Manner, *Helsinki University of Technology, Finland*

15.1 Introduction

To be useful, IP networking requires various parameters to be set up. A network node needs at least an IP address, routing information, and name services. In a fixed network this configuration is typically done with a centralized scheme, where a server hosts the configuration information and clients query the server with the Dynamic Host Configuration Protocol (DHCP). Companies, university campuses and even home broadband use the DHCP system to configure hosts. This signaling happens in the background, and users seldom need to think about it; only when things are not working properly, manual intervention is needed. The same protocol can be used in mobile networks, where the client device communicates with the access network provider and his DHCP service. The core information provided by DHCP includes a unique IP address for the host, the IP address of the closest IP router for routing messages to other networks, and the location of domain name servers.

DHCP was originally designed for IPv4 networks. In an IPv6 network, there is also the option to rely on router advertisements. These proactive messages from the routers tell the client devices where their first hop router is, and what is the network prefix. In addition, a recent effort proposes to add also DNS information into the router advertisements, thus replacing DHCP in most of the typical scenarios; routers effectively become small DHCP servers themselves. DHCP still has other features unavailable in the router-based configuration, for example, a DHCP server can control more efficiently the access to a network and the use of the IP address, and it can be used to load a boot image for a network node.

In addition, there is an increasing amount of work done to also design service discovery mechanisms. A client device could automatically look for services, say a printer, and available service providers would make themselves known. In essence, there are two types of service discovery methods, proactive and reactive. The former means that service providers make themselves known by periodically broadcasting a service announcement of themselves into the network, without prior contact by clients. Client nodes listen for the these announcements and gather service information in due time. For example, the IPv6 router advertisements operate just like this, messages are broadcast periodically regardless of whether or not there are clients in the network. In the latter reactive method, clients actively query for services. They either contact a well-known server that stores service availability information, or they broadcast a query about available services, and the service providers then reply.

In the mobile environment, the same schemes that were originally designed for fixed networks can also be used. Yet, one of the major challenges is whether broadcast and multicast are supported in the mobile access network. Moreover, the schemes designed for fixed networks typically do not consider the use of the link bandwidth, since in fixed access networks bandwidth is not such a limited resource. In wireless networks bandwidth is much more limited, and the fixed network schemes can therefore create a relatively large amount of messages in a mobile access network, and cause serious problems.

In other mobile and wireless environments, for example, wireless ad hoc and sensor networks, there typically are no clear centralized places for getting configuration information, as in DHCP, or for looking up services. Thus, totally new kinds of schemes are needed, methods that take into account the limited bandwidths of wireless links, and also support the much more dynamic environment of ad hoc networks.

A hybrid setting of fixed and ad hoc networks is also very realistic. Say there is a public WLAN service at an airport. IP information can be provided by a centralized DHCP server, but users can then announce and look for services using a distributed approach. A user could be looking for someone to play a game of chess with, someone in the same airport network. Or people could set up a chat session on the fly without the need for centralized servers. Thus, from a networking point of view, the scenario is clearly a fixed wireless network, but the services are available in an ad hoc (people come and go) and distributed way. So, ad hoc networks should be thought about as distributed, on the fly, networking in a broad sense, not necessarily as the more traditional view of stand-alone independent networks.

This chapter looks into how autoconfiguration works in a mobile environment, and how a client node can look for, and provide itself, services, in Sect. 15.2 and 15.3, respectively. There are many ways to implement such functionality, and it depends on the deployment scenarios whether a centralized scheme can be used, or whether a distributed protocol is the best solution.

This chapter starts with autoconfiguration, where the main challenge is configuration of unique values in the network, i.e., the configuration of unique IP addresses to hosts; all other information is shared by the nodes, and, therefore, the challenge is simply information dissemination. The chapter then continues to service discovery, where the goals are also in efficient information dissemination. With all information, there is the additional question of how valid the information is. Information can be stale, old, but it can as well be intentionally fake, for example, an attacker could announce a false service on someone's mobile device and cause other users to overload the victim with requests for a service he doesn't host; in a centralized world trust is much easier to manage.

15.2 Autoconfiguration of IP Networking

This section discusses ways a mobile client can gather the necessary information to make use of the Internet. This basic information is composed of three parts, getting an IP address, finding a gateway towards other networks, and how to map DNS names to IP addresses.

15.2.1 IP Autoconfiguration in Fixed Networks

In fixed networks, host computers typically rely on DHCP to configure the interfaces and the networking stack. When a node notices that an interface has link layer connectivity, for example, the Ethernet cable was plugged in, or a WLAN radio has connectivity to an access point, it can start searching for DHCP servers by broadcasting a DHCP query on the link layer. This does not require a configured IP address, because the receiving hosts can see the MAC address of the sending host. The DHCP architecture allows multiple DHCP servers to offer configuration information. Thus, the initial query sent by an end host can trigger multiple responses. The client can then choose which offer it wants, and commits to this offer through a second two-way exchange of messages. DHCP can be used in IPv4 and IPv6 networks.

A DHCP offer can carry a variety of information; it can even be used to indicate a boot image a dumb terminal can use to boot the device. Typically a DHCP offer provides a unique IP address, router IP address and DNS information.

In IPv4, a host can only have one IP address assigned to a host. IPv6 on the other hand was designed from the very beginning to support a concept of multiple addresses per interface. A networking device in an IPv6 network typically has two addresses configured, a *link local* address for communicating with nodes on the same link and a *global address* used for Internet-wide networking.[1] This is possible because the IPv6 address space can more easily support address prefixes and thus enable much more flexibility in the address configuration. IPv4 address space is more limited in size and does has not have this luxury, it can therefore only support one address per interface. There used to be a third scope called *site-local* addresses, but this has been deprecated some years ago.

The IPv6 specifications include RFC 2462, which defines the Stateless Address Autoconfiguration. The scheme is tied to single links and for a network, where each link has a unique prefix advertised by an IPv6 router. The basic concept is that a node chooses a 64-bit interface identifier, typically the MAC address, but any random set of bits is fine. It then adds the 64-bit link-local prefix *FE80::0* and forms a 128-bit *link-local* address. Now, this address is tentative, until the node can be sure with some high probability that no other node on the same link, or subnet, has the exact

---

[1] A "link" in IP terminology is a network segment, which does not require routing, for example, a network where nodes are connected through Ethernet; all nodes are directly reachable through the link layer without a need to route packets on the IP layer.

same address. Essentially, the question is about the interface address, since all nodes use the same prefix.

To validate the address, the node starts sending neighbour solicitation multicast messages. The interval and number of messages can be configured per link. If the node does not get any replies to the solicitations, it can be relatively sure that no other node on the link has used the same interface address. If a node receives a solicitation for the same address it uses itself, the node will reply with a neighbour advertisement. This messages tells the sender that the tentative address was being used already, and forces the sender to restart the procedure with a new tentative address, i.e., a new tentative interface identifier.

Once the node is sure that it has a unique interface identifier, it can configure routing. This is accomplished through the IP Router Advertisements. Each IPv6 router broadcasts periodically Router Advertisements, which inform nodes on the link about the availability of gateway out of the link, or subnetwork. They also indicate the network prefix the nodes can use to form a globally routable IP address. If the node does not receive the router advertisements, it can request one by sending a Router Solicitation.

Finally, once IP addresses are configured, and a node knows how to route packets, one important function is still missing, namely mapping names to IP address using DNS. There are two distinct scopes, Internet-wide and link-local name resolution. The former is achieved through an extension to the IPv6 router advertisements enabling them to carry DNS information (RFC 5006). Thus, an IPv6 Router advertisement can be used to (1) configure a globally routable IPv6 address, (2) set up routing to external networks, and (3) configure DNS lookups.

The latter second scope for DNS configuration consists of link-local name resolution. The idea is that a node can multicast a DNS query on the link, and a node holding the information will reply. There exists two major implementations of this concept, the Microsoft-driven Link-Local Multicast Name Resolution (LLMNR) defined in an informational RFC 4795, and the Multicast DNS, one of the features of the Apple Bonjour (formerly "Rendezvous"), included in Mac OS X. These link-local schemes support also IPv4.

The Stateless Address Autoconfiguration was originally developed for IPv6. It has since then also found its way into IPv4 as Dynamic Configuration of IPv4 Link-Local Addresses specified in IETF RFC 3927. The scheme uses different messages but the protocol logic is the same. The prefix used for communicating with local nodes is 169.254/16. Thus, there are only about 65000 addresses available for automatic IPv4 address configuration without DHCP. The more there are nodes on the same link, the higher the probability is that a tentative address will collide with an existing one.

Router advertisements can also be used in IPv4 networks (RFC 1256). However, IPv4 routers only tell that they are available, no prefix information is distributed. Thus, unless the IPv4 node already has a globally reachable, topologically correct, address, the availability of routers has very limited use; having a link local address only does not allow direct communication with Internet hosts.

In both versions of IP, there is the same fundamental challenge of address auto-configuration: how many messages should a node send and how long it should wait before concluding that it's chosen address is unique? This is a simple question of speed vs. reliability. For example, Bohnenkamp et al. [100] have studied this optimization question and show that at the same time it is not possible to get quick and reliable information about whether a tentative address is unique; either of these two can be achieved but not both.

Thus, as a summary, without a centralized DHCP service, an IPv6 node, with the help of the network routers, can configure all the important information for IP networks. An IPv4 node can also get IP connectivity without DHCP, but currently there is no scheme to that would provide Internet-wide name resolution (only link-local) and the router advertisements have limited use. The next section presents schemes for configuring the IP parameters in an ad hoc network. Essentially, the schemes are quite similar to the ones presented above, with only slight modifications to make the operation fit multihop networks.

15.2.2 IP Configuration in Ad Hoc Networks

Making IP-based nodes operate in an ad hoc network requires essentially the same three parameters as presented above. The very fundamental issue is how each node can configure a unique IP address for communication within the ad hoc network. When a unique IP address has been acquired, a node might want to search for a party using a DNS name, which, without special techniques, may be impossible within a stand-alone network of homogeneous, equal, nodes. If the ad hoc network has a connection to the Internet, nodes within the network would need to find a gateway router connecting to the Internet, and direct packets to this gateway. Moreover, the nodes within the ad hoc network would now need to have a globally reachable IP address, and, most probably information about DNS servers. The key difference between the techniques proposed for fixed and ad hoc networks is that all major schemes for fixed networks only consider a link-local environment, so IP packets are not routed between subnetworks. Ad hoc networks are typically multihop, so the schemes also need to consider routing.

This section discusses various ways a mobile node (MN) in an ad hoc network can set up IP connectivity parameters for communication within the ad hoc network, and towards the Internet. The specific type of the ad hoc network also affects how the nodes may be configured. Figures 15.1, 15.2, and 15.3 show different types of ad hoc network;. The Type I network is a basic stand-alone ad hoc network; there is no need to configure gateway routers, but the node could make use of DNS lookups within this limited environment. Type II networks have one or more nodes connected to the Internet through an access point (AP). Type III networks are similar to Type II, but may move relative to the fixed access network, a mobile network.

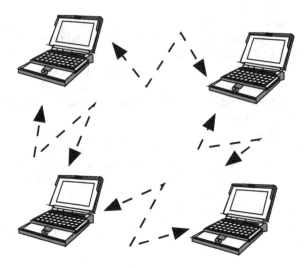

Fig. 15.1 Type I ad hoc network

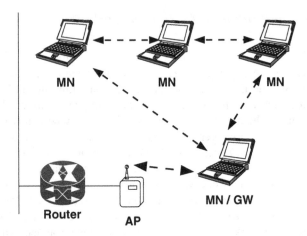

Fig. 15.2 Type II ad hoc network

Routing in Ad Hoc Networks

The IETF Mobile Ad hoc Networks (MANET) Working Group has lead the work on routing protocols for mobile and wireless ad hoc networks. Some of the most well-known ad hoc network routing protocols are AODV [664], OLSR [181], DSR [429], and TBRPF [636]. The current situation is that the working group is converging into two protocols, a reactive on-demand protocol DYMO (Dynamic MANET On-Demand Routing) and a proactive protocol, OLSRv2 (Optimized Link State Routing); the research community has developed tens of other protocols, too, so the total number of ad hoc routing protocols is probably close to a hundred nowadays.

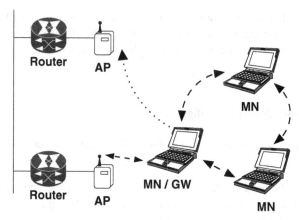

Fig. 15.3 Type III ad hoc network

Ad hoc network routing protocols try to solve the problem of routing in a highly dynamic environment. The Internet, and fixed networks in general, have very stable routes and changes in a local area are very in frequent. Mobile and wireless ad hoc networks can have very unstable topologies as nodes come and go at will, and move around. This environment is very challenging for providing a constant connectivity between nodes, and it is common that in simulations the packet loss exceeds sensible values and in practice IP-based communication is barely usable.

Ad hoc routing protocols are typically designed for multihop ad hoc networks with potentially hundreds or thousands of nodes. However, currently there are many deployment scenarios for ad hoc networks, which would not require such complex protocols, or even multihop routing at all. For example, an ad hoc network would be set up with an 802.11 wireless LAN configured in ad hoc mode. Such a network could be used, for example, in a cafe, a meeting, a lecture, or a conference session between the participants. In simple ad hoc networks, for example, when all nodes are located in the same room and can hear each other directly over a WLAN network operating in ad hoc mode, there is no need to employ a MANET-specific routing protocol. Instead, simple IP routing will be enough, as all nodes are connected to the same link and are only one hop away from each other.

There exists work on more simple protocols, for example, LUNAR [816] is a recent protocol targeted at small-scale ad hoc networks, in the range of a few tens of nodes, and a maximum of a few hops between any two nodes. The design is based on the Address Resolution Protocol (ARP), where a node looking for another sends an ARP query. This query is encapsulated into a LUNAR message and broadcast further in the network. Eventually, the answer is also encapsulated in a LUNAR message. The basic operation of the protocol is simple, but the use of broadcast implies that the network should be relatively small.

Most ad hoc routing schemes do not consider autoconfiguration at all. They expect the network nodes to have unique IP addresses, at most gateway routers can be discovered using the routing protocol. Furthermore, name resolution is seldom discussed. Yet, there exists supplementary work on ways to perform IP address configuration, DNS and gateway discovery, and IP routing. The schemes differ in their applicability to various environments, there does not exist a "one size fits all" solution.

Ad Hoc Network Address Configuration

In multihop ad hoc networks, the challenges in IP address autoconfiguration are related to

- overloading the network through flooding messages all over the network,
- unreliable nature of connectivity and node availability, e.g., a node can not rely on centralized information repositories, and
- network partitioning and merging, where two ad hoc networks having nodes with same IP addresses merge to form a larger network, leading to conflicts in addressing.

Moreover, the same consideration of speed vs. reliability discussed earlier is even more central in ad hoc networks than it is in fixed networks. If a node is eager to decide that its chosen address is unique, it leaves more chances for this decision to be incorrect, and later on lead to problems with node identification.

Auto-configuration in an ad hoc network can happen in a centralized or distributed way. A centralized scheme could rely on DHCP to allocate IP addresses and, for example, addresses of DNS servers. In a Type II ad hoc network, a DHCP server could be set up on the router connected to the GW MN. A paper by Misra et al. [585] presents a system that tries to solve address configuration, user registration and mobility all at once. The address configuration scheme is derivative of DHCP, and is based on a hierarchy of address pools. Nodes can lease IP address pools from each other and offer these addresses in turn to new nodes in the network. Address pools have priorities that are used in renumbering in case networks merge and there are address collisions, that is, the same addresses are used by multiple nodes.

A distributed auto-configuration scheme, as, for example, the IPv6 Stateless Address Autoconfiguration, or the similar scheme for IPv4 networks, may involve more messaging but can work without a centralized server. Both schemes can be used in ad hoc networks. For example, a paper by Fan and Subramani [275] present a method of using IPv6 Stateless Address Autoconfiguration in ad hoc networks. They also consider network merging by introducing a flag in the protocol messages that tell the network nodes that a merger has happened. As a consequence, the nodes restart the process of address acquisition and duplicate address detection. A paper by Weniger [863] presents also a scheme based on the IPv6 duplicate address detection. The scheme uses a hierarchy in the routing to limit the flooding. Weniger also discusses the use of the scheme with certain existing ad hoc routing protocols. Researchers in Korea present a way similar to the above [419] but tied to the AODV ad hoc routing protocol. They also take into account network mergers and seek to resolve duplicate addresses.

Recently, the IETF has set up a Ad Hoc Network Autoconfiguration (autoconf) working group to design autoconfiguration for ad hoc networks. The work seems to target only IPv6 at this stage. The work is intended to produce a problem statement, and the design for IP address and prefix configuration for IPv6-based ad hoc networks. There are also individual submissions, for example, reviews and evaluations of current autoconfiguration schemes.

Domain Name Service Configuration

Once the MN has an IP address, there may be need to use find nodes based on their DNS name. Carrying DNS information in router advertisements, as discussed earlier, would be applicable to an ad hoc network, too. In [418], the authors discuss DNS resolution within the ad hoc network using a pre-defined multicast address. All MNs run a small-scale DNS server, DNS queries are multicast to the well-known address, and the MN in charge of the name space answer to the query. The scheme can also be extended to cover external network, in that, the MN operating also as a gateway router can answer to the DNS queries by using the DNS servers within the fixed network. The authors also discuss broadcasting the DNS server information as a service within the ad hoc network, thus, making the multicast queries unnecessary, similarly than in RFC 5006, where IPv6 router advertisements carry DNS configuration information in a fixed network.

Gateways to External Networks

Routing to the Internet in an ad hoc network is quite different from fixed networks. In a LAN, a host has a default route that points to a gateway router that can route packets towards the Internet. The IP packets are sent to this router by using the link layer MAC address, usual the Ethernet MAC address; the destination IP address is that of the target hosts on the Internet. In a multihop ad hoc network, typically a host is not on the same physical link as the gateway, that is, has direct link layer connectivity to it, and thus, it can not use a MAC address for a link layer target. Instead, the ad hoc node can only figure out a good next hop from its neighbouring ad hoc nodes, which again does the same, eventually leading to the packet getting to the gateway. How the ad hoc nodes know where the gateway is, and how to reach it, is a bit of a challenge. This is even more challenging if there are multiple gateways available, and the sending ad hoc node would like to control which gateway will be used.

A very thorough design for acquiring global connectivity from an IPv6 ad hoc network can be found in an Internet Draft by Wakikawa et al. [847]. Here the gateway could advertise itself using IPv6 router advertisements. These advertisements could also carry a global prefix that the MNs could use to form a globally routable IP address (by appending the network interface address to the prefix). MNs can also solicit an advertisement from a gateway. Routing through the gateway is done using an IPv6 routing header.

An Internet Draft by Cha et al. [152] presents two different ways to route packets from an IPv6 ad hoc network MN to the gateway. The first option would be to use the IP routing header and add the gateway as the next hop in the routing. The second option would be to tunnel the data packets to the gateway, which could then decapsulate the packets, and forward then to the Internet. In both cases, the sending ad hoc node would need to know the IP address of the gateway.

15.2.3 Summary

Configuration of IP communication parameters with a centralized scheme makes life relatively easy. Yet, in a dynamic and wireless ad hoc environment, centralized schemes are not a practical solution. A node needs to rely on distributed and decentralized mechanisms. Yet, these schemes are non-trivial to implement. The challenges are in the responsiveness and reliability of the schemes, and how they handle address collisions, in both fixed and ad hoc networks. There are many proposals, each suitable to different deployment scenarios. The fact that the IETF was a working group looking at autoconfiguration in ad hoc networks gives a hint that the problem is real, and people are getting serious about it. Similarly, the use of IPv6 router advertisements to carry not only prefixes but also DNS service information tells that also in the fixed networks people are interested in looking at new solutions separately from the evolution of the centralized DHCP architecture and protocols.

15.3 Service Discovery

Service discovery is an important functionality in a dynamic environment such as ad hoc network where applications cannot rely on fixed addresses and name services may not be readily available. However migration of any service discovery framework to this kind of environment is not straightforward. Variable connectivity of ad hoc nodes may cause non-uniform or even false service information propagation especially if service directories are used. Additionally the open nature of the ad hoc network raises security concerns since they are a fairly easy target for a number of attacks

As discussed above, information required for IP networking can be seen as generic services, for example, the availability of a gateway towards external networks can be a service provided by a third party. This section starts by first going through some well-known proposals for service discovery in fixed networks. The discussion continues to investigate the topic by looking at proposals that focus on wireless ad hoc networks. There are a number of different ways to tackle distributed service discovery and how information is stored in the network, for example

- fully distributed schemes, where every node gets to know everything,
- a network of directories or a distributed overlay network that holds the service information, or
- coupling service discovery with the particular routing protocol deployed in the ad hoc network.

The fully distributed scheme can be based on push-style information dissemination, or based on pulling information out of network when needed. Furthermore, it is possible to use some kind of controlled dissemination, where information about a certain service is not stored everywhere in the network, but at one, or a handful, of nodes. Practically all the service discovery schemes designed for ad hoc networks propose some particular way to store service availability information in the network, to ease the discovery process.

As for security, the service discovery protocol needs ways to confirm the requester and the service provider. Only authorized requesters are allowed to receive an answer to a query and the requester needs some guarantees that the service is valid.

15.3.1 Service Discovery in Fixed Networks

This section covers the most common service discovery protocols today. Brief overviews of some major schemes are given, for example, SLP, Jini, UPnP, WS-Discovery, and XSDF are presented. All of these use broadcast or multicast to provide service information. Yet, the schemes do not perfectly fit ad hoc networks, and ad hoc specific solutions will be presented later in this chapter.

Service Location Protocol

Service Location Protocol (SLP) [350, 663, 448, 449, 612, 346, 447, 347, 348, 906, 907] is one of the oldest frameworks for service discovery. It has been developed by the IETF SLP Working Group as system that can perform decentralized operation in small, unadministered networks, but also scales to large enterprise scenarios. It enables policies dictating who can discover which resources.

SLP is based on three entities that perform service discovery functions:

- User Agents (UA) are the client devices performing service discovery.
- Service Agents (SA) host and advertise the location and attributes of the available services.
- Directory Agents (DA) store and distribute service information to the clients, these are like the Yellow pages in a telephone book.

SLP has two mechanisms for SAs and UAs to discover DAs, passive and active discovery. In passive discovery, SAs and UAs listen for periodically repeated multicast advertisements from DAs. In active discovery, SAs and UAs use multicast to

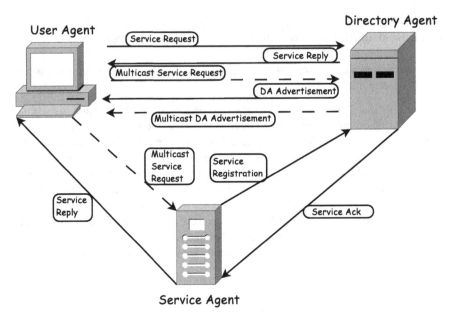

Fig. 15.4 SLP agents and core protocol messages

send SLP requests, or they can use DHCP to find DAs. When a DA is found, SAs use unicast communication to register their services to the DA and UAs use the same method to search for a service from the DA [319]. DAs are not necessarily required for the protocol to work, they can be seen as an optimization in certain deployments; their main purpose is to reduce the amount of multicast messaging in the network. DAs are designed to be very flexible in what comes to the different service types. They have no real understanding of the services they advertise; implementing new service types requires only to create new service templates of these three agents are shown in Fig. 15.4, dashed arrows are multicast messages, solid arrows are unicast messaging. UAs can send multicast service request and DAs can send multicast DA advertisements. All other messages are triggered by the reception of these multicast and sent over unicast directly to a defined recipient.

In SLP all services are advertised through a service URL, which contains all information necessary to contact the service. The services can be searched by the service type and the attributes linked to that service. Each service request has to contain a scope identifier. This mechanism enables forcing of certain service discovery policies. This can mean for example advertisement of certain services only to some group of UAs. SLP does not include any specific autoconfiguration mechanism, apart from the service queries and DA advertisements sent over multicast, but it can rely on DHCP in the network [661]. SLP works with both IPv4 and IPv6; the required modifications for IPv6 are specified in the IETF RFC 3111 [347].

SLP is very much demand-driven and suits well to active service discovery in which clients query for suitable services. However, it does not support advertisement of the services themselves, it only supports propagation of the availability of DAs. However, RFC 3082 [447] specifies experimental notification and subscription mechanisms that enables passive dissemination of SA availability over multicast, that is, SAs tell UAs when they appear in the network and when they are about to shutdown.

eXtensible Service Discovery Framework

The eXtensible Service Discovery Framework (XSDF) [827, 829] is an evolution of SLP. It was designed for dynamic mobile environments, where clients roam between multiple domains and want to look for services. Total of seven Internet Drafts were released in 2004 but seems like to work ceased there. XSDF introduces many improvements over the SLP protocol and as such was an interesting undertaking. Most important features of XSDF are:

- **Enhanced Service Model.** Services are not identified by URL as in SLP but by Universally Unique IDs (UUIDs).
- **Internet-wide location** SLP was designed for local area networks within a single administrative domain. XSDF is designed to be used for Internet-wide service discovery.
- **Load balancing and high-availability.** For the purpose of becoming Internet-wide solution for service discovery, XSDF also incorporates new features which can be used to select the best service and distribute data of registered services.

Like SLP, the XSDF architecture consists of User Agents (UA), Service Agents (SA), and Directory Agents (DA). XSDF does not change their fundamental properties, it just extends their functionality and tries to achieve architectural improvements. These *XSDF Agents* communicate with four protocols [827]:

- **eXtensible Service Location Protocol (XSLP).** XSLP is the protocol used in the actual service discovery. UAs use it to find services from DAs or directly from SAs [830].
- **eXtensible Service Register Protocol (XSRP).** XSRP is used by the SAs to register and refresh their service information to DAs [825].
- **eXtensible Service Subscription Protocol (XSSP).** XSSP is used by XSDF Agents to subscribe to get notifications of changes in service information. Typically used between DAs [826].
- **eXtensible Service Transfer Protocol (XSTP).** XSTP is used to by DAs to fetch and synchronize service information from each other, in order to share a common view of service availability to UAs [828].

Jini

The Jini architecture [787, 788] was designed by Sun Microsystems as a distributed system consisting of many different devices connected with a network to form a

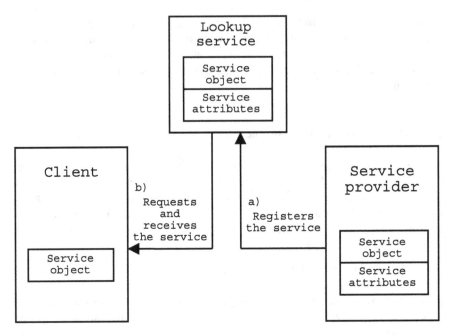

Fig. 15.5 Jini operation [319]

system like one large computer. Fundamentally, Jini is a distribution framework build on top of Java and Java/RMI. Main advantage of Jini is the support for very dynamic and flexible distributed systems, in which participating devices (hosts) can join and leave without any need of administrative work.

Jini is targeted for relatively small distributed systems, one Jini system could serve, for example, one workgroup. The devices federating in the system are also expected to be connected in a local area network; yet, it is possible to deploy Jini in larger systems over multiple hops.

A Jini system consists of three main types of components: *clients*, *service providers* and *lookup services*. These are shown in Fig. 15.5. Service providers implement and offer services (same as SAs in SLP). Service in the Jini framework is an entity that can be used by a person or a program, for example, a computation, storage, a communication channel, a filter, a hardware device, or even another user. For example, printing a document could be a service. Services can form hierarchies of services by using other services, similarly to object-oriented programming and interactions between classes. Services communicate with each other using a *service protocol*, which typically (but not necessarily always) runs on top of Java/RMI.

Lookup services act as directories (similar to DAs in SLP) in which services are registered and from which they can be found by clients or other services. Service providers find lookup services with a *discovery* protocol. Services are added to a lookup service with a *join* protocol. Clients use a different protocol, *lookup*, when they are searching for services.

The actual service discovery needed to find lookup services is based on multicast requests and announcements. Lookup services advertise themselves with announcements, or they can be also found with multicast requests. The idea behind having both is that when a new device is plugged into a Jini system, it first actively searches for lookup services with multicast requests. After a while, it ends the active search and switches to passive mode, only listening to possible advertisements of lookup services; this mode of operation is the same as in SLP.

Despite of their many functions, lookup services are not an absolute necessity in a Jini system. The Jini specifications include also a technique called *peer lookup*, which can be used to find and use services if no lookup services are available.

UPnP

UPnP is a service advertisement and discovery architecture supported by the UPnP forum [824] headed by Microsoft. UPnP standardizes the protocols used by the devices in an XML-based format. In addition to service advertisement and discovery, the UPnP specification also describes device addressing, device control, eventing, and presentation. Eventing allows clients to observe for changes in the discovered service. Presentation allows a client to obtain a generic user interface to a discovered service.

UPnP uses Simple Service Discovery Protocol (SSDP) for service discovery. It is similar to SLP, but it does not allow searching for services by their attributes [363]. SSDP does not allow currently discovery outside a single subnet [319]. Services features and capabilities are described by XML description files. The usage, invocation, of a service is expressed as Simple Object Access Protocol (SOAP) objects and their URLs in the XML file. UPnP doesn't use any kind of registry (compared to DAs of SLP) and therefore has only two basic entities Control point, and service hosted by a device.

The Control point discovers services by multicast messages. The device hosting the service responds with a URL that points to its XML description document. It gives the control point all the information required to use the service, including the messages that form the control protocol to communicate with the device. This interaction is described in Fig. 15.6. Under one Service there can be found multiple subdevices which can be controlled independently. Eventing functionality is based on the Generic Event Notification Architecture (GENA). It is a subscription-based event notification service based on HTTP. The XML description document includes four important items:

- A presentation URL allows access to device's root page, which provides a GUI for device control.
- A control URL entry points to device's control server, which accepts device-specific commands to control the device.
- An event subscription URL can be used by control points to subscribe to the device's event service.

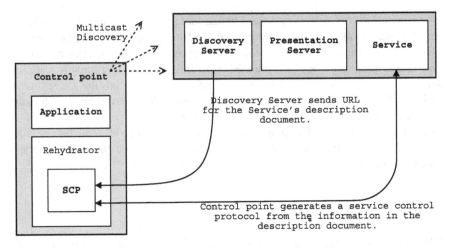

Fig. 15.6 UPnP entities and some service discovery messages [319]

- A service control protocol (SCP) definition describes the protocol which can be used to interact with the device.

Web Services Dynamic Discovery (WS-Discovery)

WS-Discovery is a multicast discovery protocol for locating Web services. It allows discovery of services in local networks that have minimum of networking services (e.g., DNS or directory services) [568].

The basic architecture of WS-Discovery has two entities: client and target service. The clients can search for type of the target service, scope of the target service (functionality, limitations, etc.), or both, by sending probe message to a multicast group. Target services that match the probe send responses directly to the client. Clients can also find services by name. This is can be done by sending a resolution request message to the same multicast group. Again the target service that matches sends a direct response to the client. When a target service joins the network it sends an announcement message to inform all the clients of its presence. This minimizes the amount of unnecessary probing by the clients because they can now only listen to the multicast group to be notified of new services. When a target service leaves the network it sends a bye message to the multicast group [568].

There is also an alternative mode for the protocol that is aimed to minimize the multicast messaging. This mode requires presence of a new entity called discovery proxy (a similar node than the SLP DA). It can work as a messenger between clients and target services. It also sends announcement messages of itself when it detects a probe or a resolution request sent by multicast. The WS-Discovery protocol specification however does not describe the discovery protocol in detail and leaves it

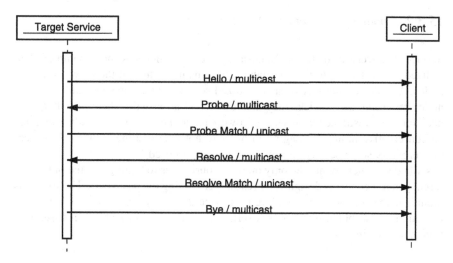

Fig. 15.7 WS-Discovery functions [568]

pretty much open. Endpoints can implement more than one of client, target service or discovery proxy entities [568].

All the messages in the WS-Discovery protocol are defined with XML descriptions in the Web services manner. Detailed descriptions of these messages can be found in the WS-Discovery specification. Messages that are sent over UDP are sent using SOAP over UDP [568]; TCP is a reliable byte streams and does not need special considerations. Messages sent by the client and target service are shown in Fig. 15.7.

Summary

Service discovery in fixed networks is mostly based on multicast. This provides some support for dynamic environments, where clients and services come and go. Multicast has also downsides, for example, messages are sent all over the network to whoever is subscribed to a particular multicast group, and currently multicast can only be used in local networks, the Internet at large does not support multicast. The focus on multicast has clearly affected the target of the schemes, since they are all, except perhaps for the XSDF undertaking, focusing on local services. The hugely popular peer-to-peer networks could also be used for service discovery: instead of looking for files, a user could also look for services. This would bring the whole Internet and its services to the users, not just the local network services. In fact, there exists some work on making use of peer-to-peer technology for service discovery. These concepts will surely lead to some great new ideas for the fixed Internet, but the dynamic and wireless mobile ad hoc networks still need other angles, as discussed in the next section.

15.3.2 Service Discovery in Ad Hoc Networks

Service discovery in distributed and ad hoc environments is more complex than in fixed networks due to the unreliable and dynamic nature of the network and its nodes. A node can not rely on a centralized server to be always available, and the nodes providing the services may also come and go at will. Therefore, there is a need for more resilience in service discovery, but at the same time the solution must be conservative in any messaging in order to not overload the network. This section presents some service discovery schemes proposed for ad hoc networks; the described schemes are not the only ones, the purpose is to highlight some of the different general concepts designed by researchers during the past years. The solutions include extensions to SLP, coupling of service discovery and routing to minimize messaging, distributed directories, overlay networks, and probabilistic dissemination of service information.

Bluetooth

Bluetooth technology (including its Service Discovery Protocol or SDP) is supervised by the **Bluetooth Special Interest Group (SIG)** [98]. Bluetooth SDP is a protocol designed to solve service discovery problem between Bluetooth enabled devices. It does not provide access to services, brokering of services, service advertisements, or service registration and there is no event notification either. SDP supports searching by service class, search by service attributes and service browsing [363]. Methods of invoking the found services are outside the scope of SDP [82]. The SDP API also supports stop rules that limit the duration of searches or the number of devices returned [319]. Devices are identified by Universally Unique Identifiers (UUID) which are generated once at the time a service is generated. They are used to avoid service definition collisions.

Bluetooth SDP is basically a client-server protocol. The server maintains a list of service records that describe the characteristics of services associated with the server. Each record contains information of a single service. Clients can retrieve information from service record by sending a SDP request. A client has to open a separate connection to actually use the service since SDP does not provide any mechanisms to do this. One Bluetooth device may contain exactly one server. A single device can operate as a server and as a client at the same time [754].

A service record maintained by the SDP server consists entirely of a list of service attributes. Each service attribute describes a single characteristic of a service. Service attributes consist of two components: an attribute ID and an attribute value. Each service is an instance of a service class. The service class definition provides the definitions of all the service's attributes. A service record contains service attributes that are specific to a service class and some attributes that are common to all services. Service searching is based on UUIDs and service browsing is based on attributes of the services [754]. The basic operation of SDP is described in Fig. 15.8.

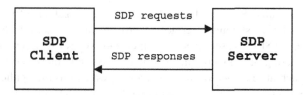

Fig. 15.8 Bluetooth SDP functions [754]

Distributed Service Discovery

One of the earliest works in this area was conducted by IBM Research in the DEAPspace project [631]. The solution used a push model where all devices hold a list of all known services. Periodically each device broadcasts its view of the services to it's neighbours, which update their own cache.

The SESSI project investigated, among other topics, service discovery in ad hoc networks. The solution was based on SLP and introduced several extensions to the standard protocol; all of these were implemented on Linux-based mobile devices. First of all, to make SLP suitable for the highly dynamic target environment, the researchers decentralized it by disabling all DA functionality; SAs communicate directly with UAs. Secondly, SLP was enhanced with support for two-way authentication and authorization as well as integrity protection and protection against replay attacks by adding signatures with logical timestamps to all messages. Furthermore, the design also enabled confidentiality protection of the messages by applying combined symmetric and asymmetric encryption to their contents [439].

Further extensions to the networking scenarios included a service discovery proxy on a gateway connecting ad hoc networks to the Internet. The proxy enables service discovery between the two networks. The SLP scopes were used to define whether a node in the ad hoc network is interested in local services or services in the access network or even Internet. Also a new operation mode, passive discovery of SAs, that is especially suitable for ad hoc networks, was further developed from the experimental extension discussed in RFC 3082 and analyzed in the scope of an ad hoc network. It allows nodes to discover services by listening to specific advertisements. The aim was to minimize network transmissions and conserve power. Passive monitoring is much better suited to dynamic networks than constant requests sent over multicast and replies coming back over unicast; request-reply suits well inactive networks.

The set of available services is bound to change constantly in a highly dynamic network. Therefore the use of traditional UA initiated service discovery model may result in a constantly high traffic especially in multihop networks. The assumption was that in an ad hoc network mobile nodes are more likely to search for services than to provide them. As a solution, in the new SLP operating mode service providers are able to advertise their services instead of just responding to service requests. This enables on average more efficient bandwidth usage as well as prolonged battery life for most nodes compared to the traditional service discovery model. This new model was called Passive SLP, and enables UAs to gather information about services by passively listening to service advertisements broadcast by SAs. The application using the SLP UA services may indicate the type of services it is interested in to enable filtering of the incoming advertisements. Only those service advertisements that match at least one of the filters are forwarded to the application [552]. Rather than being a substitute to the traditional "active" discovery described earlier, the new method is intended to augment it. UA could first get a quick overview of all available services in the network by using the traditional active discovery method and then keep track which services are emerging and disappearing by listening to the service advertisements.

Lenders at al. [506] try to find an analogy in physics and electric fields, and service discovery. In their vision, nodes broadcast their service information to all nodes in the network. Their ideas are based on the expectation that there may be multiple nodes hosting the same service. Thus, when a node requests a certain service, the request is sent to the provider that is closest to the client; "closest" here is a combination of physical proximity and *capacity of service*. If there is only one provider for a certain service, the system performs a simple shortest path forwarding.

Yang et al. [879] discuss a similar concept of full distribution and pull-based queries as the works above, but they propose a subtle difference. Instead of sending a service reply directly to a client, the reply is sent as multicast. This has the benefit that other nodes in the network also get to know about the service, and can cache the information. If these nodes later also want to use the service, they already have information about it and don't need to send a service query. Services are described in XML, which creates verbose documents high in size, thus, not very optimal from the bandwidth consumption point of view.

Service Discovery and Routing

While a certain amount of research looks at service discovery decoupled from the routing protocols, other researchers have investigated the interworking of these two functionality. The benefit is that by combining route lookup and service discovery in multihop networks, the solution can optimize the amount of signaling messages in the network, and as a result send less bytes and consume less energy. The drawback is that such a system ties together service discovery and the specific routing solution, and makes the solution less flexible, for example, from deployment perspective. Also, coupling the routing and service discovery is only useful in large multihop networks; small networks do not necessarily require a dedicated ad hoc routing protocol, for example, in an 802.11 WLAN network in ad hoc mode all nodes are listening on the same radio channel and hear each other directly.

Fan and Ho [274] have studied SLP in multihop ad hoc networks. They integrate SLP with the reactive AODV routing protocol and simulate using ns2 various scenarios. They also discuss how Quality of Service information could be used in the service discovery and how to choose the best service among multiple offerings. A similar work with an implementation of SLP and AODV for PDAs was presented by Kettunen et al. [452]. Engelstad et al. studied the same topic using simulations with GloMoSim [253].

Service advertisements and queries within proactive routing was studied by Jodra et al. [426]. They analyze both push and pull style service discovery carried by a Service Discovery Message in an ad hoc network running the proactive OLSR routing protocol. A similar proposal was studied by Li and Lamount [513]. All these papers show great benefits in merging routing and service discovery; the cost is higher complexity of the protocol architecture and loss of modularity when routing and service discovery become one.

A good overview of service discovery and multihop routing was presented by Hoebeke et al. [372]. They analyzed different routing protocols, proactive and reactive, and how service discovery should be combined with them. They conducted simulations studies and analytical analysis, and came to the conclusion that there is no simple answer as to which protocol is based, the deployment environment dictates in great deal which approach is most useful.

Directory-Based Service Discovery

A very different method for service discovery is the use of a distributed directory. The idea is that with an overlay network of directories, searching for services will result in less traffic, compared to fully distributed schemes. Key issues in this type of schemes is the algorithms and protocols employed for setting and managing the reconfiguration of the nodes managing the distributed directory, for example, when nodes arrive and disappear.

Sailhan and Issarny [723] discuss in their paper one such scheme. Their main target is a service-oriented architecture based on Web services. Services are described in XML and WSDL, and carry additional Quality of Service information for further finetuning the use of various services. A key concept in their work was the use of Bloom filters as an efficient way to describe the services; this enables hashing the service descriptions and minimize the overall traffic. Similar concepts about directory-based service discovery was presented by Juszcxyk et al. [436]. The benefits of directories depend heavily on the environment, in some environments the solutions work well, in others they may lead to increased signaling compared to other approaches.

Overlay-Based Service Discovery

Klein and König-Ries [465] discuss a concept of clustering the service availability information. Their idea is that services are organized into multilayer clusters based on their physical and semantic proximity. Physical proximity means that a node is within the radio coverage of the querying node, and semantic proximity is measured by how close the service descriptions are in some common ontology. When a client sends a query, the request is forwarded between clusters in order to find a good service provider.

Yoon et al. [887] propose the use of a distributed hash table (DHT) and a peer-to-peer overlay network for storing service information in an ad hoc network. They employ a two-phase approach where the nodes physically closer, one hop away, within the radio coverage, are preferred. If no such node exists, the DHT is used to find the service. The challenge in this work is in the combination of DHTs and the physical topology of the network, such that the search operation does not cause excessive signaling all over the ad hoc network. A somewhat similar concept was presented by Kozat and Tassiulas [475]. The idea is also about setting an overlay network for service discovery. The authors use the term *virtual backbone* and present the formation of a virtual backbone and how services registered, queried and replied.

Probabilistic Dissemination

Instead of disseminating all service availability to the entire network, a node can employ algorithms that seek to store the data in a given subset of the nodes. The algorithms are evaluated based on how well they spread the information into the network. "Well" here means that the information, the data items, should be evenly distributed in the network with a minimal number of messages.

The data dissemination algorithm should balance the need to provide data replication (to cope with failures) with the need to avoid excessive data redundancy (as nodes may have limited storage capability). Furthermore, data items should be distributed as evenly as possible among all the nodes forming the network, avoiding clustering of information in sub-areas; even dissemination of data items should leverage lower access latency to any item from any node in the network. An even distribution of information in the network implies that whenever a data item is requested by a node S, the distance to the node that provides the reply is approximately the same, regardless of the location of S. Naturally, the actual distance depends on multiple parameters, such as the number of nodes in the network, the size of the cache where data items are stored, and the number of data items. From a latency point of view, a solution should aim at minimizing distance (i.e., ideally, any data item should be available from one of the 1-hop neighbours of S). Finally, since in wireless ad hoc networks both bandwidth and battery power are precious resources, the algorithm should also minimize the amount of signaling data.

PCache [582] is one algorithm for information management in ad hoc networks, including data dissemination and efficient caching. The algorithm provides two separate operations: dissemination and retrieval of cached data items. The implementation of these operations is orchestrated such that a limited number of messages is required for retrieving any data item from the network, independently of the addition, removal or movement of nodes. This goal is achieved by a combination of four different complementary mechanisms: an efficient best-effort probabilistic broadcast mechanism, a distributed algorithm for deciding which nodes replicate a given data item, a data shuffling mechanism to improve the distribution of data replicas, and an expanded ring-search mechanism to support queries.

Each node in a PCache system has a cache of a limited and predefined size. The cache is used to store a fraction of all the data items advertised. Each data item is composed of a key, a value, an expiration time and a version number with application dependent semantics. Nodes continuously pursue a better distribution of the items, by varying the content of their caches. The goal of PCache is to provide an adequate distribution of data items so that each node is able to find a significant proportion of the total items in its cache or in the cache of the neighbours within its transmission range. Items in PCache are distinguished into those owned (that is, advertised) by a node, and so called complementary items, which are items held by a node, but not owned (that is, items advertised by other nodes).

PCache is a reactive protocol in the sense that it only generates packets to satisfy the requests of applications. Data dissemination and data retrieval operations are implemented using three types of messages. In the dissemination process, nodes cooperate to provide an adequate distribution of the replicas of new or updated versions of data items. Dissemination messages are broadcast following a probabilistic algorithm. The retrieval process is triggered by applications requesting from PCache the value associated with a key. The protocol first verifies if the value is stored in its local cache and if it is not, it broadcasts query messages. Nodes having in their cache the corresponding value address a reply message to the source of the query. Data Gathering messages are a particular type of messages used to perform queries for items satisfying a particular set of conditions, unlike query messages that are used to retrieve the value of an item indicated by its key as search parameter.

Security Issues Specific to Ad Hoc Networks

One important and very difficult challenge is the authentication of parties involved in the service discovery, and the confidentiality of the service signaling. Even if the content of a service is public information, a user still might want to confirm the source of the information and make sure the content is valid. For example, a news service or announcements at an airport are public information, but users probably want to be sure that the source is truly the news company or the airport authority, and not someone fooling around on purpose, or even trying to send unsolicited spam.

The issue here is about having the right certificates and encryption keys at the receiver to verity or decrypt some information. In ad hoc networks, there are basically three options for getting this keying material:

1. Keys are shared between parties, downloaded before entering an ad hoc network. This is simple, but the nodes are limited to the keying material previously have stored.
2. The nodes can use a chain-of-trust in the ad hoc network to get the missing keys. This requires that some keys are stored that verify certain nodes and services, and the node can then use this secure channel to query the know entities for the

missing material. This concept only works if the node can trust the entities whose keys it has stored.

3. A node ca use an alternative connection to the Internet, and use fixed network servers to aid in the authentication. This can be done by the clients, for example, they are in a WLAN-based ad hoc network, and use a 3G connection to the Internet for the authentication, or the service provider in the ad hoc network could have a secondary Internet connection for authentication of the clients.

The first two options were investigated by the SESSI project mentioned earlier. Service and user authentication was achieved through pre-shared keys, but if some keys are missing, the nodes can use a chain-of-trust implemented using the SIP protocol to exchange keying material. The length of the chain is in theory unlimited, i.e., a node receiving a query for keying material from a trusted node can send the query further to its trusted nodes. Zhu et al. [910] have investigated the third option. Their work studies both mechanisms, service- and client-driven authentication using an alternative connection to the Internet.

Almenarez and Campo [17] have also studied the security of service discovery. Their target environment is a simple single hop ad hoc network and services are advertised by broadcast, i.e., using push for information dissemination. Services have different trust degrees, and a client selects the service with the biggest trust degree. The network nodes build trust using past experience and nodes can send each other their recommendations. The scheme includes a separate request-reply protocol for this exchange of trust information.

Summary

Service discovery for wireless and mobile ad hoc networks has received somewhat more attention than IP autoconfiguration. First, this is clear from the larger set of research papers one can find. Secondly, the range of concepts and ideas used is wider. IP autoconfiguration seems to circle around the fundamental concepts used in IPv6 networking. Service discovery schemes are much more interesting and range from simple multicast-based schemes to overlay networks, coupling with routing to probabilistic dissemination. Many of the proposals are interesting on a conceptual level and look good in papers, but actual implementation and deployment will give surprises.

15.4 Summary and Outlook

IP autoconfiguration and service discovery in wireless environments is largely affected by the number and size of messages. Any solution must be conservative in how many messages are sent at various times, for example, disseminating various information, finding it, or building a structure about the information (some overlay network). The deployment scenario is also very meaningful, since a simple one hop ad hoc network does not need a complex system: if a node sends a packet over a radio link, all nodes in the vicinity hear it, regardless of whether it was meant as unicast, broadcast or multicast, thus, the messaging could as well be broadcast from the start.

All the different solutions presented above are good in some scenario, but fail in others; that is typical in research, "one size fits all" solutions very seldom exist. The solutions also differ in their applicability to Internet-wide communications, say, if the ad hoc network is connected to the Internet, how would a given proposal operate then? Moreover, some of the proposals have been analyzed using only analytical methods or simulations; these research methods are relatively easy to use but lead to simplified networking scenarios and the results on real-life testbeds can be very different. This is one of the major challenges in the research on ad hoc wireless networking: how to experiment with the protocols in a real ad hoc network composed of hundreds or even thousands of nodes? The various proposals presented have given some insight to autoconfiguration and service discovery, but can they really be deployed, it is impossible to say.

Chapter 16
Dynamic Resource Management and Cyber Foraging

Jari Porras, *Lappeenranta University of Technology, Finland*
Oriana Riva, *ETH Zürich, Switzerland*
Mads Darø Kristensen, *University of Aarhus, Denmark*

16.1 Introduction

Mobile devices such as PDAs and mobile phones are rapidly advancing to become full-fledged personal computing devices. In particular, besides supporting phone calls, mobile phones nowadays provide storage, computing, communication, and multimedia capabilities thus to be considered the primary personal computing devices of the future [68]. However, although relatively powerful, mobile devices will always be constrained in terms of physical size, thus leading to limitations in their computing and communication capabilities, battery lifetime as well as screen and keyboard size. These constraints inhibit mobile devices from fully supporting increasingly demanding mobile applications. Furthermore, although processing capabilities have followed Moore's law for the last 30 years, the more critical resource on mobile devices is battery energy density, which has shown the slowest trend in mobile computing [651].

Current trends in the field of mobile and ubiquitous computing, such as advances in sensor technology, wireless sensor networks, and mobile ad hoc networks, enable and promote the usage of networked resources to augment resource-constrained mobile devices. According to the ubiquitous computing vision, embedding computation into the surrounding environment enables people to exploit available computing capabilities in an unobtrusive manner, so that ubiquitous computing systems ultimately become an invisible technology and interactions with computers become natural [858]. Computing utilities of such ubiquitous environments, often called smart spaces, include traditional desktop devices, wireless mobile devices, digital assistants, game devices, wrist watches, clothing, sensors, RFIDs, cars, consumer electronics (e.g., TV, microwave), etc. Mobile devices entering smart spaces probe their surroundings to look for devices offering resources such as processors and storage repositories, or input/output devices (e.g., displays, microphones, and video cameras). They opportunistically use such resources and, every time any of such devices becomes unavailable or new ones appear, they adapt accordingly. This process is usually referred to as *cyber foraging* defined as "living off the land" [729].

Cyber foraging is not the only possible approach to accomplish dynamic resource management. Both parallel and distributed computing systems have used similar approaches. Parallel computing environments take advantage of several processing elements by partitioning the computational problem and executing multiple parts simultaneously. In traditional parallel computing, the execution environment is usually fixed although dynamic processes such as dynamic scheduling, load balancing, and process migration are commonly used to make efficient use of parallel resources. Distributed or grid computing represents a special type of parallel computing which assumes the availability of actual machines with CPU, memory storage, power, network interface, etc. connected to a wired network, instead of multiprocessors connected to a single computer bus. Distributed systems aim to unify multiple system resources (e.g., servers, storage systems, and networks) into a single large system and hide the distributed nature of the environment to the end user. In contrast to traditional parallel and distributed approaches, cyber foraging primarily targets wireless-enabled mobile devices operating in dynamic mobile environments.

This chapter starts with a description of example scenarios where cyber foraging can be applied. It then gives an overview of the cyber foraging process with a special focus on the challenges that arise in every step of the process. The discussion continues with an overview of some existing prototype systems that support cyber foraging.

16.2 Scenarios

Cyber foraging can be applied in several mobile computing scenarios. This section outlines three examples with a special focus on the type of problems cyber foraging can solve.

Wearable computing. In wearable computing systems, a common goal is usually to minimize the size of the computing equipment while retaining the necessary device functionality. An example scenario, taken from [480], is a doctor wearing a small microphone while doing home visits. Using the microphone the doctor is able to enter information about his patients into an electronic journal. To enable this functionality the microphone must be able to do speech recognition and send the recognized sentences over a secure connection to some central server. While the latter is indeed possible the former of these operations is not—a mobile device of such small dimensions will simply not have the raw processing power to perform proper speech recognition. On the other hand, if a more powerful computing device is currently available in its vicinity, the task of speech recognition can be forwarded to such a machine. If no powerful device is currently available, the microphone will store the recorded sound for later processing, when a computing server becomes available.

Image processing. During crowded events such as a political convention or a football match at the Olympic stadium, policemen can use mobile cameras to identify suspicious entities. Face recognition algorithms usually rely on relatively powerful computing capabilities and on the availability of large databases of facial images. Policemen's cameras collect photos and can directly process part of the collected images through a verification application where an algorithm verifies that a certain face corresponds to a claimed identity by using a locally available database of a small number of mugshots. Alternatively, in the presence of powerful computing servers, photos of unverified identities can be processed by more advanced algorithms that are capable of identifying unknown faces by relying on much larger mugshot databases.

Region monitoring. After an earthquake it is necessary to monitor the state of precarious buildings in the immediate proximity of a disaster area, so as to help with the rescue operations. Using the sensors of computing devices situated within the disaster area, a monitoring client can carry out visual or sensorial analysis of buildings and streets in the affected region. A mobile ad hoc network can be formed on-the-fly using all available devices and observations can be provided to the remote monitoring client. In spite of device mobility and battery exhaustion of the devices hosting the monitoring task, the remote client would like to receive a continuous stream of observations. Consequently, after a certain device has left the region of observation or has failed, a new device capable of hosting the monitoring task should be quickly discovered and the task migrated to the new device.

These usage scenarios demonstrate how cyber foraging can be useful in many different situations and for different purposes. In the first scenario, cyber foraging permits augmenting the limited speech recognition capabilities of a small wearable microphone with the processing capabilities of a powerful server. In the second scenario, it allows policemen equipped with small portable camera devices to carry out advanced face recognition by occasionally interacting with larger databases of mugshots and the available infrastructure. Finally, the remote monitoring client presented in the last scenario is able to collect observations from the immediate proximity of a disaster area by delegating the monitoring task to a network of collaborating devices currently located in the disaster area.

16.3 The Cyber Foraging Process

Cyber foraging aims to dynamically augment the capabilities of client applications running on resource-constrained mobile devices. Cyber foraging can be defined as *the opportunistic use of resources and services provided by computing devices available in the surrounding environment.* These nearby computing devices are generally referred to as *surrogates*.

As Fig. 16.1 illustrates, to fully accomplish cyber foraging a multi-step process needs to successfully take place. First of all, client applications that want to delegate computing tasks to external resources must *discover* surrogates available in their vicinity. Once the surrogate discovery has completed, the application needs to

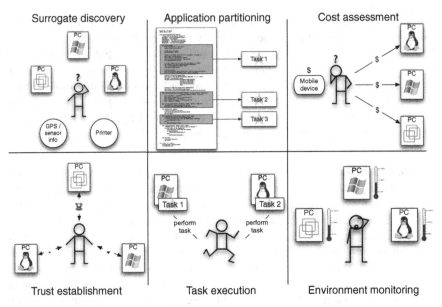

Fig. 16.1 The cyber foraging process in six steps

be *partitioned* into locally executable tasks and remotely executable tasks that may
be assigned to surrogates. The best *execution strategy* specifying "where" to execute "which" tasks is determined by trading off cost and performance of the cyber
foraging process if performed in the current execution environment. In addition, an
important requirement of many applications and surrogates is that before cyber foraging can occur a *trust* relationship must be established among the two parties, for
instance, to prevent malicious code from harming the surrogate and vice versa. Once
all prerequisites are guaranteed, cyber foraging can occur. The actual *execution* can
take different forms. For example, application tasks can be pre-installed on surrogates or can be migrated on-demand. The communication paradigm between client
and surrogates (e.g., message-based, publish/subscribe, client-server paradigm) can
also vary a lot depending on the execution environment and the type of task. Finally,
as surrogates available in mobile environments can quickly appear and disappear
as well as change their offered resources, clients must constantly *monitor* changes
of their execution environment and adapt the cyber foraging process accordingly.
Consequently, client applications that rely on cyber foraging are also required to
integrate a certain degree of adaptability and be able to adjust their execution to the
current level of resource availability.

Depending on the execution conditions, all or some steps of the cyber foraging
process may need to be repeated multiple times. For instance, when a more powerful surrogate appears in the environment the cyber foraging tasks can be moved
from the old surrogate to the new one. This implies that the steps of application
partitioning, cost assessment, and trust establishment must be re-executed. Another
possibility is that at some point a currently employed surrogate becomes overloaded

thus decreasing the quality of the provided services. In this case, a new application partitioning strategy can move some of the tasks to another surrogate while the other tasks continue uninterrupted on the current surrogate.

In this section, we describe the process of cyber foraging and highlight the challenges that arise in each step of such a process together with possible solutions.

16.3.1 Surrogate Discovery

The first step of any cyber foraging process is the discovery of surrogates. Client applications need to discover surrogates available in the surrounding environment that are capable of providing the necessary resources. Surrogates thus need to be discoverable by means of some of the available networking communication protocols. Some technologies (e.g., Bluetooth) permit limiting the discoverability of the device in order to avoid misuses of the device's resources. In the context of this chapter, we assume that devices are discoverable through the used networking technology. However being discoverable does not mean that the surrogate is willing to share its resources. A surrogate willing to make some of its resources available to others needs to advertise its availability and capabilities as well as provide means to be accessed.

Several middleware architectures and protocols for service discovery exist [708, 553]. These can be employed to support surrogate discovery as well. In static environments, cyber foraging can employ centralized discovery protocols designed for fixed local area networks where the number of participants is limited and the devices are relatively static. Examples of these types of protocols include Salutation [724] and Jini [786]. Surrogates can register with the protocol registry by specifying the offered resources. A client can then contact the service discovery server and submit its request for surrogates. The service discovery server matches the client's request against the registered surrogates and, in case of many available surrogates, the best matching surrogate is selected based on surrogate-specific attributes. The surrogate's address together with other qualifying properties are then returned to the client that can therefore contact and use its resources. A good example of centralized discovery model is the Web services approach. Web services use the Universal Description, Discovery and Integration (UDDI) standard [822] for registering and discovering services.

A large part of the solutions proposed for fixed networks rely on centralized registries. This makes the surrogate discovery process dependent on the availability of the central directory, which constitutes a bottleneck. To achieve larger scalability, distributed approaches such as VIA [145] permits sharing data among several discovery domains. Decentralized solutions such as Universal Plug and Play (UPnP) [824] and Service Location Protocol (SLP) [350] represent a better fit for dynamic environments such as mobile ad hoc networks, where mobility and failures are common and it is therefore not possible to rely on any centralized server. Some proposed solutions are Bluetooth SDP [97], GSD [156], Konark [364], and

SSD [723]. Peer-to-peer technologies have also been used for distributed resource discovery. Hoschek has proposed a peer-to-peer based approach for distributed databases that was applied to his distributed Web Service Discovery Architecture [378, 379]. Another Distributed Web Service Discovery Architecture is presented by Sapkota et al. in [725].

There exist also hybrid solutions that integrate both infrastructure-based and infrastructure-less approaches. PeerHood [680] combines different networking technologies (e.g., Bluetooth, WLAN, GPRS) and discovery protocols under one interface thus providing a unified view of the available surrogates and offered services. Clients use PeerHood to discover surrogates either in their close proximity (using Bluetooth and WLAN connectivity) or further on in the network (using fixed service registries accessible using GPRS). Clients may also register their own sharable resources to PeerHood thus allowing others to use them.

A first challenge with existing service discovery protocols is interoperability. Even though these protocols share the same basic principles, they all have different origins and employ different technologies. Due to incompatible data representation and communication formats, service discovery protocols do not interoperate with each other. Hence, in general, clients are able to discover only services that are advertised with the protocol(s) they support. Since it is very unlikely that in the future one service discovery protocol will dominate or that device manufactures will offer service discovery technologies on low-cost devices, proxy-based and middleware architectures to enable service discovery interoperability have been proposed [349, 693, 284].

A second challenge to be considered is mobility. In highly dynamic environments, it is hard to constantly maintain up-to-date information on the number and location of surrogates as well as on the type and quality of provided resources. Update mechanisms are usually built using a proactive or reactive approach. In the reactive approach, updates are exchanged only when an event occurs, for instance, a surrogate leaves the network. In the proactive approach, update messages are constantly exchanged and a consistent view of the network is maintained. In addition, mobility information can be used to adjust the surrogate advertisement rate and the range of dissemination [156, 692].

A third challenge of the discovery process is how to describe the offered resources. Surrogates need to provide sufficient information describing their resources so that clients can select the most suitable surrogates for their execution. Web services use the Web Services Description Language (WSDL) [871] and Resource Description Framework (RDF) [698] to describe services and resources. Sihvonen presents in his PhD thesis [755] an approach called Personal Service Environment (PSE) that changes its configuration depending on the available resources or capabilities. Capabilities are described with Composite Capability Preference Profile (CC/PP) and RDF. In the field of grid computing, Liny and Raman [529] present the Classad mechanism that allows description of both resource and task requirements. This approach supports the matchmaking procedure needed in the placement decision.

16.3.2 Application Partitioning

To enable the use of cyber foraging an application has to be split into *locally executable code*, such as the application GUI and its model data, and *remotely executable code* that may benefit from remote execution. An application may contain multiple tasks that are good candidates for remote execution depending on the amount of computing, storage, and communication resources they require to be performed. As these resources are relatively constrained on small mobile devices, such as PDAs and mobile phones, the more resource-intensive a task is, the more beneficial the remote execution process will be. On the other hand, as we will explain in the next subsection, cyber foraging implies some overhead, for example in terms of the communication bandwidth required to migrate tasks, transfer control messages, and receive task responses. Therefore, in principle, a task should be executed remotely only when this overhead can be amortized by the gain brought by utilizing cyber foraging.

A very important question to answer is whether the application partitioning should be a manual process performed by the application developer, or whether it is possible to automate such a process. There exist several tradeoffs in deciding between manual and automatic task partitioning. An automatic partitioning system sounds alluring, since it implies less work on the application developer's side. An example of such a system is Coign [394]. Coign makes partitioning and distribution of tasks possible without altering the source code of the application. Coign works on applications consisting of distributable COM components. It constructs a graph model of the application's inter-component communication where the nodes represent the COM components, the edges represent the communication between such components, and the edge weights represent the amount of communication. Given this graph a graph-cutting algorithm is employed to partition the application by cutting at the edges with the smallest weights. This information paired with a network profile may then be used to make positioning decisions at runtime.

While automated partitioning has its merits, in many cases, manual partitioning can be more effective because, as also noted by Flinn et al. in [283], a little application-specific knowledge can go a long way when preparing an application for distribution. The inclusion of distribution into the original application logic often alters entirely the actual program structure. For instance, it may become more convenient to execute an initially linear program as a number of parallel tasks—thus better utilizing the available surrogates. Such optimizations would be hard to detect for an automated distribution algorithm. Furthermore, an automated partitioning approach will always solve tasks to the same fixed degree. It thus does not offer any variability in the task execution which, as we will discuss shortly, can be of high importance in a cyber foraging setting. Ideally, a cyber foraging framework should cater for both kinds of distribution, falling back to automatic distribution when no instrumented version of an application is available.

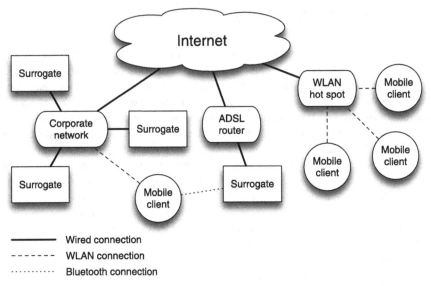

Fig. 16.2 An example environment where cyber foraging may be performed

16.3.3 Placement Decision and Cost Assessment

Once an application has been partitioned into locally and remotely executable tasks, the basis for utilizing cyber foraging has been built. The next step is to decide on an *execution plan*. The execution plan specifies where a given task is to be executed. Execution plans are dynamic in nature since they must take factors such as the current network environment, device resource levels, and surrogate availability into consideration.

For example, the network depicted in Fig. 16.2 shows a number of possible network settings that a mobile device may encounter during its operation. In the left side of the image, there is a corporate network where a number of computers (surrogates) connected to the wired network are offering several types of services and resources. These surrogates may be discovered and accessed by a mobile client through the corporate network infrastructure. In the middle of the image, there is a personal computer in a home network connected to the Internet by an ADSL connection. Its surrogate services may be reached through a direct Bluetooth connection. Finally, on the right side of the image, there is a WLAN hotspot, such as the ones found in many cafés today. Using this WLAN mobile clients can act as surrogates for each other and perform tasks in co-operation. The heterogeneity of the environment depicted in Fig. 16.2 points out many of the factors that the task placement and cost assessment processes need to take into account when selecting an execution plan. Some of these factors are the following:

- Network bandwidth and latency between:

- the client and the surrogate(s),
- the surrogate(s) and the Internet, if the remotely executed task requires Internet connectivity, and
- the client and the Internet.

- Full capacity and current utilization of resources such as battery, CPU, and storage of:

  - the client device, and
  - the surrogate(s).

- Resources consumed by:

  - data transfer over the available wireless links, and
  - CPU-intensive tasks.

- Task-specific properties:

  - Resource demands—either as defined by the author of the task or based on history-driven profiling,
  - input and output data/code sizes, and
  - adaptability in face of changing resource levels.

These factors interrelate in different ways. For instance, when considering whether to remotely execute a given task, the time and energy necessary for transferring input and output data between the client and its surrogate must be taken into account. This computation involves task-specific properties (sizes of input and output data), information on the execution environment (current network latency and bandwidth), and device-specific properties (energy consumed by wireless communication). Taking all these factors into consideration is not straightforward. However, there are systems [283, 48] that have made progress into this direction.

In choosing an execution plan, a scheduler needs to decide on what should be optimized: Is the execution time the most important aspect? Or is the energy consumption more relevant? Or should both factors be equally weighted in selecting the most appropriate plan? If cyber foraging is employed primarily to reduce the overall execution time, then the decision of remotely executing a task should be based entirely on the latency estimation. If, instead, reducing energy consumption has higher priority it may be necessary to remotely execute some tasks at the cost of a higher latency.

16.3.4 Security and Trust

Security and trust are essential requirements to enable sharing of resources in cyber foraging scenarios. When a client's task is remotely executed on a surrogate, the task code needs to be protected from a malicious surrogate and the surrogate needs to be protected from a malicious client's tasks. More specifically, a surrogate needs to prevent a malicious client's task from making an excessive use of its resources. For example, a malicious task may exhaust the energy of the hosting surrogate or corrupt some of its data, launch an attack on the device, or infect other tasks running on the surrogate. A potential solution to this problem is to perform admission control of incoming tasks and require the migrating tasks to specify upper bounds on the amount of necessary resources.

On the other hand, a malicious surrogate may alter the data of a migrating task and compromise its correct execution. Or it may even alter the task's code and propagate viruses to other surrogates and to the client itself. This can be solved by transferring the task's code to surrogates in an encrypted form. In this case, each client and surrogate will need to carry a pair of public/private keys [324]. Another security threat is represented by the situation in which a malicious surrogate may attract client's tasks by pretending to possess false resources (i.e., fake surrogates). In a distributed setting, this issue can be partly solved by using redundancy—when operating in an untrusted environment the same task could be placed on multiple surrogates. Otherwise, if available, a trusted third party can certify that surrogates claim to have resources that they actually own. Redundancy could also be employed to hamper the damages possibly caused by hostile surrogates. The same task is executed on more than one surrogate. When the results are collected, if a surrogate returns a result different from the results provided by the majority of the surrogates, this surrogate is deemed as hostile and will not be used in any further execution. This approach is employed in the Slingshot system [778].

Integrating support for trust in a cyber foraging system is generally a complex task. If a centralized authority is available, this can authenticate clients and surrogates, monitor the environment, and detect malicious entities. If a centralized authority is not present or simply too expensive to be maintained, as in the case of mobile ad hoc networks, an entity can establish trustworthy relationships by relying on its direct experiences with the same entities as well as by relying on others' recommendations. A recommendation is generally defined as "the perception that a node creates through past actions about its intentions and norms" [596]. To isolate malicious entities, reputation mechanisms such as [565, 525, 909] can be used.

The issues met when considering security and trust in a cyber foraging setting vary depending on the code distribution approach that is used. This is closely related to the execution strategy adopted by the cyber foraging system, and is therefore discussed in the following subsection.

16.3.5 Task Execution

Once it has been decided which surrogates are available (surrogate discovery), which tasks can be executed remotely (application partitioning), and what cost is associated with every possible distribution strategy (cost assessment), it is the responsibility of the execution phase to decide on 1) how the tasks should be physically allocated to surrogates, 2) how the communication with surrogates should be implemented, and *3)* how the task executions on multiple surrogates should be co-ordinated.

In the distributed computing field, CORBA and Java RMI offer a means for supporting remote execution in distributed computing environments. CORBA uses object references to hide the location of the executing service. Java RMI implements methods for remote procedure calls and thus enables the use of distributed resources. Both approaches present some weaknesses when applied to mobile environments such as failure detection, high latency, and communication overhead. Another approach to distributed computing is offered by cluster computing and grid computing. Both originating from the parallel computing field, they are solely based on the use of shared resources. Although earlier approaches exist, Beowulf [78] clustering and the SETI@home [744] applications started an era in which extra-computing resources are efficiently used for common purposes. However, in these distributed computing approaches the discovery of resources is usually based on static structures, e.g. host files, centralized registers, resource brokers, and portals. The use of static structures hinders their applicability to a mobile environment, even though current trends in grid computing have started focusing also on more dynamic behaviors by looking at the dynamic service creation and discovery issues [10].

In the mobile context, the selection of the best mechanism for physically allocating tasks over the available surrogates depends on the execution environment. If it is the case of mobile devices that perform cyber foraging in known environments, such as within a corporate network, tasks may be pre-installed on fixed surrogates and clients can then use simple RPC to migrate tasks. If mobile devices are performing cyber foraging in unknown, and thus not previously configured, environments a solution based on mobile code where task code is pushed onto surrogates for execution may be more feasible.

The advantages of using pre-installed tasks in a trusted environment are obvious: *1)* there is a smaller overhead involved in performing tasks remotely, *2)* surrogates can trust the code that they are executing since only their administrators may install tasks, and *3)* clients can trust the results returned by surrogates. The disadvantage of such a scheme is the lack of flexibility and in particular the poor support for mobility—a mobile device using this kind of cyber foraging is bound to a specific physical location, and when this known environment is left behind all tasks must be performed locally. Approaches based on mobile code require surrogates to only offer a generic cyber foraging service to clients. Clients can use this service to directly install their tasks and remove them once the execution is completed. However, by using mobile code in unknown environments some serious trust and security issues arise—issues that are well-known within the field of mobile agents, see e.g., [169]. A more in-depth discussion of the different execution possibilities can be found in Sect. 16.4.

16.3.6 Environment Monitoring and Application Adaptability

Pervasive computing environments present a variable level of resource availability [729], such as computing power, storage capabilities, and network bandwidth. Applications need to exploit computing, storage, and communication opportunities whenever available, and must be able to survive when such resources are not available anymore. In addition, with portable wireless devices, disconnections and device failures have to be treated as part of normal operation. Disconnections can occur either accidentally due to loss of wireless connectivity or voluntarily to save battery or reduce connection costs. Wireless communication can also be temporarily degraded due to signal interference and thus cause packet losses, variable bandwidth, and high error rates.

Therefore, applications should constantly monitor their execution environment while consuming almost no power [257]. Every time the execution environment changes, the change must be detected and if it is permanent enough to trigger a reconfiguration, then the behavior of the application must change accordingly. Adapting the application's fidelity to fluctuating resource levels has been shown to be effective in coping with such dynamism [632, 283, 218, 50, 49]. Fidelity defines the degree to which service results returned to the application matches the expected service quality. For instance, fidelity can be temporarily degraded to the minimal level acceptable by the user in order to minimize the resource consumption in a resource-poor environment.

How the application fidelity can be varied strongly depends on the data the application manages. For instance, in the case of an application performing speech recognition, when no surrogates are available the lowest level of fidelity is provided: the application simply stores the speech audio files for later processing. When a "weak" surrogate is available a medium level of fidelity is provided: the application uses a "1.5-way" task-directed speech recognition where the speaker adheres to a certain grammar and only utterances within that grammar can be recognized. Finally, when "powerful" surrogates are available the highest level of fidelity is provided using a more resource-consuming approach: the application uses a "2-way" general speech recognition where the recognizer tries to recognize every single word without assuming any grammar.

In addition, learning about the resource usage of an application, predicting resource impoverishment, and anticipating changes of the application's requirements are all important challenges to fully accomplish efficient reconfiguration.

16.3.7 Summary

In this section, we have illustrated the steps that a system needs to implement in order to successfully perform cyber foraging. Table 16.1 summarizes these steps and includes a description of their main characteristics.

Table 16.1 Summary of challenges posed when designing a cyber foraging system

Step	Description
Surrogate discovery	• Surrogates must be discoverable by means of some wireless network technology. • Services and resources offered by each surrogate device must be described and advertised.
Application partitioning	• Applications have to be partitioned into locally executable tasks and remotely executable tasks. • Both automatic and manual partitioning schemes may be considered.
Cost assessment	• Before placing tasks on external surrogates the cost of using remote execution must be assessed. • A placement decision aims to minimize an application-specific cost function assessing power, communication, and processing resources consumed by cyber foraging. • Local execution should *always* be a possibility so that the application is not dependent on the availability of surrogates.
Trust establishment	• Clients must be able to trust surrogates and vice versa. • Clients must be sure that a surrogate does not: *1)* alter the code or state of their tasks, *2)* return false results, *3)* access private data, and *4)* pretend to own resources that it does not own. • Surrogates must be sure that clients do not: *1)* perform DoS attacks on the surrogate, *2)* use the surrogate to launch attacks on other peers, *3)* access private data stored on the surrogate, *4)* modify the code or state of other tasks currently performed on the surrogate.
Task execution	• It must be decided how tasks are physically allocated to surrogates (pre-installed RPCs, mobile code, virtual machines). • A means for communicating between clients and surrogates must be chosen (Java RMI, CORBA, etc.) • Optionally, the possibility of utilizing multiple surrogates in parallel – for improving the execution's performance or security reasons – should be considered too.
Environment monitoring	• Monitoring of the execution environment is necessary to dynamically adapt to changes in the level of resource availability. • Several resource parameters such as CPU utilization, power consumption, and network latency/bandwidth need to be monitored. • The surrogate availability must also be monitored.

16.4 Cyber Foraging Approaches

Several approaches have been proposed to accomplish cyber foraging in mobile computing environments. In this section we review three well-known mechanisms that have been used in this domain and present the corresponding implemented systems that integrate such approaches. As depicted in Fig. 16.3, cyber foraging mechanisms can be based on remote procedure call (RPC), virtual machine (VM) techniques, or mobile code.

RPC-based approaches assume the environment to be pre-configured. RPC functions are pre-installed on available surrogates such that mobile clients entering the computing environment can remotely invoke functions offered by the surrogates.

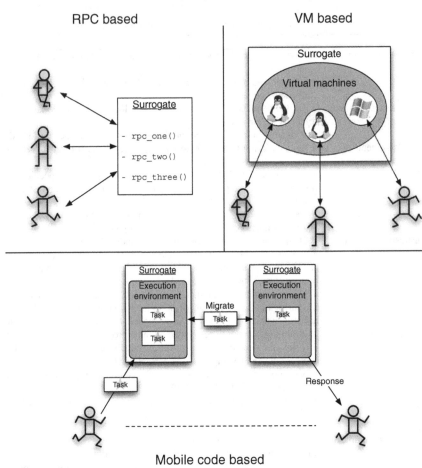

Fig. 16.3 Three classes of cyber foraging approaches

Due to the need to pre-configure the environment, RPC-based approaches do not provide high flexibility and therefore suits only static environments. VM-based approaches can provide higher flexibility by allowing clients and surrogates to install arbitrary code on the surrogate machines. Approaches based on mobile code provides the highest flexibility. The application task can migrate from one surrogate machine to another thus guaranteeing to always execute on the best surrogate available in terms of communication latency, computing resources, reliability, etc. On the other hand, the overhead for migrating and installing the task on the surrogate node can affect the application's responsiveness depending on the system implementation.

16.4.1 RPC-Based

Two examples of RPC-based cyber foraging approaches are Spectra [48, 283] and Chroma [49]. In these systems client applications are partitioned into locally executable code and a number of remotely executable tasks. These tasks are preinstalled on surrogate computers that then offer RPC services for invocation. Apart from requiring the pre-installation of tasks a shared Coda [730] file system is also used to exchange data between clients and surrogates. This means that Spectra and Chroma only function in preconfigured environments that support specific functions required by each applications.

In Spectra both mobile clients and surrogates are running Spectra servers along with some resource monitors. When a client is within the range of a surrogate it decides, based on both the monitored resource levels of the current environment and the execution history of the tasks, where to place a task; *i.e.* whether to locally execute the task or invoke the corresponding RPC function at the surrogate. Spectra uses the Coda file system for exchanging data between clients and surrogates. Input and output files of the RPC functions are placed in this shared file system.

Chroma is an advanced version of Spectra. Chroma seeks to support computationally intensive interactive applications, such as speech recognition, natural language translation, and augmented-reality applications, which can operate at multiple fidelities. Since these interactive applications generally require minimal user distraction, users specify in advance their high-level preferences and then the system autonomously decides at runtime how to execute applications. Chroma has been designed to be effective in applying different cyber foraging strategies depending on the resource conditions and the level of fidelity to be guaranteed.

Chroma particularly targets the problem of application repartitioning in dynamic mobile computing environments. It makes uses of *tactics* that are high-level declarative specifications of application-specific knowledge to determine optimal application partitioning. Each tactic describes one way of combining RPCs to execute a certain task. In selecting the best tactic, Chroma measures the available resources and selects the tactic that maximizes certain user-specific utility functions. Utility functions express the user's preferences in one or more fidelity attributes.

PeerHood is another example of this type of an approach. When a PeerHood-enabled device enters a new environment it proactively scans the neighborhood and looks for PeerHood-enabled surrogates. PeerHood-enabled surrogates advertise their services and thus clients can find out about the available RPC service functions. Both mobile devices and ordinary servers may behave as PeerHood-enabled surrogates. Kallonen et al. present in [438] an image processing application on top of the PeerHood middleware. The surrogate carries out image processing and returns the results to the mobile device. PeerHood also provides a remote monitoring service that allows the transfer of monitoring actions from one surrogate to another surrogate. In this way, a mobile client can be constantly informed about changes occurring in its vicinity while saving battery.

16.4.2 VM-Based

Where the RPC-based approaches presented required fully prepared environments, both with regards to individual application support and the use of a shared file system, VM-based approaches provide increased flexibility. The usage of virtual machines makes it possible to allow users to "install" their own functionality on the surrogates on-demand. Apart from the flexibility gained by utilizing virtual machines VM-based approaches also allow clients to fairly share resources of surrogate machines by using the load balancing functionality of virtual machine managers. In addition, they enable easy clean-up once the client's execution has completed as all execution has taken place within an easily replaceable virtual machine image. Two systems using VM-based mechanisms to support cyber foraging are Slingshot [778] and the system described by Goyal and Carter in [324].

Slingshot replicates services on surrogate machines located at hotspots. A primary replica of each services runs on a remote server owned by the mobile user. Secondary replicas are instantiated on surrogates at or near the hotspot where the user is currently located. A proxy running on the resource-constrained mobile device broadcasts each service requests to all replicas. The first received response is passed to the application. The advantage is that secondary replicas located at the user's hotspot can improve the response time while the primary replica serves as a stable repository for the application state in case of surrogate crashes. Each replica runs within its own VM, which encapsulates all-application specific state such as a guest OS, shared libraries, executables, and data files. The surrogate machines simply consists of the host OS (i.e., Linux), the VM monitor (i.e., VMware), and Slingshot.

In the infrastructure proposed by Goyal and Carter [324], surrogate managers maintain root partition images for each available OS. In response to a client request, the manager initializes a pre-allocated root partition with the appropriate root image and starts a new virtual server. The IP address of the virtual server is returned to the client. To submit a task to the surrogate, the client sends a request to the virtual server manager running on the server. The request consists of a URL pointing to the program the server is requested to run on the client's behalf. The server manager can download and install the necessary software packages from such a URL. Once the execution is completed or the client's allocated time slot expires, the server removes the installed software and restores the original clean state.

16.4.3 Mobile Code

A third approach to supporting cyber foraging is based on mobile code. Due to its flexibility, this approach particularly suits highly mobile environments where clients and surrogates move and surrogates change their resource capabilities over time. This is the case of an ad hoc network of collaborating devices, where nodes dynamically join and leave the environment and fail or voluntarily switch themselves off to save resources.

Context-aware migratory services [709] is a framework designed to support service execution in highly volatile mobile ad hoc networks. A migratory service is a service capable of migrating to different nodes in the network in order to effectively accomplish its task. The service executes on a certain surrogate node as long as it is able to provide semantically correct results to the client; when this is not possible anymore, it migrates through the network until it finds a new surrogate node where the execution can be resumed.

For instance, if we consider the region monitoring scenario previously described (see Sect. 16.1), the migratory service monitors the disaster area by executing on a surrogate device available in such a region. When such a device moves away from the region of observation, the service migrates to a new surrogate device currently located in the region of interest. The monitoring service periodically transfers observations of the disaster area to the remote client. There are two main advantages of using migratory services in implementing such a scenario. First, when the current surrogate node becomes unsuitable for hosting the monitoring service, the client does not need to perform any surrogate discovery because the current service can autonomously migrate to a new surrogate that is qualified for accomplishing the current task. Second, the migratory service incorporates all the state information necessary to resume the interaction with the client upon the migration to a different surrogate has completed.

The service migration occurs transparently to the client, and except for a certain delay, no service interruption is perceived by the client. Although a migratory service is physically located on different surrogates over time, it constantly presents a single virtual end-point to the client. Hence, a continuous client-service interaction can be maintained.

The migratory services model incorporates three main mechanisms. The first monitors the dynamism of interacting entities (client and surrogates) by assessing context parameters characterizing their state of execution and available resources. The second specifies, through context rules, how the service execution is influenced and should be modified based on the variations of those context parameters. The third makes the service capable of migrating from one surrogate to another and of resuming its execution once migrated.

To support the migratory services model, the migratory services framework needs to runs on each node willing to co-operate in the ad hoc network. This framework was built on top of the Smart Messages [106, 441] distributed computing platform which provides support for execution migration, naming, routing, and security. Smart Messages are similar to mobile agents, which also use migration of code in the network. A mobile agent can be seen as a task that explicitly migrates from node to node. However, mobile agents typically name nodes by fixed addresses and know the network configuration a priori. Instead, Smart Messages are responsible for their own routing at each node in the path between two nodes of interest. This feature makes approaches such as Smart Messages capable of more quickly adapting to changes that may occur in the network topology and resource distribution.

Another system based on mobile code is Locusts [480]. In this system mobile code is used to enable higher client mobility. As discussed earlier, using mobile code makes a cyber foraging system more flexible, because it makes it possible for a client to utilize cyber foraging in unknown environments, as long as the cyber foraging framework itself is installed on the surrogates. Using tasks specified as graphs of interconnected, mobile-code based services the Locusts framework is capable of dynamically installing and performing tasks in unprepared environments.

Finally, an important requirement for systems based on mobile code is portability. The more independent the approach is from the underlying hardware and software platform and the more usable it is. On the other hand, guaranteeing portability may affect the overall performance of the system. The Smart Messages platform, for instance, was initially implemented by modifying the Java virtual machine running on the surrogate nodes in order to provide efficient migration. Later on, the Portable Smart Messages [694] platform capable of running on unmodified virtual machines was proposed. Performing migration without having access to the VM internals required a lightweight migration approach based on Java bytecode instrumentation. This made the implementation portable, but costlier than the first implementation in terms of execution time.

16.4.4 Summary

This section introduced three different approaches towards realizing cyber foraging: RPC-based, VM-based, and mobile code based. The advantages and disadvantages of the different approaches are summarized in Table 16.2.

As Table 16.2 shows, each approach works well within a specific usage scenario. If, for example, cyber foraging is employed within the confines of a single organization/home, an RPC-based approach may be a viable solution. This would offer fast access to cyber foraging with the needed security measures already given by the existing computing infrastructure, and the maintenance of surrogates and installation of tasks may be performed by the administrative staff. In a less static setting, where the user of the cyber foraging service must be able to perform cyber foraging in unprepared environments, using a VM-based or mobile code based approach is necessary. In this case the rate of mobility becomes important; if the user is static for long periods of time (hours instead of minutes) a VM-based approach may be viable. The main drawback of a VM-based approach is the very high setup time; when the client initially enters a new area, it will take on the order of minutes to establish a "connection" to a surrogate, and therefore the VM-based approach is mainly useful for low mobility. If, on the other hand, the client is highly mobile, a mobile code approach may be suitable. Using mobile code, the "setup" time between client and surrogate is minimal and migration of running tasks (e.g., a task is moved from one surrogate to another because of client mobility) becomes possible—even in the face of high client mobility.

Table 16.2 Advantages and disadvantages of the different cyber foraging approaches

Approach	Advantages	Disadvantages
RPC-based	+ Small overhead when performing tasks. + Security and trust is easier to establish. + Portable—can be language agnostic.	– Not very mobile—tasks must be pre-installed. – No obvious support for task migration.
VM-based	+ Flexible—no preparation of surrogates needed. + After initial setup the overhead when performing tasks can be kept small. + Portable—Completely language agnostic. + Migration of the entire state is trivial.	– Very heavyweight with regard to initialization. – Not very mobile because of the high cost of initialization.
Mobile code	+ Very flexible—no preparation of surrogates needed. + High mobility—low initialization times means that unprepared surrogates can be utilized almost immediately. + Migration of tasks is possible—even autonomous migration if needed (mobile agents). + Migration can be done quickly—with very little overhead.	– Not always portable—typically the mobile code has to be expressed in a specific language. – Security issues with regards to using mobile code must be addressed.

16.5 Summary and Outlook

This chapter has introduced cyber foraging and showed how it can be employed to support mobile applications running on resource-constrained mobile devices. The core principle behind cyber foraging is to make opportunistic use of resources and services provided by nearby computing devices, called surrogates. The first part of the chapter has focused on describing the cyber foraging process consisting of six main processes: surrogate discovery, application partitioning, cost assessment, trust establishment, task execution, and environment monitoring. Problems and solutions to accomplish each single step have been described. The second part of the chapter presented three different approaches to supporting cyber foraging and described existing systems that implement such approaches.

There are still numerous challenges that need to be solved to make cyber foraging a widespread technique. First of all, security and trust issues can hinder the applicability of cyber foraging to unknown environments. This has been demonstrated by several systems based on code mobility which did not manage to go beyond small-scale research prototypes. Pre-configuring computing environments to support cyber foraging is a viable solution but lack the flexibility required by mobile users. Another challenge that current research is facing is how to provide reasonable performance in highly volatile environments such as ad hoc networks. In these environments not only nodes are mobile, but they also offer heterogeneous resources and are subject to frequent failures.

Another major challenge that must be considered to make cyber foraging a viable solution outside of research laboratories, is ease of the development of cyber foraging enabled applications. It is a well-known fact that application developers find it hard to develop parallel programs. Cyber foraging applications are not only parallel but they are also distributed and must work in *very* unstable environments. Hence adequate frameworks that provide abstractions for distribution, task placement, concurrency, error-correction, etc. are needed. In principle, a developer should only be responsible for defining the tasks that the application would like to have performed and the rest should be handled transparently by the cyber foraging system.

Chapter 17
Vehicular Networks and Applications

Aline Senart, *Department of Computer Science, Trinity College Dublin, Ireland*
Mélanie Bouroche, *Department of Computer Science, Trinity College Dublin, Ireland*
Vinny Cahill, *Department of Computer Science, Trinity College Dublin, Ireland*
Stefan Weber, *Department of Computer Science, Trinity College Dublin, Ireland*

17.1 Introduction

Vehicular networks represent a particularly challenging class of mobile (ad hoc) networks that enable vehicles to communicate with each other and/or with roadside infrastructure. This chapter describes potential applications, middleware approaches and communication protocols proposed for vehicular networks.

17.1.1 Background

Historically, drivers have used their voice, gestures, horns, and observation of each other's trajectory to coordinate their behavior. When the proliferation of vehicles made this insufficient, in the second half of the 19th century, traffic police were put in charge of controlling traffic using hand signals, semaphores and colored lights [699]. The 1930s saw the automation of traffic lights, and the 1940s the widespread deployment of car indicators. Variable-message signs were introduced in the 1960s, providing drivers with information that could be adapted to current circumstances. The information communicated via all of these means is, however, fairly limited: road infrastructure typically provides the same information to all cars, and the amount of information that drivers can share directly with one another is restricted. More recently, drivers can communicate more information, such as directions and traffic information, to each other via citizen band radio or car phones.

Wireless communication allows more personalized and complete information to be exchanged. There is a body of on-going research addressing wireless communication in VANETs, as well as on-going standardization, notably the Wireless Access for the Vehicular Environment (WAVE) standards [402] based on the emerging IEEE 802.11p specification [400]. Wireless communication essentially enables *infrastructure-to-vehicle (I2V)*, *vehicle-to-infrastructure (V2I)*, and *vehicle-*

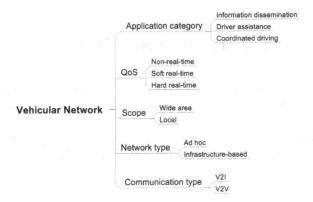

Fig. 17.1 Vehicular networks taxonomy

to-vehicle (V2V) communication. In the following we use the term V2I to refer to both I2V and V2I communication.

17.1.2 A Vehicular Networks Taxonomy

Vehicular networks can typically support three categories of applications (see Fig. 17.1): information dissemination, driver assistance, and coordinated driving. Applications in the *information dissemination* category focus on providing information to both drivers and their passengers. This information might be adapted to the context of the users, such as the identification of a petrol station in a vehicle's vicinity, and might be updated in real-time, for example, while playing games. *Driver assistance* applications support drivers in their maneuvers, such as emergency vehicle braking warning, where vehicles braking sharply to avoid an accident warn surrounding vehicles, allowing them to slow down even if they cannot yet see the vehicle that is braking. Finally, *coordinated driving* applications control the trajectory of vehicles, such as cruise control that adapts the speed of a vehicle depending on the distance to the vehicle that it is following.

As illustrated by the aforementioned examples, vehicular networks applications range from entertainment to safety-critical operations. Consequently, the Quality of Service (QoS) required varies from *non-real-time*, to *soft real-time* where a timing failure might compromise service quality, up to *hard real-time* where a timing failure might lead to a catastrophe. These applications can also be characterized by their scope, i.e., whether they provide communication over a *wide area*, or are *local* only. Finally, such applications can vary in their networking approach: *ad hoc*, where vehicles communicate spontaneously, or *infrastructure-based*, where communication is managed by fixed base stations; and in their communication type: *V2I* and *V2V*.

17.1.3 Issues and Requirements

Communication, and in particular real-time communication, in wireless networks is highly unreliable and the achievable timeliness varies greatly over time and location [300]. In addition, vehicular networks exhibit a number of unique issues. Because no central coordination can be assumed, a single shared control channel is required at the MAC layer (the so-called one channel paradigm) [360]. Mobility patterns of vehicular networks are also very specific, e.g. vehicles move along roads, in predefined directions, and this requires new specific mobility models to be devised. In addition, cars have very high mobility rates, with relative speeds of up to 300 km/h that makes wireless transmission very challenging [99].

Furthermore, the vehicle density exhibits spatio-temporal variations: it might be very scarce, with no vehicle or only few, up to very dense, with over 500 vehicles per kilometer [616]. Both ends of the density spectrum are particularly challenging.

The applications of vehicular networks should also fulfill a number of nonfunctional requirements, such as potentially very high reliability, but also security to ensure that potentially safety-critical applications cannot be tempered with. Vehicles range over very large geographical areas (cities or countries), and therefore require potentially large-scale networks, and especially a very extensive deployment of equipment if infrastructure-based networks are used.

In addition to technical challenges, socio-economic challenges have to be solved. The benefits of V2V communication only become significant when there is a sufficient large number of vehicles using the technology. Vehicular applications must therefore be able to operate and be useful despite initial low penetration.

17.2 Applications

This section reviews existing and future applications for vehicular networks as identified in Sect. 17.1: information dissemination, providing services for all passengers of a vehicle; driver assistance, offering support for driving maneuvers; and coordinated driving, controlling the trajectory of vehicles.

17.2.1 Information Dissemination

Passengers in a vehicle are potentially interested by many different kinds of information: the area in which they are currently traveling, the facilities in their vicinity, or might want to access the Internet or watch television to pass time. In addition, traffic authorities are interested in obtaining information about road users for traffic monitoring, i.e., obtaining real-time information about traffic flows to detect potential congestion. Finally, information dissemination can also be used to manage traffic, both in normal conditions and when an accident occurs.

Local Information Systems. A significant application domain for vehicular networks is the provision of information to road users on their surroundings. For example, AdTorrent [611] provides wireless digital billboards on the roadside that can be used for advertisement.

A more interactive application is FleaNet [503], which supports opportunistic interaction between road users, pedestrians and business in urban areas via a virtual flee market where users can disseminate both requests for and offers of goods.

Other applications are targeted to a more specific usage. [421], for example, predicts parking lot occupancy using information broadcast by parking lots and estimated arrival time. This application minimizes the effort required to discover a free parking space, hence saving driver time and money and decreasing pollution. Another possible application is the provision of regional alerts about accidents or road conditions [779]. Such notifications are maintained for a certain duration to allow cars arriving in the area to be informed.

Entertainment. A number of applications aim to entertain passengers who spend a very long period in transit. One such project is FleetNet [291] that provides Internet access, as well as communication between passengers in cars in the same vicinity, allowing them to play games for example. Pavan [312] is another application that deals with the discovery and delivery of video and audio content to car passengers.

Traffic Monitoring. As infrastructure-based solutions are costly, vehicular networks are now being used for traffic monitoring. For example, in 2004, a number of cars in Atlanta were fitted with on-board facilities including GPS units and cellular modems allowing the tracking of the cars' position and speed. Such data was used to deduce current congestion levels [341]. This approach could be extended by the use of audio and video devices, that could be used for terrorist activities monitoring [502]. In SOTIS [867], each vehicle monitors the locally observed traffic situation and shares it with surrounding vehicles. StreetSmart [232] improves this approach by sharing information only when detected traffic levels are different from some prediction, hence allowing a significant reduction in the data exchanged. Another project in the same area [421] aims at providing faster delivery of traffic information, and improving the efficiency and accuracy of traffic detection by allowing collaborative processing of information between vehicles.

Traffic Management. Vehicular networks have been shown to be particularly useful for traffic management. For instance, V2I solutions for road tolling are widely deployed. In Singapore, all vehicles are equipped with on-board units to pay for road usage on congested roads, with prices fixed depending on congestion levels [880]. In the future, vehicular networks could enable pervasive pricing, where drivers would be charged for their specific usage of the road network [193]. Another field of application is the optimization of traffic flow. If junctions were equipped with a controller that can either listen to communication between vehicles or receive messages from arriving vehicles, then the controller would be able to build an accurate view of the traffic at the junction and could therefore adapt its behavior to optimize the throughput [328]. Vehicular networks can also be used to allow emergency vehicles to preempt the lights at signalized intersection [832], and clear the path of emergency vehicles in real-time, hence saving crucial time [743].

17.2.2 Driver Assistance

Vehicular networks can also be used to support driving maneuvers by providing drivers with information that they might have missed or might not yet be able to see, or by helping their decision making.

Increased Horizon. Vehicular networks can be used to extend the drivers' horizons. For example, Virtual Warning Signs [546], displayed on the dashboard of a vehicle, inform drivers about conditions in their surroundings, for instance to warn of an accident that has occurred around a blind corner. Similarly, TrafficView [608] displays information on the traffic beyond what a driver can actually see. While these applications are very similar to traffic monitoring, they are concerned with immediate surroundings of a given vehicle, rather than an aggregate view of traffic conditions in an area.

Collision Warning. By having vehicles exhibiting abnormal driving patterns, such as a dramatic change of direction, send messages to inform cars in their vicinity, drivers can be warned earlier of potential hazards, and therefore get more time to react and avoid accidents [89]. This approach has also been applied at intersections where cars communicate their current position and speed, making it possible to predict collisions between cars [573].

Other Driver Assistance Applications. Other applications of vehicular networks to driver assistance include supporting decision making. For example, SASPENCE determines current road conditions using a long-range radar, maps of the area, lane recognition cameras and messages from other vehicles, and provides drivers with recommendations on proper velocity and headway for these conditions [795]. Another application is a lane positioning system that uses inter-vehicle communication to improve GPS accuracy and provide lane-level positioning [210]. Such detailed positioning allows the provision of services such as lane departure warning, as well as lane-level navigation systems.

17.2.3 Coordinated Driving

In the last decade, the idea of autonomous vehicles has moved from the domain of pure science fiction to a vision achievable in a not-too-distant future. The DARPA Urban Challenge, which took place in November 2007, has demonstrated that prototype autonomous vehicles are able to operate in urban areas while obeying traffic regulations, negotiating other traffic and obstacles, and merging into traffic. The performance and safety of such vehicles could be improved if they would collaborate with each other [467]. Proposed coordinated driving applications center around three scenarios: adaptive cruise control, platooning and intersection management.

Adaptive Cruise Control. The simplest coordination scenario for vehicular networks is Adaptive Cruise Control that maintains a safe distance for each vehicle to the vehicle in front using forward sensors. Cooperative Adaptive Cruise Control exploits wireless communication to allow vehicles to cooperatively perform control maneuvers [874].

Platooning. In a platoon, a leader traditionally coordinates platoon members using V2V communication [752]. A more sophisticated collaboration model is to coordinate the vehicles through teamwork models [356] in which autonomous vehicles maintain the platoon in a decentralized fashion. Platoons are expected to enable increased road capacity and efficiency, reduced congestion, energy consumption and pollution, and enhance safety and comfort [564]. Demonstrations of cars traveling in platoons [445, 252] have proven the feasibility of the approach in protected settings, but they have yet to be used on public roads.

Intersection Management. Another application scenario is collaborative collision avoidance [750], and more particularly collaborative intersection crossing of autonomous cars. A slot-based method has been suggested, whereby each vehicle approaching a junction requests and receives slots from an intersection manager, during which it may pass. This reservation-based system improves both delay on vehicle journey and throughput of the intersection in comparison to traffic lights or stop signs [238]. Another approach consists of vehicles sharing their intended path and adapting their velocity to avoid collisions [749]. Finally, the Cybercar project defined an algorithm for collaborative intersection crossing that relies on refining partial trajectories [113]. While Cybercar was tested on real vehicles, none of the intersection management applications have yet been deployed in real-life settings.

17.3 Middleware

Traditional middleware architectures such as Microsoft's .NET framework [569], the Object Management Group's Common Object Request Broker Architecture [642] or Sun's Java Remote Method Invocation [677] have successfully accommodated a wide range of diverse distributed applications. However, these architectures are inadequate for vehicular networks as they do not enable loosely-coupled integration between communication parties, which is an inherent requirement of vehicular applications. Therefore, middleware for vehicular applications are based on (extension to) paradigms that provide such loose coupling including publish/subscribe [561], peer-to-peer [534] and service-oriented computing [759].

17.3.1 Middleware Requirements

From the applications presented in Sect. 17.2, this section identifies a set of specific requirements that middleware for vehicular networks should support.

Context Acquisition.   Context information can be obtained from a number of different sources including on-board and infrastructure-based sensors, as well as through the use of wireless communication. Examples of on-board sensors in vehicular applications range from widely-used GPS positioning systems [341], to RFID tags for reservation enforcement on highways [695], to impact sensors used to generate emergency calls [245]. Imaging sensors using visible light or infrared radiation can also be used but are relatively expensive and involve a substantial amount of processing [484]. Cheaper approaches use time-of-flight sensors mounted on vehicles to detect obstacles but suffer from interference [431]. For all of these vehicle-based systems, the area in which context can be acquired is restricted to the surroundings of the vehicles. Hence it is useful to additionally deploy sensors in the infrastructure and to relay information between vehicles to obtain remote context information.

Infrastructure-based sensors are embedded in the road (*e.g.*, in "cat's eyes"), or the surroundings of the road (*e.g.*, in signs or buildings). For example, road conditions [563] and weather information systems [587] are traditionally deployed in the environment. Inductive loops embedded in the road are used to count vehicles as they pass over the loop [301]. Cameras are mounted on poles or gantries above or adjacent to the roadway to detect incidents or measure lane-by-lane traffic flow [366], but only at busy or dangerous places due to their high cost.

Various forms of wireless communication technologies have been proposed for vehicular networks allowing new cheap ways to sense context. Radio devices identifying a vehicle as it passes under a gantry over the roadway can be used in electronic toll collection [533]. Switched-on mobile phones can become traffic probes by leveraging existing mobile phone networks. Wireless communication has significant advantages over conventional sensing methods as no infrastructure or hardware is needed to be built in cars or along the road.

Context Aggregation and Representation.   With the vast amount of sensor information available, combining multiple sensor information is essential. For example, in TrafficView [607], context data is aggregated based on their relative distances, or when a vehicle has more than one record describing the same phenomenon in its data storage.

The most common way to represent the aggregated context information is via predicates (*e.g.*, RDFS) or onlogogies (*e.g.*, OWL) [337]. Freshness of context information is ensured through periodic update [337], revocation messages [248] or timers [607]. Further, in order to adapt to new situations, vehicular applications require additional services to reason about contexts, for example to stop if there is an road accident. Rule-based [175] or semantic-based reasoning [337] can be employed.

Data Diffusion. Three main mechanisms for data dissemination in vehicular networks can be used: flooding, forwarding and dissemination. Flooding, in which each vehicle re-broadcasts packets it receives, is the most common approach to broadcasting, but causes broadcast storms [810]. A more scalable alternative is periodic dissemination. Dissemination can either transmit information to vehicles in all directions, or perform a directed transmission to vehicles behind. Moreover, communication can be relayed using only vehicles traveling in the same direction, vehicles traveling in the opposite direction, or vehicles traveling in both directions [608]. Furthermore, if messages are relayed in a store-and-forward fashion, it is possible to improve successful message delivery in spite of partitions [494].

In VMesh [162], vehicles dynamically form an ad hoc mobile network to relay data between different clusters of static nodes that are otherwise disconnected. One example application is to interconnect buildings that are equipped with wireless sensors to monitor gas, water and electricity usage to a backbone wide-area network infrastructure such as the Internet. In Hull et al., each vehicle collects data from local sensors and uses opportunistic wireless connectivity to deliver them to a central portal [393]. Finally, in [905], traffic information is sent from data centers to vehicles, and buffering and rebroadcasting at road intersections are used as techniques to reduce the amount of data sent.

Safety. While significant activities are underway to address wireless communication in VANETs (see Sect. 17.4), in general this work only addresses best-effort communication suitable for non-safety-critical applications. Future safety-critical automotive applications such as adaptive cruise control will, however, require guaranteed real-time message delivery. To address this issue, RTSTEAM [562] middleware has been designed for real-time communication in wireless (vehicular) networks. This middleware exploits proximity-based event propagation to guarantee periodic real-time communication only within a dynamic (varying over time) proximity.

Security. Authentication is the main challenge for securing VANETs. A vehicle could falsely report that a road is jammed with traffic or impersonate another vehicle or roadside infrastructure to trigger safety hazards. Networks of trust could be used but broadcast authentication is really challenging in such dynamic environments as vehicles might not have met before. Electronic license plates, tamper-proof GPS and symmetric keys exchange for vehicle coordination are possible solutions [697].

17.3.2 Extensions to Existing System Architectures

In this section, we review system architectures that have been extended to meet the specific requirements of vehicular networks. As noted earlier, most middleware represent adaptations to or extensions of middleware architectures that feature loose coupling between systems, notably the event-based, peer-to-peer, and service-oriented architectures.

Publish-Subscribe. [293] presents an event-based middleware that enriches events and subscriptions with context information including the notions of space and time. This allows publishers to constrain the diffusion of events and to define persistent events that should remain available for a specified time after their publication. Similarly, subscribers can subscribe to events that are relevant in specified contexts and originate from publishers belonging to a particular context.

The middleware designed in [507] additionally exploits the vehicles' navigation system (i) to automatically generate interests and to filter incoming notifications, (ii) to exploit contextual information contained in the database (location, destination, points of interests, etc.), and (iii) to efficiently perform geographical routing by using the suggested routes of the navigation system.

Peer-to-Peer. Mobile ad hoc peer-to-peer (P2P) networking is yet another style for organizing vehicular networks that assures flexibility of interconnections, expansion of bandwidth and enhancement of computing and storage capabilities [173].

C2P2 [311] is a car-to-car P2P network that provides on-demand delivery of continuous audio and video clips to moving vehicles. Vehicles traversing a road segment broadcast their belief on the state of the segment, *e.g.*, a belief that the segment is congested. Their navigation system can then make route planning decisions depending on available information about traffic conditions turning the transportation network into a controlled network.

Rather than focusing on infotainment, [174] proposes a P2P Vehicular Communication Platform (VCP) dedicated to supporting the exchange of safety-relevant information between traffic participants (*e.g.*, road accidents). Vehicles and fixed roadside entities can use services and resources offered by other traffic participants. Participants are organized in dynamic communication zones named Peer Spaces (PS), in which they share a common interest.

Service-Oriented Computing. The architectural principle underlying service-oriented computing is to separate functions into loosely-coupled distinct services which can be distributed over a network and can be combined and reused to create applications.

MYCAREVENT is a European Project of the 6th Framework Program that has developed service-oriented V2I middleware for vehicular applications directed to the support of the roadside assistance, the workshops, and the driver in a breakdown situation [860]. The proposed architecture is divided into two layers that interact with vehicular applications through Web service interfaces.

EmergentITS [701] has demonstrated the use of the Jini technology for providing a V2V service-based architecture to remote in-vehicle systems and Palm devices. The middleware is organized into two service groups: (i) lightweight application services that deliver information to remote vehicles (e.g., navigation service) and (ii) management services that monitor application services and provide configuration and fault-tolerance capabilities.

The VMTL project [309] proposes to minimize communication by putting the evaluation logic to monitor vehicle data in the vehicles themselves. The middleware is structured in two parts: (i) a protocol part based on SOAP [586], and (ii) a service part including basic services (e.g., service publishing or monitoring) and application services that can be plugged in at runtime.

To tolerate partial disconnections frequent in vehicular networks, JNomad integrates service-oriented computing (Jini technology) with mobile agents [343]. The framework essentially allows a service to relocate itself to another vehicle to overcome the fluctuations in its connectivity.

17.4 Internet Connectivity

The connectivity of applications in VANETs encompasses mechanisms that enable the transfer of information between nodes and ensure the reliability of communication between applications. These mechanisms have been integrated in a number of protocols that provide functionality such as the determination of routes for information, the handling of errors and the control of information flow. The communication protocols that implement this functionality have been developed for wired networks over the last 30 years. However, VANETs exhibit specific characteristics (see Sect. 17.1.3) that often conflict with the assumptions made in the development of traditional communication protocols. In the following we will discuss proposed solutions for VANETs in three categories: the routing of packets, the provision of IP connectivity and the transport of data between applications.

17.4.1 Routing

A number of solutions have been derived from routing protocols for generic MANETs. These approaches aim at specializing these routing protocols by adapting them to the characteristics of VANETs. In the following sections, we first discuss a number of mobility models that have been suggested as the basis for the development of routing protocols and then examine a number of routing protocols that have been developed to provide the forwarding of traffic in VANETs.

Mobility Modeling. Nodes in VANETs are restricted in their movement by physical phenomena such as lanes, junctions, and access ramps. These limitations have led to the development of mobility models that aim to represent more accurately the movement of nodes in VANETs [34]. Two categories of mobility models can be distinguished: mobility models that derive their movement patterns from randomized decisions and models that derive their topology restrictions from maps such as TIGER [831].

Solutions are generally evaluated using mobility models such as Random Walk [250] or the Random Waypoint [120]. The Random Walk model specifies that a node moves in a given direction at a given speed. Broch et al. [120] modified this model by introducing a pause time that a node wait when it reaches a destination before continuing to the next waypoint. In order to adapt this model to the behavior of vehicles, Bettstetter et al. [79] proposed the Smooth Random Mobility model that introduces times for acceleration and slow down. A vehicle that leaves a waypoint accelerates towards the target velocity over a given time and slows down to a stop before reaching a destination.

A variety of mobility models for VANETs argue against the use of random trajectories for nodes and propose the use of map-based topologies through which nodes can navigate. The City Section mobility model [216] assumes a grid of streets. The center of the grid represents the main thorough fair of a city with mid-speed streets while rows and columns further from the center represent streets in the suburbs with low speeds. Similarly, the STRAW mobility model [177] is based on street maps with a simple car-following model.

The above mobility models are closely connected to routing protocols. Fuessler et al. [299] suggest a classification of routing protocols following two criteria: the *method* that is used to construct a route and the *time* of the construction of a route. In topology-based methods, a route is established according to neighboring nodes, whereas in position-based approaches a route is established according to the availability of location information. Both methods are employed either proactively (as an ongoing process) or reactively (as reaction to the need of a specific route).

Topology-Based Routing Approaches. A variety of routing protocols that have been developed for generic MANETs have been adapted to the specific characteristics of VANETs. Singh et al. [758] discuss an adaptation of the Optimized Link State Routing protocol (OLSR) that takes into account the history of Signal-to-Noise Ratio (SNR) and the development of affinity over time. Sommer and Dressler [764] evaluate DYMO in a number of VANET scenarios.

Position-Based Routing Approaches. Takano et al. [793] classify position-based routing approaches into approaches based on next-hop forwarding or directed flooding.

Karp and Kung [444] have proposed the Greedy Perimeter Stateless Routing (GPSR) protocol, that forwards a packet to the node closest to the destination of the packet. Festag et al. [277] proposed a modification to GPSR that controls the transmission power in order to increase the number of possible transmissions in a given area. The Greedy Perimeter Coordinator Routing (GPCR) [531] protocol uses a greedy forwarding mechanism similar to GPSR, but forwards packets based on the distribution of nodes and junctions (based on a map).

Lochert et al. [530] suggest a Geographic Source Routing approach that is based on the availability of a map and a Reactive Location Service. The location service uses a flooding approach to determine the location of a destination. The routing protocol then computes the junctions that a packet has to traverse in order to travel from the source to the destination. The Greedy Traffic Aware Routing Protocol [420] takes a similar approach with the availability of a map and a location service such as the Grid Location Service [512]. The junctions that a packet should traverse on its way to the destination are determined on a junction-by-junction basis, depending on the traffic density and distance from the destination. The REACT protocol by Van de Velde et al. [835] uses the trajectories of nodes derived from their positions to elect the next forwarder of a packet.

17.4.2 Internet Protocol

The internet protocol (IP) is one of the major protocols at the network layer of the OSI stack. This protocol provides the connectivity between two nodes in possibly different local area networks (LANs). Maihoefer et al. [545] discuss a number of implementation alternatives for the provision of IP in VANETs.

Mobile IP and Proxies. In the FleetNet project [291] a combination of NAT and protocol translation are used to connect nodes in VANETs to infrastructure networks. Nodes in a VANET are assigned local IP addresses and communicate with gateways outside the VANET. The MOCCA architecture [72] represents an IPv6-like implementation. Bechler et al. [73] discuss the discovery of access points in VANETs in detail.

17.4.3 Transport Protocols

A number of approaches have been proposed to provide TCP or TCP-like support for applications in MANETs; for example, ATCP [526], ATP [780], Split-TCP [470], SCTP over DSR [30]. In the following, we will discuss new approaches suitable for VANETs that either adapt traditional TCP implementations to VANETs or devise alternative transport protocols to exploit the specific characteristics of VANETs.

TCP in V2V and V2I. TCP-Vegas exhibits the best performance when compared with TCP-Reno and TCP-NewReno in dynamic environments [650]. However, the use of TCP-Vegas in MANETs is flawed because a route change can render the round trip time, BaseRTT, that has been used over a previous route, obsolete. TCP-Vegas-ad hoc [456] addresses this issue by introducing a feedback mechanism that informs the sender about route changes and allows the round trip time to be adjusted.

Alternative Transport Protocols. Two categories of alternative protocols have been proposed: protocols that provide streamed communication from a single source application to a single destination application, and protocols that provide packet distribution in a geographically restricted area.

Schmitz et al. investigate connectivity in VANETs and propose a preliminary design for a Vehicular Transport Protocol [734]. This protocol takes into account the distance between communication endpoints and statistical knowledge about connectivity between vehicles on highways.

The Vehicle Information Transfer Protocol (VITP) [229] falls into the second category as it forwards messages between nodes in a geographic area during a request phase until a return condition is fulfilled. The messages are defined as URIs that may include an expiration time, a time-to-live or a number of expressions that define the return condition. The Robust Data Transfer Protocol [179] provides a similar flooding approach as VITP. In this protocol, however, nodes at the end of a query region as well as intermediate nodes generate replies that are send back to the source of a request.

17.5 Summary and Outlook

Inter-vehicle communication enables a wide sprectrum of new applications. However, work on vehicular infrastructure and middleware to support these applications is still in early stage and there is not much deployment experience in practice. Major research in the area focuses on wireless communication, whose feasibility has now been established. Currently, vehicle-to-vehicle communication is mostly used to extend the horizon of drivers and on-board information systems. Collaboration/self-organisation of multiple vehicles has not been thoroughly investigated.

Glossary

Adaptivity: the degree to which a service adapts to a change in its context.

Application contract: a formal definition of the worst case security-related behavior of an application.

Atomicity: the ability to perform multiple operations on a diverse set of processes in a single atomic step.

Availability: the measure of the delivery of correct service with respect to the alternation between correct and incorrect service.

Broadcast: the process of disseminating a piece of information to all nodes of the system.

Broker: entity that mediates the interaction among two or more participants in a distributed computation; the broker can be, itself, a distributed component (brokers typically form an overlay network).

Buffer overflow: writing past the bounds of an allocated region in memory, overwriting adjacent data.

Configuration: a complete description of the state of all components of a distributed system at some moment in time.

Content-based pub/sub: a publish/subscribe model that supports the routing of messages or packets based on their content.

Context: information about the execution environment and the user profile that can be used to improve the performance of a service.

Context-aware protocol: a protocol that takes context into consideration.

Coordination: a programming paradigm whose goal is to separate the definition of the individual behavior of application components from the mechanics of their interaction.

Cyber foraging: a process by which a mobile device opportunistically probes its surroundings to look for resources offered by other devices, e.g., processing power and storage space.

Dependability: the measure in which reliance can justifiably be placed on the service delivered by the system.

Deterministic algorithm: an algorithm where the next step for a component to perform is uniquely determined by that component's current state.

Disconnected operation: the ability to provide service during network partitions. Since coordination among nodes residing in different partitions is impossible, disconnected operation usually requires the execution of reconciliation procedures when the network is re-connected.

Event flooding: a simple technique for propagating information by flooding it in the distributed environment.

Event matching: the process of associating an event message with a number of subscriptions. For an event to match a subscription, it must satisfy the constraints defined in that subscription.

Event Service: a logically centralized service that provides the publish/subscribe functionality.

Execution: one possible sequence of steps that a distributed system can take while components follow their assigned algorithms.

Fairness: in situations where schedules are unpredictable, a fairness condition is a restriction on what schedules are possible, usually ensuring that different components all get to take steps.

Filter: a constraint on a message. Typically filters are defined using stateless Boolean functions.

Filter-based routing: a routing technique that uses filters to selectively propagate information towards subscribers. Typically, messages are not delivered to those parts of a network that have not expressed prior interest in receiving the messages.

Filter Merging: an optimization technique that aggregates routing information, namely filters. The aggregation is typically based on logical rules or probabilistic techniques.

Flooding: a brute-force method of disseminating information in a network, where each node forwards the data to each and every one of its neighbors.

Gossip: an interaction style where each node periodically perform an exchange of information with one or more of its neighbours, typically chosen at random (see Chap. 8).

Gossiping event routing: a probabilistic event routing technique that does not typically require a particular network structure, and aims to guarantee the delivery of messages in a highly dynamic environment.

Interaction graph: a graph where nodes represent processes in a mobile system. An edge between a pair of processes is included in the graph if it is possible for those two processes to meet and communicate directly with each other.

Machine model: abstract high-level representation of a platform.

Managed platform: platform in which code is compiled into an intermediate language, independent from any processor architecture. Code in the intermediate language is then executed on a virtual machine.

Matching: in pub/sub systems, the process of checking an event against a subscription (see Chap. 10).

Mobile ad hoc network: a network whose components move, and which has no infrastructure set up initially.

Native platform: platform in which code is compiled into native machine code and is executed directly on the CPU.

Opportunistic network: a network where the mobility of nodes is exploited to provide connectivity in scenarios where the source and destination nodes might never be connected to the same network (see Chap. 6)

Overlay network: a logical network constructed on top of an underlying (typically TCP/IP) network.

Platform: combination of architecture, operating system, compiler and programming language.

Population protocol model: a computational model where a collection of very limited devices interact in unpredictable patterns.

Publish/Subscribe: interaction paradigm where each participant can take on the role of a publisher or a subscriber of information. Publishers produce information in form of events, which is consumed by subscribers issuing subscriptions representing their interest only in specific events. Pub/sub introduces indirection by decoupling publishers and subscribers from each other (see Chap. 10).

Publish/Subscribe Mobility Support: a family of mobility support protocols that aim to support the network mobility of either subscribers, publishers, or both. In this context, network mobility means that a subscribing or publishing endpoint relocates from one physical network attachment point to another. In order to support seamless information delivery, the publish/subscribe event routing topology needs to be adjusted due to this mobility.

Pull based protocol: a protocol where the recipient of the information has the onus of requesting its transmission from potential owners.

Push based protocol: a protocol where the owner of (new) information has the onus of forwarding it to the (potential) recipients.

Random walk model: a mathematical model of a trajectory (derived from Brownian motion), which consists in taking successive random steps.

Random waypoint model: a commonly used model for mobility in ad hoc networks, where each node moves along a line from one waypoint to the next, the waypoints being uniformly distributed; a variant exists for vehicular ad hoc networks known as STRAW (street random waypoint).

Register machine: an abstract model of a sequential computer where the basic steps are instructions that can access an elements of an array.

Reliability: the measure of the continuous delivery of correct service.

Rendezvous node: a third party, known to two or more participants, that is used to establish communication among these participants when they do not known each other a priori.

Rendezvous-based Routing: a routing technique that uses special network nodes, called Rendezvous Points, to coordinate event routing table updates and message propagation. Rendezvous-based routing is typically implemented using an overlay network.

Schedule: the order in which different components of a system execute steps of their algorithms.

Selective event routing: a process that aims to reduce the size of the distributed routing tables and the protocol overhead by selectively propagating routing information and messages.

Self-stabilizing algorithm: an algorithm that eventually begins exhibiting correct behavior even when all components of the system are initialized in arbitrary states.

Semilinear: a set of vectors is called semilinear if it is a union of finitely many sets of the form $\{\mathbf{b} + k_1\mathbf{a}_1 + k_2\mathbf{a}_2 + \cdots + k_m\mathbf{a}_m \mid k_1, \ldots, k_m \in \mathbb{N}\}$, for some m and vectors $\mathbf{b}, \mathbf{a}_1, \ldots, \mathbf{a}_m$. A predicate on vectors is semilinear if the set of vectors for which it is true is semilinear.

Stack cookie: protection against buffer overflows by placing a random secret value before the return address on function entry and verifying the number on function exit.

System policy: a formal definition of the worst case security-related behavior of an application that is allowed by the system.

SxC: a mechanism to enforce system policies upon applications, using different enforcement techniques. Applications typically come with contracts, which can be used in the enforcement process.

Topic-based Routing: a publish/subscribe model that supports the routing of messages or packets based on a particular topic. This topic has to be agreed beforehand by the communicating entities.

Tuple: a sequence of typed fields, as in ⟨"foo", 9, 27.5⟩, containing the information being communicated in a tuple space.

Tuple space: a globally accessible, persistent, content-addressable data structure containing elementary data structures called tuples.

Type-based Routing: a publish/subscribe model that supports the routing of messages or packets based on a type definition.

References

1. M. Abadi, M. Budiu, Ú. Erlingsson, and J. Ligatti. Control-flow Integrity. In *Proceedings of the 12th ACM Conference on Computer and Communications Security*, pages 340–353, Alexandria, Virginia, U.S.A., November 2005. The ACM Press.
2. T. Abdelzaher, B. Blum, Q. Cao, Y. Chen, D. Evans, J. George, S. George, L. Gu, T. He, S. Krishnamurthy, L. Luo, S. Son, J. Stankovic, R. Stoleru, and A. Wood. EnviroTrack: Towards an Environmental Computing Paradigm for Distributed Sensor Networks. In *Proceedings of the 23th International Conference on Distributed Computing Systems (ICDCS'04)*, 2004.
3. G. Abowd, A. Dey, P. Brown, N. Davies, M. Smith, and P. Steggles. Towards a Better Understanding of Context and Context-Awareness. In *Proceedings of the 1st International Symposium on Handheld and Ubiquitous Computing (HUC'99)*, pages 304–307, London, UK, 1999. Springer-Verlag.
4. Ad Hoc On-Demand Distance Vector Routing. Available at http://moment.cs.ucsb.edu/AODV.
5. A. Adya, P. Bahl, J. Padhye, A. Wolman, and L. Zhou. A Multi-Radio Unification Protocol for IEEE 802.11 Wireless Networks. In *Proceedings of the First International Conference on Broadband Networks (BROADNETS'04)*, pages 344–354. IEEE Press, 2004.
6. I. Aekaterinidis and P. Triantafillou. PastryStrings: A Comprehensive Content-Based Publish/Subscribe DHT Network. In *Proceedings of the 26th IEEE International Conference on Distributed Computing and Systems (ICDCS'06)*, July 2006.
7. D. Aguayo, J. Bicket, S. Biswas, G. Judd, and R. Morris. Link-level Measurements from an 802.11b Mesh Network. In *Proceedings of the 2004 Conference on Applications, Technologies, Architectures, and Protocols for Computer Communications (SGCOMM 04)*, volume 34, pages 121–132. The ACM Press, October 2004.
8. M. Aguilera, R. Strom, D. Sturman, M. Astley, and T.D. Chandra. Matching Events in a Content-Based Subscription System. In *Proceedings of The ACM Symposium on Principles of Distributed Computing (PODC 1999)*, pages 53–61, 1999.
9. H. Ailisto, P. Alahuhta, V. Haataja, V. Kyllönen, and M. Lindholm. Structuring Context Aware Applications: Five-Layer Model and Example Case. In *Workshop at UbiComp 2002*, 2002.
10. Akogrimo—Access to Knowledge through the Grid in a Mobile World. Available at http://www.akogrimo.org.
11. I. Akyildiz and I. Kasimoglu. Wireless Sensor and Actor Networks: Research Challenges. *Ad Hoc Networks Journal*, 2(4), 2004.

12. I. Akyildiz, W. Su, Y. Sankarasubramaniam, and E. Cayirci. Wireless Sensor Networks: A Survey. *Computer Networks*, 38(4):393–422, March 2002.

13. S. Alamouti. A Simple Transmit Diversity Technique for Wireless Communications. In *IEEE Journal on Selected Areas in Communications*, pages 1451–1458, 1998.

14. R. Albert and A.-L. Barabasi. Statistical Mechanics of Complex Networks. *Review of Modern Physics*, 74:47–97, 2002.

15. S. Alexander. Defeating Compiler-level Buffer Overflow Protection. *;login: The USENIX Magazine*, 30(3), June 2005.

16. M. Allani, B. Garbinato, F. Pedone, and M. Stamenkovic. Scalable and Reliable Stream Diffusion: A Gambling Resource-Aware Approach. In *Proceedings of 26th IEEE Symposium on Reliable Distributed Systems (SRDS'2007)*, pages 288–297, Beijing, China, 2007.

17. F. Almenarez and C. Campo. SPDP: A Secure Service Discovery Protocol for Ad-Hoc Networks. In *Proceedings of EUNICE 2003 9th Open European Summer School and IFIP Workshop on Next Generation Networks*, 2003.

18. M. Altherr, M. Erzberg, and S. Maffeis. iBus—A Software Bus Middleware for the Java Platform. In *Proceedings of the International Workshop on Reliable Middleware Systems*, pages 49–65, October 1999.

19. E. Anceaume, A. Datta, M. Gradinariu, and G. Simon. Publish/Subscribe Scheme for Mobile Networks. In *Proceedings of the 2nd Workshop on Principles of Mobile Computing (POMC 2002)*, pages 74–81, 2002.

20. D. Angluin, J. Aspnes, Z. Diamadi, M. Fischer, and R. Peralta. Urn Automata. Technical Report YALEU/DCS/TR-1280, Yale University Department of Computer Science, November 2003.

21. D. Angluin, J. Aspnes, M. Chan, M. Fischer, H. Jiang, and R. Peralta. Stably Computable Properties of Network Graphs. In *Proceedings of the 1st IEEE International Conference on Distributed Computing in Sensor Systems*, pages 63–74, 2005.

22. D. Angluin, J. Aspnes, D. Eisenstat, and E. Ruppert. On the Power of Anonymous One-way Communication. In *Proceedings on 9th International Conference on Principles of Distributed Systems*, pages 396–411, 2005.

23. D. Angluin, J. Aspnes, M. Fischer, and H. Jiang. Self-stabilizing Population Protocols. In *Proceedings Principles of Distributed Systems, 9th International Conference*, pages 103–117, 2005.

24. D. Angluin, J. Aspnes, Z. Diamadi, M. Fischer, and R. Peralta. Computation in Networks of Passively Mobile Finite-State Sensors. *Distributed Computing*, 18(4):235–253, March 2006.

25. D. Angluin, J. Aspnes, and D. Eisenstat. Stably Computable Predicates are Semilinear. In *Proceedings of the 25th Annual ACM Symposium on Principles of Distributed Computing*, pages 292–299, 2006.

26. D. Angluin, M. Fischer, and H. Jiang. Stabilizing Consensus in Mobile Networks. In *Proceedings of the 2nd IEEE International Conference on Distributed Computing in Sensor Systems*, pages 37–50, 2006.

27. D. Angluin, J. Aspnes, D. Eisenstat, and E. Ruppert. The Computational Power of Population Protocols. *Distributed Computing*, 20(4):279–304, 2007.

28. D. Angluin, J. Aspnes, and D. Eisenstat. Fast Computation by Population Protocols with a Leader. *Distributed Computing*, 21(3):183–199, September 2008.

29. D. Angluin, J. Aspnes, and D. Eisenstat. A Simple Population Protocol for Fast Robust Approximate Majority. *Distributed Computing*, 21(2):87–102, July 2008.

30. A. Argyriou and V. Madisetti. Using a New Protocol to Enhance Path Reliability and Realize Load Balancing in Mobile Ad Hoc Networks. *Ad Hoc Networks*, 4(1), January 2006.

31. G. Asada, M. Dong, T. Lin, F. Newberg, G. Pottie, W. Kaiser, and H. Marcy. Wireless Integrated Network Sensors: Low Power Systems on a Chip. In *Proceedings of the 24th European Solid-State Circuits Conference (ESSCIRC)*, pages 9–16, Hague, Netherlands, September 1998.

32. J. Aspnes and E. Ruppert. An Introduction to Population Protocols. *Bulletin of the EATCS*, 93:98–117, October 2007.

33. D. Athanasopoulos, A. Zarras, V. Issarny, E. Pitoura, and P. Vassiliadis. CoWSAMI: Interface-Aware Context Gathering in Ambient Intelligence Environments. *Pervasive and Mobile Computing Journal (PMC)*, 4(3):360–389, June 2008.

34. E. Atsan and O. Oezkasap. A Classification and Performance Comparison of Mobility Models. *5th International Conference on AD-HOC Networks and Wireless (ADHOC-NOW'06)*, August 2006.

35. D. Austin, W. Bowen, and J. McMillan. Intraspecific Variation in Movement Patterns: Modeling Individual Behaviour in a Large Marine Predator. *Oikos*, 105(1):15–30, 2004.

36. M. Avvenuti, D. Pedroni, and A. Vecchio. Core Services in a Middleware for Mobile Ad-Hoc Networks. In *Proceedings of the 9th IEEE Workshop on Future Trends of Distributed Computing Systems (FTDCS'03)*, pages 152–158. IEEE Press, May 2003.

37. A. Awan, S. Jagannathan, and A. Grama. Macroprogramming Heterogeneous Sensor Networks Using Cosmos. *SIGOPS Operating Systems Review*, 41(3):159–172, 2007.

38. S. Baehni, C. Chhabra, and R. Guerraoui. Mobility Friendly Publish/Subscribe. In *EPFL Technical Report 200488*, 2004.

39. S. Baehni, P. Eugster, and R. Guerraoui. Data-Aware Multicast. In *Proceedings of the 2004 International Conference on Dependable Systems and Networks (DSN 2004)*, pages 233–242, 2004.

40. S. Baehni, C. Chhabra, and R. Guerraoui. Frugal Event Dissemination in a Mobile Environment. In *The 7th International Middleware Conference*, pages 205–224, 2005.

41. R. Baeza-Yates and B. Ribeiro-Neto. *Modern Information Retrieval*. ACM Press, Addison Wesley, 1999.

42. R. Baeza-Yates, J. Culberson, and G. Rawlins. Searching in the Plane. *Information and Computation*, 106(2):234–252, 1993.

43. R. Bagrodia, R. Meyer, R. Takai, Y. Chen, X. Zeng, J. Martin, B. Park, and H. Song. A Parallel Simulation Environment for Complex Systems. *IEEE Computer*, pages 77–85, October 1998.

44. F. Bai, N. Sadagopan, and A. Helmy. IMPORTANT: A Framework to Systematically Analyze the Impact of Mobility on Performance of RouTing Protocols for Adhoc NeTworks. In *INFOCOM'03*, 2003.

45. A. Bakshi, V.K. Prasanna, J. Reich, and D. Larner. The Abstract Task Graph: A Methodology for Architecture-Independent Programming of Networked Sensor Systems. In *Workshop on End-to-end Sense-and-respond Systems (EESR)*, 2005.

46. A. Balachandran, G. Voelker, P. Bahl, and P. Rangan. Characterizing User Behavior and Network Performance in a Public Wireless LAN. In *Proceedings of SIGMETRICS'02*, pages 195–205, New York, NY, USA, 2002. ACM Press.

47. H. Balakrishnan, V. Padmanabhan, and R. Katz. The Effects of Asymmetry on TCP Performance. In *Mobile Computing and Networking*, pages 77–89, 1997.

48. R. Balan, J. Flinn, M. Satyanarayanan, S. Sinnamohideen, and H. Yang. The Case for Cyber Foraging. In *Proceedings of the 10th Workshop on ACM SIGOPS European Workshop: Beyond the PC (EW10)*, pages 87–92. The ACM Press, 2002.

49. R. Balan, M. Satyanarayanan, S. Park, and T. Okoshi. Tactics-based Remote Execution for Mobile Computing. In *Proceedings of the 1st International Conference on Mobile Systems, Applications and Services (MobiSys'03)*, pages 273–286. The ACM Press, 2003.

50. R. Balan, D. Gergle, M. Satyanarayanan, and J. Herbsleb. Simplifying Cyber Foraging for Mobile Devices. In *Proceedings of the 5th International Conference on Mobile Systems, Applications and Services (MobiSys'07)*, pages 272–285. The ACM Press, 2007.

51. M. Balazinska and P. Castro. Characterizing Mobility and Network Usage in a Corporate Wireless Local-Area Network. In *1st International Conference on Mobile Systems, Applications, and Services (MobiSys'03)*, San Francisco, CA, May 2003.

52. M. Baldauf, S. Dustdar, and F. Rosenberg. A Survey on Context-Aware Systems. *International Journal of Ad Hoc and Ubiquitous Computing (IJAHUC)*, 2(4):263–277, June 2007.

53. R. Baldoni, M. Contenti, and A. Virgillito. The Evolution of Publish/Subscribe Systems. In André Schiper, Alexander A. Shvartsman, Hakim Weatherspoon, and Ben Y. Zhao, edi-

tors, *Future Trends in Distributed Computing, Research and Position Papers*, volume 2584. Springer-Verlag, 2003.

54. R. Baldoni, R. Beraldi, G. Cugola, M. Migliavacca, and L. Querzoni. Structure-less Content-Based Routing in Mobile Ad Hoc Networks. In *Proceedings of the International Conference on Pervasive Services (ICPS'05)*, Santorini, Greece, July 2005, 2005.

55. R. Baldoni, R. Beraldi, S. Tucci Piergiovanni, and A. Virgillito. On the Modelling of Publish/Subscribe Communication Systems. *Concurrency and Computation: Practice and Experience*, 17(12):1471–1495, 2005.

56. R. Baldoni, C. Marchetti, A. Virgillito, and R. Vitenberg. Content-based Publish-Subscribe over Structured Overlay Networks. In *International Conference on Distributed Computing Systems (ICDCS'05)*, 2005.

57. R. Baldoni, R. Beraldi, V. Quema, L. Querzoni, and S. Tucci-Piergiovanni. TERA: Topic-Based Event Routing for Peer-to-Peer Architectures. In *Proceedings of the 2007 Inaugural International Conference on Distributed Event-Based Systems (DEBS'07)*, pages 2–13. The ACM Press, 2007.

58. D. Balzarotti, P. Costa, and G. Picco. The LighTS Tuple Space Framework and its Customization for Context-Aware Applications. *International Journal on Web Intelligence and Agent Systems (WIAS)*, 5(2):215–231, June 2007.

59. G. Banavar, T. Chandra, B. Mukherjee, J. Nagarajarao, R. Strom, and D. Sturman. An Efficient Multicast Protocol for Content-based Publish-Subscribe Systems. In *Proceedings of International Conference on Distributed Computing Systems (ICDCS'99)*, 1999.

60. S. Banerjee, B. Bhattacharjee, and C. Kommareddy. Scalable Application Layer Multicast. In *SIGCOMM '02: Proceedings of the 2002 Conference on Applications, Technologies, Architectures, and Protocols for Computer Communications*, New York, NY, USA, July 2000. The ACM Press.

61. S. Banerjee, C. Kommareddy, K. Kar, B. Bhattacharjee, and S. Khuller. OMNI: An Efficient Overlay Multicast Infrastructure for Real-Time Applications. *Computer Networks*, 50(6):826–841, 2006.

62. A. Baratloo, N. Singh, and T. Tsai. Transparent Run-Time Defense Against Stack Smashing Attacks. In *USENIX 2000 Annual Technical Conference Proceedings*, pages 251–262, San Diego, California, U.S.A., June 2000. USENIX Association.

63. E. Barrantes, D. Ackley, S. Forrest, T. Palmer, D. Stefanović, and D. Zovi. Randomized Instruction Set Emulation to Disrupt Binary Code Injection Attacks. In *Proceedings of the 10th ACM Conference on Computer and Communications Security (CCS2003)*, pages 281–289, Washington, District of Columbia, U.S.A., October 2003. The ACM Press.

64. L. Barrière, P. Fraigniaud, L. Narayanan, and J. Opatrny. Robust Position-Based Routing in Wireless Ad Hoc Networks with Irregular Transmission Ranges. *Wireless Communications And Mobile Computing Journal*, 2002.

65. J. Barros and S. Servetto. Network Information Flow with Correlated Sources. *IEEE Transactions on Information Theory*, 52(1):155–170, January 2006.

66. M. Bartlett. *Stochastic Population Models in Ecology and Epidemiology*. Methuen, London, 1960.

67. J. Barton and T. Kindberg. The Cooltown User Experience. Technical Report, Hewlett Packard, 2001. http://cooltown.hp.com/mpulse/1103-computon.asp.

68. J. Barton, S. Zhai, and S. Cousins. Mobile Phones Will Become The Primary Personal Computing Devices. In *Proceedings of the 7th IEEE Workshop on Mobile Computing Systems and Applications (WMCSA'06)*, pages 3–9, 2006.

69. R. Baumann, F. Legendre, and P. Sommer. Generic Mobility Simulation Framework (GMSF). In *Proceedings of First ACM SIGMOBILE International Workshop on Mobility Models for Networking Research (MobilityModels'08)*, Hong Kong S.A.R., China, May 2008.

70. J. Beauquier, J. Clement, S. Messika, L. Rosaz, and B. Rozoy. Self-Stabilizing Counting in Mobile Sensor Networks with a Base Station. In *Proceedings of the 21st International Symposium on Distributed Computing (DISC)*, volume 4731 of *LNCS*, pages 63–76, 2007.

71. S. Bechhofer, F. van Harmelen, J. Hendler, I. Horrocks, D. McGuinness, P. Patel-Schneider, and L. Stein. OWL Web Ontology Language Reference. http://www.w3.org/TR/owl-ref, 2004.

72. M. Bechler, W. Franz, and L. Wolf. Mobile Internet Access in FleetNet. In *13. Fachtagung Kommunikation in Verteilten Systemen (KiVS'03)*, Leipzig, Germany, February 2003.

73. M. Bechler, L. Wolf, O. Storz, and W. Franz. Efficient Discovery of Internet Gateways in Future Vehicular Communication Systems. In *57th IEEE Vehicular Technology Conference (VTC'03)*, Jeju Island, Korea, April 2003.

74. BelAir Networks. Available at http://www.belairnetworks.com.

75. E. Belding-Royer. Multi-level Hierarchies for Scalable Ad Hoc Routing. *Wireless Networks*, 9:461–478, 2003.

76. P. Bellavista, A. Corradi, R. Montanari, and C. Stefanelli. Context-Aware Middleware for Resource Management in the Wireless Internet. *IEEE Transactions on Software Engineering*, 29(12):1086–1099, 2003.

77. N. Bencomo, P. Grace, C. Flores Cortes, D. Hughes, and G. Blair. Genie: Supporting the Model Driven Development of Reflective, Component-based Adaptive Systems. In *Proceedings of the 30th International Conference on Software Engineering (ICSE'08)*, pages 811–814, New York, NY, USA, 2008. The ACM Press.

78. Beowulf.org. Available at http://www.beowulf.org.

79. C. Bettstetter. Smooth is Better than Sharp: A Random Mobility Model for Simulation of Wireless Networks. In *Proceedings of the 4th ACM International Symposium on Modeling, Analysis and Simulation of Wireless and Mobile Systems*, pages 19–27. ACM, 2001.

80. C. Bettstetter, G. Resta, and P. Santi. The Node Distribution of the Random Waypoint Mobility Model for Wireless Ad Hoc Networks. *IEEE Transactions on Mobile Computing*, 2(3):257–269, 2003.

81. C. Bettstetter, H. Hartenstein, and X. Pérez-Costa. Stochastic Properties of the Random Waypoint Mobility Model. *Wireless Networks (Special Issue on Modeling and Analysis of Mobile Networks)*, 10(5):555–567, 2004.

82. P. Bhagwat. Bluetooth: Technology For Short-range Wireless Apps. *IEEE Internet Computing*, 5:83–93, 2001.

83. M. Bhardwaj, A. Chandrakasan, and T. Garnett. Upper Bounds on the Lifetime of Sensor Networks. In *Proceedings of the IEEE International Conference on Communications*, pages 785–790, 2001.

84. D. Bhattacharjee, A. Rao, C. Shah, M. Shah, and A. Helmy. Empirical Modeling of Campus-wide Pedestrian Mobility Observations on the USC Campus. In *Proceedings of the Vehicular Technology Conference, 2004 (VTC2004-Fall)*, volume 4, pages 2887–2891, 2004.

85. S. Bhatti, J. Carlson, H. Dai, J. Deng, J. Rose, A. Sheth, B. Shucker, C. Gruenwald, A. Torgerson, and R. Han. MANTIS OS: An Embedded Multithreaded Operating System for Wireless Micro Sensor Platforms. *ACM/Kluwer Mobile Networks & Applications (MONET), Special Issue on Wireless Sensor Networks*, 10(4):563–579, August 2005.

86. S. Bhola, R. Strom, S. Bagchi, Y. Zhao, and J. Auerbach. Exactly-once Delivery in a Content-based Publish-Subscribe System. In *Proceedings of The International Conference on Dependable Systems and Networks*, 2002.

87. J. Bicket, D. Aguayo, S. Biswas, and R. Morris. Architecture and evaluation of an unplanned 802.11b mesh network. In *Proceedings of the 11th Annual International Conference on Mobile Computing and Networking (MobiCom'05)*, pages 31–42. The ACM Press, 2005.

88. K. Birman, M. Hayden, O. Ozkasap, Z. Xiao, M. Budiu, and Y. Minsky. Bimodal Multicast. *ACM Transactions on Computer Systems*, 17(2):41–88, May 1999.

89. S. Biswas, R. Tatchikou, and F. Dion. Vehicle-to-Vehicle Wireless Communication Protocols for Enhancing Highway Traffic Safety. *Communications Magazine*, 44(1):74–82, January 2006.

90. S. Bittner, A. Hinze. A Classification of Filtering Algorithms in Content-based Publish/Subscribe Systems. In *Proceedings of COOPIS 2005*, 2005.

91. S. Bittner and A. Hinze. On the Benefits of Non-Canonical Filtering in Publish/Subscribe Systems. In *Proceedings of the International Workshop on Distributed Event-Based Systems (ICDCS/DEBS'05)*, 2005.

92. S. Björk, J. Falk, R. Hansson, and P. Ljungstrand. Pirates! Using the Physical World as a Game Board. In *Proceedings of the Human-Computer Interaction Conference (Interact'01)*. IOS Press, 2001.

93. G. Blair, G. Coulson, A. Andersen, L. Blair, M. Clarke, F. Costa, H. Duran-Limon, T. Fitzpatrick, L. Johnston, R. Moreira, N. Parlavantzas, and K. Saikoski. The Design and Implementation of Open ORB 2. *IEEE Distributed Systems Online*, 2(6), September 2001.

94. K. Blakely and B. Lowekamp. A structured group mobility model for the simulation of mobile ad hoc networks. In *Proceedings of MobiWac'04*, pages 111–118. The ACM Press, 2004.

95. S. Blostein and H. Leib. Multiple Antenna Systems: Their Role and Impact in Future Wireless Access. In *IEEE Communications Magazine*, pages 94–101, 2003.

96. D. Blough and P. Santi. Investigating Upper Bounds on Network Lifetime Extension for Cell-Based Energy Conservation Techniques in Stationary Ad Hoc Networks. In *Proceedings of the 8th ACM International Conference on Mobile Computing and Networking (MOBICOM'02)*, 2002.

97. Bluetooth Consortium. Specification of the bluetooth system core version 1.0b: Part e, service discovery protocol (SDP), November 1999. Available at http://www.bluetooth.com.

98. Bluetooth Special Interest Group (SIG) trade association website. Available at http://www.bluetooth.com, April 2004.

99. J. Blum, A. Eskandarian, and L. Hoffman. Challenges of Intervehicle Ad Hoc Networks. *IEEE Transactions on Intelligent Transportation Systems*, 5(4), December 2004.

100. H. Bohnenkamp, P. Stok, H. Hermanns, and F. Vaandrager. Cost-optimisation of the IPv4 zeroconf protocol. In *Proceedings of International Performance and Dependability Symposium (IPDS 2003)*, pages 531–540, 2003.

101. C. Boldrini, M. Conti, I. Iacopini, and A. Passarella. HiBOp: A History Based Routing Protocol for Opportunistic Networks. *Proceedings of the IEEE International Symposium on a World of Wireless, Mobile and Multimedia Networks (WoWMoM'07)*, pages 1–12, 2007.

102. C. Boldrini, M. Conti, and A. Passarella. Users Mobility Models for Opportunistic Networks: The Role of Physical Locations. *Proceedings of the IEEE Wireless Rural and Emergency Communications Conference (WRECOM'07)*, 2007.

103. C. Boldrini, M. Conti, and A. Passarella. Exploiting Users' Social Relations to Forward Data in Opportunistic Networks: The HiBOp Solution. *Elsevier Pervasive and Mobile Computing*, 2008.

104. J. Bondy and U. Murty. *Graph Theory with Applications*. Elsevier North-Holland, 1976.

105. J. Boner. AspectWerkz Dynamic AOP for Java. In *Proceeding of the 3rd International Conference on Aspect-Oriented Software Development (AOSD 2004)*, March 2004.

106. C. Borcea, D. Iyer, P. Kang, A. Saxena, and L. Iftode. Cooperative Computing for Distributed Embedded Systems. In *Proceedings of the 22nd International Conference on Distributed Computing Systems (ICDCS 2002)*, pages 227–236, July 2002.

107. E. Borgia, M. Conti, F. Delmastro, and L. Pelusi. Lessons from an Ad-Hoc Network Test-Bed: Middleware and Routing Issues. *Ad Hoc & Sensor Wireless Networks, An International Journal*, 1(1–2), 2005.

108. P. Bose and P. Morin. Online Routing in Triangulations. In *Proceedings of the 10th Annual International Symposium on Algorithms and Computation (ISAAC'99)*, 1999.

109. P. Bose, P. Morin, I. Stojmenovic, and J. Urrutia. Routing with Guaranteed Delivery in ad hoc Wireless Networks. In *International Workshop on Discrete Algorithms and Methods for Mobile Computing and Communications (DIALM'99)*, pages 48–55, 1999.

110. S. Bouckaert, J. Bergs, D. Naudts, J. Baekelmans, E. De Kegel, N. van den Wijngaert, C. Blondia, I. Moerman, and P. Demeester. A Mobile Crisis Management System for Emergency Services: From Concept to Field Test. In *Proceedings of the 1st International Workshop on Wireless Mesh: Moving Towards Applications (WiMeshNets'06)*. The ACM Press, 2006.

111. G. Boudol. Asynchrony and the Pi-calculus. Technical Report 1702, Institut National de Recherche en Informatique et en Automatique (INRIA), 1992.

112. A. Boulis, C. Han, and M. Srivastava. Design and Implementation of a Framework for Efficient and Programmable Sensor Networks. In *Proceedings of the 1st International Conference on Mobile Systems, Applications and Services (MobiSys'03)*, pages 187–200. The ACM Press, 2003.

113. L. Bouraoui, S. Petti, A. Laouiti, T. Fraichard, and M. Parent. Cybercar Cooperation for Safe Intersections. In *Conference on Intelligent Transportation Systems*, Toronto, Ontario, September 2006.

114. A. Bozdog, R. van Renesse, and D. Dumitriu. SelectCast—A Scalable and Self-Repairing Multicast Overlay Routing Facility. In *Proceedings of the 1st ACM Workshop on Survivable and Self-Regenerative Systems*, 2003.

115. B. Bray. Compiler Security Checks In Depth. Available at http://msdn.microsoft.com/library/en-us/dv_vstechart/html/vctchCompilerSecurityChecksInDepth.asp, February 2002.

116. B. Bray. Security Improvements to the Whidbey Compiler. Available at http://weblogs.asp.net/branbray/archive/2003/11/11/51012.aspx, November 2003.

117. T. Bray, J. Paoli, and C. Sperberg-McQueen. Extensible Markup Language (XML) 1.0. *World Wide Web Consortium Recommendation*, February 1998.

118. S. Brin and L. Page. The Anatomy of a Large-Scale Hypertextual Web Search Engine. In *Proceedings of the 7th International Conference on World Wide Web*, pages 107–117, April 1998.

119. Broadband and Wireless Network Laboratory. Available at http://www.ece.gatech.edu/research/labs/bwn/mesh.

120. J. Broch, D. Maltz, D. Johnson, Y.-C. Hu, and J. Jetcheva. A Performance Comparison of Multi-hop Wireless Ad Hoc Network Routing Protocols. In *Proceedings of the 4th Annual ACM/IEEE International Conference on Mobile Computing and Networking (MOBICOM'98)*, pages 85–97. The ACM Press, 1998.

121. P. Brown. The Stick-e Document: A Framework for Creating Context-aware Applications. In *Proceedings of EP'96, Palo Alto*, volume 8, pages 259–272. Electronic Publisher, January 1996.

122. R. Bruno, M. Conti, and E. Gregori. Mesh Networks: Commodity Multihop Ad Hoc Networks. *IEEE Communications Magazine*, pages 123–131, March 2005.

123. R. Bruno, M. Conti, and A. Passarella. Opportunistic Networking Overlays for ICT Services in Crisis. In *Proceedings ISCRAM*, May 2008.

124. I. Burcea, H.-A. Jacobsen, E. de Lara, V. Muthusamy, and M. Petrovic. Disconnected Operation in Publish/Subscribe Middleware. In *Mobile Data Management*, 2004.

125. J. Burgess, B. Gallagher, D. Jensen, and B. Levine. MaxProp: Routing for Vehicle-Based Disruption-Tolerant Networks. *Proceedings of the 25th IEEE Annual Joint Conference of the IEEE Computer and Communications Societies (INFOCOM'06)*, 2006.

126. B. Burns, O. Brock, and B. Levine. MV Routing and Capacity Building in Disruption Tolerant Networks. *Proceedings of the 24th IEEE Annual Joint Conference of the IEEE Computer and Communications Societies (INFOCOM'05)*, 2005.

127. N. Busi and G. Zavattaro. Publish/Subscribe vs. Shared Dataspace Coordination Infrastructures: Is It Just a Matter of Taste? In *Proceedings of the 10th Int. Wkshp. on Enabling Technologies (WETICE)*, 2001.

128. R. Calegari, M. Musolesi, F. Raimondi, and C. Mascolo. CTG: A Connectivity Trace Generator for Testing the Performance of Opportunistic Mobile Systems. In *Proceedings of the European Software Engineering Conference and the International ACM SIGSOFT Symposium on the Foundations of Software Engineering (ESEC/FSE07)*, Dubrovnik, Croatia, September 2007. The ACM Press.

129. F. Cali, M. Conti, and E. Gregori. Dynamic tuning of the IEEE 802.11 Protocol to Achieve a Theoretical Throughput Limit. *IEEE/ACM Transaction on Networking*, 8(6):785–799, 2000.

130. T. Camp, J. Boleng, and V. Davies. A Survey of Mobility Models for Ad Hoc Network Research. *Wireless Communication and Mobile Computing Special Issue on Mobile Ad Hoc Networking: Research, Trends and Applications*, 2(5):483–502, 2002.

131. A. Campailla, S. Chaki, E.M. Clarke, S. Jha, and H. Veith. Efficient Filtering in Publish-Subscribe Systems Using Binary Decision Diagrams. In *Proceedings of the International Conference on Software Engineering*, pages 443–452, 2001.

132. F. Cao and J. Singh. Efficient Event Routing in Content-based Publish-Subscribe Service Networks. In *Proceedings of the 23rd Conference on Computer Communications (IEEE IN-FOCOM 2004)*, 2004.

133. F. Cao and J. Singh. MEDYM: Match-Early with Dynamic Multicast for Content-based Publish-Subscribe Networks. In *Proceedings of the ACM/IFIP/USENIX 6th International Middleware Conference (Middleware 2005)*, 2005.

134. G. Cao, L. Yin, and C. Das. Cooperative Cache-Based Data Access in Ad Hoc Networks. *IEEE Computer*, 37(2):32–39, February 2004.

135. M. Caporuscio, A. Carzaniga, and A. Wolf. Design and Evaluation of a Support Service for Mobile, Wireless Publish/Subscribe Applications. *IEEE Transactions on Software Engineering*, 29(12):1059–1071, December 2003.

136. L. Capra, W. Emmerich, and C. Mascolo. A Micro-Economic Approach to Conflict Resolution in Mobile Computing. In *Proceedings of the 10th ACM SIGSOFT Symposium on Foundations of Software Engineering (SIGSOFT'02/FSE-10)*, pages 31–40, New York, NY, USA, 2002. The ACM Press.

137. L. Capra, W. Emmerich, and C. Mascolo. CARISMA: Context-Aware Reflective Middleware System for Mobile Applications. *IEEE Transactions on Software Engineering*, 29(10):929–945, October 2003.

138. M. Capruscio, P. Inverardi, and P. Pellicione. Formal Analysis of Client Mobility in the Siena Publish-Subscribe Middleware. Technical Report, University of L'Aquila, 2002.

139. J. Carroll, I. Dickinson, C. Dollin, D. Reynolds, A. Seaborne, and K. Wilkinson. Jena: Implementing the Semantic Web Recommendations. Technical Report, Hewlett Packard, 2003.

140. J. Cartigny and D. Simplot. Border Node Retransmission Based Probabilistic Broadcast Protocols in Ad-Hoc Networks. *Telecommunication Systems*, 22(1–4):189–204, 2003.

141. A. Carzaniga, A. Wolf. Forwarding in a Content-Based Network. In *Proceedings of ACM SIGCOMM 2003*, pages 163–174, 2003.

142. A. Carzaniga, D. Rosenblum, and A. Wolf. Achieving Scalability and Expressiveness in an Internet-Scale Event Notification Service. In *Proceedings of the ACM Symposium on Principles of Distributed Computing*, pages 219–227, 2000.

143. A. Carzaniga, D. Rosenblum, and A. Wolf. Design and Evaluation of a Wide-Area Notification Service. *ACM Transactions on Computer Systems*, 3(19):332–383, August 2001.

144. A. Carzaniga, M. Rutherford, and A. Wolf. A Routing Scheme for Content-Based Networking. In *Proceedings of IEEE INFOCOM 2004*, 2004.

145. P. Castro, B. Greenstein, R. Muntz, C. Bisdikian, P. Kermani, and M. Papadopouli. Locating Application Data Across Service Discovery Domains. In *Proceedings of the 7th Annual International Conference on Mobile Computing and Networking (MobiCom'01)*, pages 28–42. The ACM Press, 2001.

146. M. Castro, P. Druschel, A.-M. Kermarrec, and A. Rowston. SCRIBE: A Large-Scale and Decentralized Application-Level Multicast Infrastructure. *IEEE Journal on Selected Areas in Communications*, 20(8), October 2002.

147. M. Castro, P. Druschel, A.-M. Kermarrec, A. Nandi, A. Rowstron, and A. Singh. SplitStream: High-Bandwidth Multicast in Cooperative Environments. In *SOSP '03: Proceedings of the Nineteenth ACM Symposium on Operating Systems Principles*, pages 298–313, New York, NY, USA, 2003. The ACM Press.

148. D. Cavin, Y. Sasson, and A. Schiper. On the Accuracy of MANET Simulators. In *Proceedings of the Workshop on Principles of Mobile Computing (POMC'02)*, pages 38–43, October 2002.

149. S. Cen, P. Cosman, and G. Voelker. End-to-End Differentiation of Congestion and Wireless Losses. In *Proceedings of ACM Multimedia Computing and Networking*, pages 785–799, 2002.

150. M. Ceriotti, A. Murphy, and G. Picco. Data Sharing vs. Message Passing: Synergy or Incompatibility? An Implementation-driven Case Study. In *Proceedings of the 23rd Symposium on Applied Computing (SAC)*, 2008.

151. A. Cerpa, N. Busek, and D. Estrin. SCALE: A Tool for Simple Connectivity Assessment in Lossy Environments. Technical Report 0021, UCLA Center for Embedded Network Sensing (CENS), September 2003.

152. H.-W. Cha, J.-S. Par, and H.-J. Kim. Support of Internet Connectivity for AODV. Internet draft (work in progress), Internet Engineering Task Force, February 2004.

153. A. Chaintreau, P. Hui, J. Crowcroft, C. Diot, R. Gass, and J. Scott. Pocket Switched Networks: Real-World Mobility and Its Consequences for Opportunistic Forwarding. Technical Report ucam-cl-tr-617, University of Cambridge, Computer Laboratory, February 2005.

154. A. Chaintreau, P. Hui, J. Crowcroft, C. Diot, R. Gass, and J. Scott. Impact of Human Mobility on Opportunistic Forwarding Algorithms. *IEEE Transactions on Mobile Computing*, 6(6):606–620, 2007.

155. I. Chakeres and C. Perkins. Dynamic MANET On-demand (DYMO) Routing. Internet draft, June 2008. Work in progress.

156. D. Chakraborty, A. Joshi, T. Finin, and Y. Yesha. GSD: A Novel Groupbased Service Discovery Protocol for MANETS. In *Proceedings of 4th International Workshop Mobile and Wireless Communications Networks (MWCN'02)*, pages 140–144. IEEE Press, September 2002.

157. A. Chan and S.-N. Chuang. MobiPADS: A Reflective Middleware for Context-Aware Mobile Computing. *IEEE Transactions on Software Engineering*, 29(12):1072–1085, 2003.

158. R. Chand and P. Felber. XNET: A Reliable Content-Based Publish/Subscribe System. In *23rd International Symposium on Reliable Distributed Systems (SRDS 2004)*, pages 264–273, 2004.

159. R. Chand and P. Felber. Semantic Peer-to-Peer Overlays for Publish/Subscribe Networks. In *Parallel Processing, 11th International Euro-Par Conference (Euro-par 2005)*, pages 1194–1204, 2005.

160. T. Chandra and S. Toueg. Unreliable Failure Detectors for Reliable Distributed Systems. *Journal of the ACM*, 43(2):225–267, 1996.

161. K. Chandran, S. Raghunathan, and S. Prakash. A Feed-Back-Based Scheme for Improving TCP Performance in Ad Hoc Wireless Networks. *IEEE Personal Communications*, 8(1):34–39, 2001.

162. H. Chang, H. Du, J. Anda, C.-N. Chuah, D. Ghosal, and H. Zhang. Enabling Energy Demand Response with Vehicular Mesh Networks. In *International Conference on Mobile and Wireless Communication Networks*, volume 162, Paris, France, October 2004.

163. J. Chang and L. Tassiulas. Routing for Maximum System Lifetime in Wireless Ad-Hoc Networks. In *Proceedings of the 37th Annual Allerton Conference on Communication, Control, and Computing*, September 1999.

164. J.-H. Chang and L. Tassiulas. Energy Conserving Routing in Wireless Ad-Hoc Networks. In *Proceedings of the 19th Annual Joint Conference of the IEEE Computer and Communications Societies INFOCOM '00*, pages 22–31, March 2000.

165. Y. Chawathe, S. McCanne, and E. Brewer. RMX: Reliable Multicast for Heterogeneous Networks. In *INFOCOM*, pages 795–804, Tel Aviv, Israel, 2000. IEEE Press.

166. H. Chen, T. Finin, and A. Joshi. An Ontology for Context-Aware Pervasive Computing Environments. *Knowl. Eng. Rev.*, 18(3):197–207, September 2003.

167. W. Chen, M. Hiltunen, and R. Schlichting. Constructing Adaptive Software in Distributed Systems. In *Proceedings of the The 21st International Conference on Distributed Computing Systems (Middleware '04)*, pages 635–643, Washington, DC, USA, 2001. IEEE Computer Society.

168. Y. Chen, K. Schwan, D. Zhou. Opportunistic Channels: Mobility-Aware Event Delivery. In *Middleware*, pages 182–201, 2003.

169. D. Chess. Security Issues in Mobile Code Systems. In *Mobile Agents and Security*, pages 1–14. Springer, Berlin/Heidelberg, 1998.

170. S. Chetan, J. Al-Muhtadi, R. Campbell, and M. Mickunas. Mobile Gaia: A Middleware for Ad-Hoc Pervasive Computing. In *Proceedings of Consumer Communications and Networking Conference (CCNC'05)*, pages 223–228. IEEE Computer Society, January 2005.

171. P. Chevillat, J. Jelitto, A. Barreto, and H. Turong. A Dynamic Link Adaptation Algorithm for IEEE 802.11a Wireless LANs. In *Proceedings of the IEEE International Conference on Communications (ICC)*, pages 111–145, 2003.

172. S. Chiba. Javassist—A Reflection-based Programming Wizard for Java. In *Proceedings of OOPSLA'98 Workshop on Reflective Programming in C++ and Java*, October 1998.

173. I. Chisalita and N. Shahmehri. A Novel Architecture for Supporting Vehicular Communication. In *IEEE International Vehicular Technology Conference (VTC'02)*, volume 2, pages 1002–1006, Vancouver, British Columbia, Canada, September 2002.

174. I. Chisalita and N. Shahmehri. A Peer-to-Peer Approach to Vehicular Communication for the Support of Traffic Safety Applications. In *International Conference on Intelligent Transportation Systems*, Singapore, September 2002.

175. I. Chisalita and N. Shahmehri. A Context-based Vehicular Communication Protocol. In *IEEE International Symposium on Personal, Indoor and Mobile Radio Communications*, volume 4, September 2004.

176. G. Chockler, R. Melamed, Y. Tock, and R. Vitenberg. SpiderCast: A Scalable Interest-Aware Overlay for Topic-Based Pub/Sub Communication. In *Proceedings of the 2007 Inaugural International Conference on Distributed Event-Based Systems (DEBS'07)*, pages 14–25. The ACM Press, 2007.

177. D. Choffnes and F. Bustamante. An Integrated Mobility and Traffic Model for Vehicular Ad Hoc Networks. In *2nd ACM International Workshop on Vehicular Ad Hoc Networks (VANET'05)*, pages 69–78, Cologne, Germany, September 2005.

178. Y. Chu, S. Rao, and H. Zhang. A Case For End System Multicast. In *Proceedings of ACM Sigmetrics, June 2000*, pages 1–12, June 2000.

179. M. Chuah and F. Fu. Performance Study of Robust Data Transfer Protocol for VANETs. *Lecture Notes in Computer Science*, 4325, November 2006.

180. M. Cilia. *An Active Functionality Service for Open Distributed Heterogeneous Environments*. PhD thesis, Department of Computer Science, Darmstadt University of Technology, August 2002.

181. T. Clausen and P. Jacquet. Optimized Link State Routing Protocol. Request for Comments 3626, Internet Engineering Task Force, October 2003.

182. T. Clausen, P. Jacquet (editors), C. Adjih, A. Laouiti, P. Minet, P. Muhlethaler, A. Qayyum, and L. Viennot. Optimized Link State Routing Protocol (OLSR). RFC 3626, October 2003. Network Working Group.

183. T. Clausen, P. Jacquet, A. Laouiti, P. Muhlethaler, A. Qayyum, and L. Viennot. Optimized Link State Routing Protocol. In *Proceedings of the 5th IEEE Multi Topic Conference (INMIC'01)*, December 2001.

184. C. Cleveland, editor. *Encyclopedia of Energy*. Elsevier Academic Press, Amsterdam, Netherlands, 2004.

185. M. Conti and S. Giordano. Multihop Ad Hoc Networking: The Reality. *IEEE Communications Magazine*, 45(4):88–95, 2007.

186. M. Conti and S. Giordano. Multihop Ad Hoc Networking: The Theory. *IEEE Communications Magazine*, 45(4):78–86, 2007.

187. P. Costa, and D. Frey. Publish-Subscribe Tree-Maintenance over a DHT. In *Proceedings of the International Workshop on Distributed Event-Based Systems (ICDCS/DEBS'05)*, 2005.

188. P. Costa, M. Migliavacca, G. Picco, and G. Cugola. Introducing Reliability in Content-Based Publish-Subscribe through Epidemic Algorithms. In *Proceedings of the 2nd International Workshop on Distributed Event-Based Systems (DEBS'03)*, 2003.

189. P. Costa, G. Picco, and S. Rossetto. Publish-Subscribe on Sensor Networks: A Semiprobabilistic Approach. In *Proceedings of the 2nd IEEE International Conference on Mobile Ad Hoc and Sensor Systems (MASS 2005)*, 2005.

190. P. Costa, L. Mottola, A.L. Murphy, and G. Picco. TeenyLIME. Transiently Shared Tuple Space Middleware for Wireless Sensor Networks. In *Proceedings of the 1st Int. Wkshp. on Middleware for Sensor Networks (MidSens)*, 2006.

191. P. Costa, L. Mottola, A. Murphy, and G. Picco. Programming Wireless Sensor Networks with the TeenyLIME Middleware. In *Proceedings of the 8th ACM/USENIX Int. Middleware Conf.*, 2007.

192. P. Costa, C. Mascolo, M. Musolesi, and G. Picco. Socially-Aware Routing for Publish-Subscribe in Delay-Tolerant Mobile Ad Hoc Networks. *IEEE Journal on Selected Areas in Communications*, 26(5):748–760, June 2008.

193. D. Cottingham, A. Beresford, and R. Harle. A Survey of Technologies for the Implementation of National-Scale Road User Charging. *Transport Reviews*, 27(4):499–523, July 2007.

194. G. Coulson, G. Blair, P. Grace, F. Taiani, A. Joolia, K. Lee, J. Ueyama, and T. Sivaharan. A Generic Component Model for Building Systems Software. *ACM Transactions on Computer Systems*, 26(1):1–42, 2008.

195. C. Cowan, C. Pu, D. Maier, H. Hinton, J. Walpole, P. Bakke, S. Beattie, A. Grier, P. Wagle, and Q. Zhang. StackGuard: Automatic Adaptive Detection and Prevention of Buffer-Overflow Attacks. In *Proceedings of the 7th USENIX Security Symposium*, pages 63–78, San Antonio, Texas, U.S.A., January 1998. USENIX Association.

196. C. Cowan, S. Beattie, R. Day, C. Pu, P. Wagle, and E. Walthinsen. Protecting Systems from Stack Smashing Attacks with StackGuard. In *Proceedings of Linux Expo 1999*, Raleigh, North Carolina, U.S.A., May 1999.

197. C. Cowan, M. Barringer, S. Beattie, G. Kroah-Hartman, M. Frantzen, and J. Lokier. Format-Guard: Automatic Protection from Printf Format String Vulnerabilities. In *Proceedings of the 10th USENIX Security Symposium*, pages 191–200, Washington, District of Columbia, U.S.A., August 2001. USENIX Association.

198. C. Cowan, S. Beattie, J. Johansen, and P. Wagle. PointGuard: Protecting Pointers from Buffer Overflow Vulnerabilities. In *Proceedings of the 12th USENIX Security Symposium*, pages 91–104, Washington, District of Columbia, U.S.A., August 2003. USENIX Association.

199. A. Crespo, O. Buyukkokten, and H. Garcia-Molina. Query Merging: Improving Query Subscription Processing in a Multicast Environment. *IEEE Trans. Knowl. Data Eng.*, 15(1):174–191, 2003.

200. G. Cugola and H.-A. Jacobsen. Using Publish/Subscribe Middleware for Mobile Systems. *ACM SIGMOBILE Mobile Computing and Communication Review*, 6(4):25–33, 2002.

201. G. Cugola and J. Munoz de Cote. On Introducing Location Awareness in Publish-Subscribe Middleware. In *Proceedings of the International Workshop on Distributed Event-Based Systems (ICDCS/DEBS'05)*, 2005.

202. G. Cugola and G. Picco. REDS: A Reconfigurable Dispatching System. In *Technical Report, Politecnico di Milano*, 2005.

203. G. Cugola, E. Di Nitto, and A. Fuggetta. Exploiting an Event-Based Infrastructure to Develop Complex Distributed Systems. In *Proceedings of the 10th International Conference on Software Engineering (ICSE'98)*, April 1998.

204. G. Cugola, E. Di Nitto, and A. Fuggetta. The JEDI Event-Based Infrastructure and Its Application to the Development of the OPSS WFMS. *IEEE Transactions on Software Engineering*, 27(9):827–850, September 2001.

205. Y. Cui, B. Li, and K. Nahrstedt. oStream: Asynchronous Streaming Multicast in Application-Layer Overlay Networks. *Selected Areas in Communications, IEEE Journal on*, 22(1):91–106, 2004.

206. D. Culler and H. Mulder. Smart Sensors to Network the World. *Scientific American*, 2004.

207. C. Curino, M. Giani, M. Giorgetta, A. Giusti, A. Murphy, and G. Picco. Mobile Data Collection in Sensor Networks: The TinyLime Middleware. *Elsevier Pervasive and Mobile Computing Journal*, 4(1), 2005.

208. CUWiN. Available at http://www.cuwireless.net.

209. E. Daly and M. Haahr. Social Network Analysis for Routing in Disconnected Delay-Tolerant MANETs. In *Proceedings of MobiHoc'07*, pages 32–40. The ACM Press, 2007.

210. T.-S. Dao, K. Leung, C. Clark, and J. Huissoon. Markov-Based Lane Positioning Using Intervehicle Communication. *IEEE Transactions on Intelligent Transportation Systems*, 8(4):641–650, December 2007.
211. S. Das, R. Castañeda, and J. Yan. Simulation-Based Performance Evaluation of Routing Protocols for Mobile Ad Hoc Networks. *Mobile Networks and Applications*, 5(3):179–189, 2000.
212. S. Das, C. Perkins, and E. Royer. Performance Comparison of Two On-Demand Routing Protocols for Ad Hoc Networks. In *Proceedings of the 9th Annual Joint Conference of the IEEE Computer and Communications Societi* (INFOCOM '00), volume 1, pages 3–12. IEEE Press, March 2000.
213. A. Datta, S. Quarteroni, and K. Aberer. Autonomous Gossiping: A Self-organizing Epidemic Algorithm for Selective Information Dissemination in Mobile Ad-Hoc Networks. In *Proceedings of the International Conference on Semantics of a Networked World*, 2004.
214. O. Davidyuk, J. Riekki, V.-M. Rautio, and J. Sun. Context-Aware Middlewar for Mobile Multimedia Applications. In *Proceedings of the 3rd International Conference on Mobile and Ubiquitous Multimedia (MUM '04)*, pages 213–220, New York, NY, USA, 2004. The ACM Press.
215. N. Davies, A. Friday, S.P. Wade, and G. Blair. L^{2}imbo: A Distributed Systems Platform for Mobile Computing. *Mob. Netw. Appl.*, 3(2):143–156, 1998.
216. V. Davies. Evaluating Mobility Models within an Ad Hoc Network. Master's thesis, Colorado School of Mines, 2000.
217. D. De Couto, D. Aguayo, J. Bicket, and R. Morris. A High-Throughput Path Metric for Multi-Hop Wireless Routing. *Wireless Networks*, 11:419–434, 2005.
218. E. de Lara, D. Wallach, and W. Zwaenepoel. Puppeteer: Component-based Adaptation for Mobile Computing. In *Proceedings of the 3rd USENIX Symposium on Internet Technologies and Systems (USITS'01)*, pages 159–170, Berkeley, CA, USA, 2001. USENIX Association.
219. J. Dedecker. Ambient-Oriented Programming in AmbientTalk: Combining Mobile Hardware with Simplicity and Expressiveness. In *Companion to the 20th Annual ACM SIGPLAN Conference on Object-Oriented Programming, Systems, Languages, and Applications (OOPSLA '05)*, pages 196–197. The ACM Press, 2005.
220. J. Dedecker, T. Van Cutsem, S. Mostinckx, T. D'Hondt, W. De Meuter. Ambient-Oriented Programming in AmbientTalk. In *Proceedings of the 20th European Conference on Object-Oriented Programming (ECOOP'06)*, pages 230–254, 2006.
221. C. Delporte-Gallet, H. Fauconnier, R. Guerraoui, and E. Ruppert. When Birds Die: Making Population Protocols Fault-tolerant. In *Proceedings of the 2nd IEEE International Conference on Distributed Computing in Sensor Systems*, pages 51–66, 2006.
222. A. Demers, D. Greene, C. Hauser, W. Irish, and J. Larson. Epidemic Algorithms for Replicated Database Maintenance. In *Proceedings of the 6th Annual ACM Symposium on Principles of Distributed Computing*, pages 1–12, August 1987.
223. A. Deshpande, C. Guestrin, and S. Madden. Resource-Aware Wireless Sensor-Actuator Networks. *IEEE Data Engineering*, 28(1), 2005.
224. L. Desmet, W. Joosen, F. Massacci, P. Philippaerts, F. Piessens, I. Siahaan, and D. Vanoverberghe. Security-by-Contract on the .NET Platform. *Information Security Technical Report*, 13(1):25–32, 2008.
225. A. Dey, G. Abowd, and A. Wood. CyberDesk: A Framework for Providing Self-integrating Context-aware Services. In *Proceedings of the Symposium on User Interface Software and Technology*, pages 75–76. The ACM Press, 1997.
226. Z. Diamadi and M. Fischer. A Simple Game for the Study of Trust in Distributed Systems. *Wuhan University Journal of Natural Sciences*, 6(1–2):72–82, March 2001. Also appears as Yale Technical Report TR-1207, Jan. 2001.
227. J. Dickman, K. Rath, and L. Kotecha. Proposal for 802.16 Connection Oriented Mesh. In *IEEE 802.16 Standard Proposal*, 2003.
228. E. Dijkstra. Self-Stabilizing Systems in Spite of Distributed Control. *Communications of the ACM*, 17(11):643–644, 1974.

229. M. Dikaiakos, S. Iqbal, T. Nadeem, and Li. Iftode. VITP: An Information Transfer Protocol for Vehicular Computing. In *2nd ACM Workshop on Vehicular Ad Hoc Networks (VANET05)*, Cologne, Germany, September 2005.

230. C. Diot, B.N. Levine, B. Lyles, H. Kassem, and D. Balensiefen. Deployment Issues for the IP Multicast Service. *IEEE Network Magazine, special issue on Multicasting*, January/February 2000.

231. S. Dolev. *Self-Stabilization*. The MIT Press, 2000.

232. S. Dornbush and A. Joshi. StreetSmart Traffic: Discovering and Disseminating Automobile Congestion Using VANET's. In *Vehicular Technology Conference (VTC'07)*, pages 11–15, Dublin, Ireland, April 2007.

233. J. Dowling and V. Cahill. The K-Component Architecture Meta-model for Self-Adaptive Software. In *Proceedings of the 3rd International Conference on Metalevel Architectures and Separation of Crosscutting Concerns (REFLECTION '01)*, pages 81–88, London, UK, 2001. Springer-Verlag.

234. J. Dowling, E. Curran, R. Cunningham, and V. Cahill. Using Feedback in Collaborative Reinforcement Learning to Adaptively Optimize MANET Routing. *IEEE Transactions on Systems, Man, and Cybernetics, Part A*, 35(3):360–372, 2005.

235. V. Drabkin, R. Friedman, and M. Segal. Efficient Byzantine Broadcast in Wireless Ad-Hoc Networks. In *Proceedings of the 6th IEEE Conference on Dependable Systems and Networks*, pages 160–169, June 2005.

236. V. Drabkin, R. Friedman, G. Kliot, and M. Segal. RAPID: Reliable Probabilistic Dissemination in Wireless Ad-Hoc Networks. In *Proceedings 26th IEEE Symposium on Reliable Distributed Systems*, October 2007.

237. R. Draves, J. Padhye, and B. Zill. Comparison of Routing Metrics for Static Multi-Hop Wireless Networks. In *Proceedings of the 2004 Conference on Applications, Technologies, Architectures, and Protocols for Computer Communications (SIGCOMM '04)*, pages 133–144. The ACM Press, 2004.

238. K. Dresner and P. Stone. Multiagent Traffic Management: An Improved Intersection Control Mechanism. In *International Joint Conference on Autonomous Agents and Multiagent Systems*, Utrecht, The Netherlands, July 2005.

239. S. Drossopoulou and S. Eisenbach. The Java Type System is Sound—Probably. In *Proceedings of the 11th European Conference of Object Oriented Programming (ECOOP'97)*, number 1241 in LNCS, pages 389–418. Springer-Verlag, 1997.

240. O. Drugan, T. Plagemann, and E. Munthe-Kaas. Building Resource Aware Middleware Services over MANET for Rescue and Emergency Applications Middleware Services for Information Sharing in Mobile Ad-hoc Networks. In *The 16th Annual IEEE International Symposium on Personal Indoor and Mobile Radio Communications, International Congress Center (ICC)*, Berlin, Germany, Sep 2005. IEEE Computer Society.

241. A. Dunkels, B. Grönvall, and T. Voigt. Contiki—A Lightweight and Flexible Operating System for Tiny Networked Sensors. In *Proceedings of the First IEEE Workshop on Embedded Networked Sensors (Emnets-I)*, Tampa, Florida, USA, November 2004.

242. A. Dunkels, O. Schmidt, and T. Voigt. Using Protothreads for Sensor Node Programming. In *Proceedings of the Workshop on Real-World Wireless Sensor Networks (REALWSN'05)*, June 2005.

243. Dynamic Source Routing Protocol. Available at http://www.cs.cmu.edu/~dmaltz/dsr.html.

244. N. Eagle, and A. Pentland. Reality Mining: Sensing Complex Social Systems. *Personal and Ubiquitous Computing*, 10(4):255–268, May 2006.

245. eCall project. http://www.e-call.at/ec/index.php?lng=e. Accessed February 2008.

246. A. Edwards, R. Phillips, N. Watkins, M. Freeman, E. Murphy, V. Afanasyev, S. Buldyrev, M. da Luz, E. Raposo, H. Stanley, and G. Viswanathan. Revisiting Lévy Flight Search Patterns of Wandering Albatrosses, Bumblebees and Deer. *Nature*, 449:1044–1048, October 2005.

247. C. Efstratiou, A. Friday, N. Davies, and K. Cheverst. A Platform Supporting Coordinated Adaptation in Mobile Systems. In *Proceedings of the 4th IEEE Workshop on Mobile Com-

puting Systems and Applications (WMCSA '02), pages 128–137, Washington, DC, USA, 2002. IEEE Computer Society.

248. S. Eichler, C. Merkle, and M. Strassberger. Data Aggregation System for Distributing Inter-Vehicle Warning Messages. In *31st IEEE Conference on Local Computer Networks (LCN'06)*, pages 543–544, Tampa, Florida, November 2006.

249. A. Einstein. *Investigations on the Theory of the Brownian Movement*. Dover Publications, 1956.

250. A. Einstein. Ueber die von der molekularkinetischen Theorie der Waerme geforderte Bewegung von in ruhenden Fluessigkeiten suspendierten Teilchen. In *Annals of Physic 17*, pages 549–560, 1905. English translation: On the motion of small particles suspended in liquids at rest required by the molecular-kinetic theory of heat, in Einstein's Miraculous Year, Princeton University Press, pp. 85–98, 1998.

251. F. Ekman, A. Keränen, J. Karvo, and J. Ott. Working Day Movement Model. In *Proceedings of 1st ACM SIGMOBILE International Workshop on Mobility Models for Networking Research (MobilityModels'08)*, Hong Kong S.A.R., China, May 2008.

252. D. Empey. PATH Vehicles Will Roll at Demo 2003. *Intellimotion*, 10(1), 2002.

253. P. Engelstad, Y. Zheng, R. Koodli, and C. Perkins. Service Discovery Architectures for On-demand Ad Hoc Networks. *Journal of Ad Hoc and Sensor Wireless Networks*, 2:27–58, March 2006.

254. D. Eppstein. Spanning Trees and Spanners. In *Handbook of Computational Geometry*, pages 425–461. Elsevier North-Holland, 2000.

255. U. Erlingsson. *The Inlined Reference Monitor Approach to Security Policy Enforcement*. PhD thesis. Cornell University, 2004.

256. D. Estrin, R. Govindan, J. Heidemann, and S. Kumar. Next Century Challenges: Scalable Coordination in Sensor Networks. In *Proceedings of the 5th Int. Conf. on Mobile Computing and Networking (MOBICOM)*, 1999.

257. D. Estrin, D. Culler, K. Pister, and G. Sukhatme. Connecting the Physical World with Pervasive Networks. *IEEE Pervasive Computing*, 1(1):59–69, January 2002. IEEE Educational Activities Department.

258. A. Eswaran, A. Rowe, and R. Rajkumar. Nano-RK: An Energy-Aware Resource-Centric Operating System for Sensor Networks. In *Proceedings of the IEEE Real-Time Systems Symposium (RTSS'05)*, December 2005.

259. H. Etoh and K. Yoda. Protecting from Stack-Smashing Attacks. Technical Report, IBM Research Division, Tokyo Research Laboratory, June 2000.

260. P. Eugster. *Type-Based Publish/Subscribe*. PhD thesis, EPFL, December 2001.

261. P. Eugster and R. Guerraoui. Probabilistic Multicast. In *Proceedings of the 3rd IEEE International Conference on Dependable Systems and Networks (DSN 2002)*, pages 313–322, 2002.

262. P. Eugster and R. Guerraoui. Distributed Programming with Typed Events. *IEEE Software*, 21(2):56–64, 2004.

263. P. Eugster, R. Guerraoui, and C. Damm. On Objects and Events. In *Proceedings of the Conference on Object-Oriented Programming Systems, Languages and Applications (OOPSLA)*, 2001.

264. P. Eugster, P. Felber, R. Guerraoui, and S. Handurukande. Event Systems: How to Have Your Cake and Eat It Too. In *Proceedings of the International Workshop on Distributed Event-Based Systems (DEBS'02)*, 2002.

265. P. Eugster, P. Felber, R. Guerraoui, and A.-M. Kermarrec. The Many Faces of Publish/Subscribe. *ACM Computing Surveys*, 35(2):114–131, 2003.

266. P. Eugster, R. Guerraoui, A.-M. Kermarrec, and L. Massoulie. From Epidemics to Distributed Computing. *IEEE Computer*, 37(5):60–67, 2004.

267. P. Eugster, B. Garbinato, and A. Holzer. Location-based Publish/Subscribe. In *Proceedings of the Fourth IEEE International Symposium on Network Computing and Applications (NCA'05)*, pages 279–282. IEEE Press, 2005.

268. F-Secure Malware Information Pages: Cabir. Available at http://www.f-secure.com/v-descs/cabir.shtml.

269. F-Secure Malware Information Pages: Commwarrior. Available at http://www.f-secure.com/v-descs/commwarrior.shtml.

270. F-Secure Virus Descriptions: Mabir.A. Available at http://www.f-secure.com/v-descs/mabir.shtml.

271. F. Fabret, A. Jacobsen, F. Llirbat, J. Pereira, K. Ross, and D. Shasha. Filtering Algorithms and Implementation for Very Fast Publish/Subscribe. In *Proceedings of the 20th Intl. Conference on Management of Data (SIGMOD 2001)*, pages 115–126, 2001.

272. P. Fahy and S. Clarke. CASS—Middleware for Mobile Context-Aware Applications. In *Mobisys Workshop on Context Awareness*, Boston, Massachusetts, USA, 2004. ACM SIGOPS.

273. K. Fall. A Delay-Tolerant Network Architecture for Challenged Internets. *Proceedings of the 2003 ACM Conference on Applications, Technologies, Architectures, and Protocols for Computer Communications (SIGCOMM 2003)*, pages 27–34, 2003.

274. Z. Fan and E. Ho. Service Discovery in Ad Hoc Networks: Performance Evaluation and QoS Enhancement. *Wireless Personal Communications*, 40:215–231, January 2007.

275. Z. Fan and S. Subramani. An Address Autoconfiguration Protocol for IPv6 Hosts in a Mobile Ad Hoc Network. *Computer Communications*, 28:339–350, March 2005.

276. L. Feeney and M. Nilsson. Investigating the Energy Consumption of a Wireless Network Interface in an Ad Hoc Networking Environment. In *Proceedings of the 20th Annual Joint Conference of the IEEE Computer and Communications Societies (INFOCOM '01)*, 2001.

277. A. Festag, R. Baldessari, and H. Wang. On Power-Aware Greedy Forwarding in Highway Scenarios. In *5th International Workshop Intelligent Transportation (WIT'07)*, Hamburg, Germany, March 2007.

278. L. Fiege, F. Gärtner, O. Kasten, and A. Zeidler. Supporting Mobility in Content-Based Publish/Subscribe Middleware. In *Proceedings of the 4th ACM/IFIP/USENIX International Middleware Conference (Middleware 2003)*, pages 103–122, 2003.

279. L. Fiege, A. Zeidler, A. Buchmann, R. Kilian-Kehr, and G. Mühl. Security Aspects in Publish/Subscribe Systems. In *Third Intl. Workshop on Distributed Event-based Systems (DEBS'04)*, May 2004.

280. G. Finn. Routing and Addressing Problems in Large Metropolitan-Scale Internetworks. Technical Report ISU/RR-87-180, Institute for Scientific Information, March 1987.

281. FireTide. Available at http://www.firetide.com.

282. M. Fischer and H. Jiang. Self-stabilizing Leader Election in Networks of Finite-state Anonymous Agents. In *Proceedings Principles of Distributed Systems, 10th International Conference*, pages 395–409, 2006.

283. J. Flinn, S. Park, and M. Satyanarayanan. Balancing Performance, Energy, and Quality in Pervasive Computing. In *Proceedings of the 22nd International Conference on Distributed Computing Systems (ICDCS'02)*, pages 217–226, 2002.

284. C. Flores-Cortés, G. Blair, and P. Grace. An Adaptive Middleware to Overcome Service Discovery Heterogeneity in Mobile Ad Hoc Environments. *IEEE Distributed Systems Online*, 8(7):1, 2007.

285. S. Floyd, V. Jacobson, S. McCanne, C.-G. Liu, and L. Zhang. A Reliable Multicast Framework for Light-Weight Sessions and Application Level Framing. *SIGCOMM Comput. Commun. Rev.*, 25(4), 1995.

286. C. Fok, G.-C. Roman, and G. Hackmann. A Lightweight Coordination Middleware for Mobile Computing. In *Proceedings of the 6th International Conference on Coordination Models and Languages (Coordination'04)*, pages 135–151. LNCS, 2004.

287. C.-L. Fok, G.-C. Roman, and C. Lu. Rapid Development and Flexible Deployment of Adaptive Wireless Sensor Network Applications. In *Proceedings of the 25th International Conference on Distributed Computing Systems (ICDCS'05)*, pages 653–662. IEEE Press, 2005.

288. C. Fournet and A. Gordon. Stack Inspection: Theory and Variants. In *29th ACM SIGPLAN-SIGACT Symposium on Principles of Programming Languages (POPL)*, pages 307–318. The ACM Press, January 2002.

289. P. Fraigniaud and C. Gavoille. A Space Lower Bound for Routing in Trees. In *Proceedings of the 19th Annual Symposium on Theoretical Aspects of Computer Science (STACS)*, LNCS. Springer-Verlag, 2002.

290. P. Francis. Yoid: Extending the Internet Multicast Architecture. Technical Report, AT&T Center for Internet Research at ICSI (ACIRI), 2000.

291. W. Franz, H. Hartenstein, and M. Mauve, editors. *Inter-Vehicle Communications Based on Ad Hoc Networking Principles—The FleetNet Project*. Universitaetsverlag Karlsruhe, June 2005.

292. L. Freeman. A Set of Measuring Centrality Based on Betweenness. *Sociometry*, 40:35–41, 1977.

293. D. Frey and G.-C. Roman. Context-Aware Publish Subscribe in Mobile Ad Hoc Networks. In *Proceedings of the 9th International Conference on Coordination Models and Languages (Coordination'07)*, pages 37–55, June 2007.

294. H. Frey and I. Stojmenovic. On Delivery Guarantees of Face and Combined Greedy-Face Routing in Ad Hoc and Sensor Networks. In *Proceedings of the 12th Annual International Conference on Mobile Computing and Networking (MOBICOM '06)*, pages 390–401. The ACM Press, 2006.

295. R. Friedman and R. van Renesse. Packing Messages as a Tool for Boosting the Performance of Total Ordering Protocols. In *Proceedings of the Sixth IEEE International Symposium on High Performance Distributed Computing*, pages 233–242, August 1997.

296. R. Friedman, D. Gavidia, L. Rodrigues, A.C. Viana, and S. Voulgaris. Gossiping on MANETs: The Beauty and the Beast. *SIGOPS Operating Systems Review*, 41(5):67–74, 2007.

297. A. Fuggetta, G.P. Picco, and G. Vigna. Understanding Code Mobility. *IEEE Trans. Softw. Eng.*, 24(5), 1998.

298. W. Fung, D. Sun, and J. Gehrke. COUGAR: The Network is the Database. In *Proceedings of the ACM International Conference on Management of Data (SIGMOD'02)*, 2002.

299. H. Füßler, M. Mauve, H. Hartenstein, C. Lochert, D. Vollmer, D. Herrmann, and W. Franz. *Position-based Routing for Car-to-Car Communication*, pages 117–143. Universittsverlag Karlsruhe, Karlsruhe, Germany, June 2005.

300. G. Gaertner and V. Cahill. Understanding Link Quality in 802.11 Mobile Ad Hoc Networks. *IEEE Internet Computing*, 8(1), January–February 2004.

301. J. Gajda, R. Sroka, M. Stencel, A. Wajda, and T. Zeglen. A Vehicle Classification Based on Inductive Loop Detectors. In *Instrumentation and Measurement Technology Conference*, Budapest, Hungary, May 2001.

302. F. Gandon and N. Sadeh. Context-Awareness, Privacy and Mobile Access: A Web Semantic and Multiagent Approach. In *Proceedings of the 1st French-speaking Conference on Mobility and Ubiquity Computing (UbiMob'04)*, pages 123–130, New York, NY, USA, 2004. The ACM Press.

303. D. Ganesan, B. Krishnamachari, A. Woo, D. Culler, D. Estrin, and S. Wicker. Complex Behavior at Scale: An Experimental Study of Low-Power Wireless Sensor Networks. Technical Report CSD-TR 02-0013, UCLA, February 2002.

304. J. Gao, L. Guibas, J. Hershberger, L. Zhang, and A. Zhu. Geometric Spanners for Routing in Mobile Networks. In *Proceedings of the 2nd ACM Symposium on Mobile Ad Hoc Networking and Computing (MOBIHOC '01)*, 2001.

305. B. Garbinato, F. Pedone, and R. Schmidt. An Adaptive Algorithm for Efficient Message Diffusion in Unreliable Environments. In *Proceedings of IEEE DSN'04*, pages 507–516, June 2004.

306. M. Garetto and E. Leonardi. Analysis of Random Mobility Models with PDE's. In *Proceedings of MobiHoc'06*, pages 73–84. The ACM Press, 2006.

307. D. Garlan, D. Siewiorek, A. Smailagic, and P. Steenkiste. Project Aura: Toward Distraction-Free Pervasive Computing. *IEEE Pervasive Computing*, 1(2):22–31, April-June 2002.

308. D. Gay, P. Levis, R. von Behren, M. Welsh, E. Brewer, and D. Culler. The nesC Language: A Holistic Approach to Network Embedded Systems. I. In *ACM SIGPLAN Conference on Programming Language Design and Implementation (PLDI'03)*, pages 1–11. The ACM Press, 2003.

309. G. Gehlen and G. Mavromatis. A Rule Based Data Monitoring Middleware for Mobile Applications. In *IEEE International Vehicular Technology Conference (VTC'05)*, volume 5, Stockholm, Sweden, June 2005.

310. D. Gelernter. Generative Communication in Linda. *ACM Transactions on Programming Languages and Systems (TOPLAS)*, 7(1):80–112, January 1985.

311. S. Ghandeharizadeh and B. Krishnamachari. C2P2: A Peer-to-Peer Network for On-Demand Automobile Information Services. In *First International Workshop on Grid and Peer-to-Peer Computing Impacts on Large Scale Heterogeneous Distributed Database Systems (GLOBE'04)*, Zaragoza, Spain, August 2004.

312. S. Ghandeharizadeh, S. Kapadia, and B. Krishnamachari. PAVAN: A Policy Framework for Availability in Vehicular Ad-Hoc Networks. In *First ACM Workshop on Vehicular Ad Hoc Networks (VANET'04)*, Philadelphia, PA, October 2004.

313. P. Gibbons, B. Karp, Y. Ke, S. Nath, and S. Seshan. IrisNet: An Architecture for a World-Wide Sensor Web. *IEEE Pervasive Computing*, 2(4), October-December 2003.

314. D. Gillespie. Exact Stochastic Simulation of Coupled Chemical Reactions. *Journal of Physical Chemistry*, 81(25):2340–2361, 1977.

315. D. Gillespie. A Rigorous Derivation of the Chemical Master Equation. *Physica A*, 188:404–425, 1992.

316. S. Ginsburg and E. Spanier. Semigroups, Presburger Formulas, and Languages. *Pacific Journal of Mathematics*, 16:285–296, 1966.

317. I. Glauche, W. Krause, R. Sollacher, and M. Greiner. Continuum Percolation of Wireless Ad Hoc Communication Networks. *Physica A*, 325:577–600, 2003.

318. G. Golden. Detection Algorithm and Initial Laboratory Results Using V-BLAST Space-time Communication Architecture. *IEE Electronics Letters*, 35(1):14–16, 1999.

319. R. Golden. Service Advertisement and Discovery. *IEEE Internet Computing*, 4:18–26, 2000.

320. A. Goldsmith and S. Wicker. Design Challenges for Energy-Constrained Ad Hoc Wireless Networks. *IEEE Wireless Communications*, 9(4):8–27, August 2002.

321. M. Gonzalez, C. Hidalgo, and A.-L. Barabasi. Understanding Individual Human Mobility Patterns. *Nature*, 453(7196):779–782, 2008.

322. K. Gough and G. Smith. Efficient Recognition of Events in Distributed Systems. In *Proceedings of the ACSC-18*, 1995.

323. O. Goussevskaia, M. Machado, R. Mini, A. Loureiro, G. Mateus, and J. Nogueira. Data Dissemination Based on the Energy Map. *IEEE Communications Magazine*, 43(7):134–143, July 2005.

324. S. Goyal and J. Carter. A Lightweight Secure Cyber Foraging Infrastructure for Resource-Constrained Devices. In *Proceedings of the 6th IEEE Workshop on Mobile Computing Systems and Applications (WMCSA'04)*, pages 186–195, 2004.

325. P. Grace, G. Blair, and S. Samuel. A Reflective Framework for Discovery and Interaction in Heterogeneous Mobile Environments. *ACM SIGMOBILE Mobile Computing and Communication Review*, 9(1):2–14, 2005.

326. P. Grace, G. Coulson, G. Blair, and B. Porter. Deep Middleware for the Divergent Grid. In *ACM/IFIP/USENIX, 6th International Middleware Conference*, volume 3790 of *Lecture Notes in Computer Science*, pages 334–353. Springer-Verlag, November 2005.

327. P. Grace, B. Lagaisse, E. Truyen, and W. Joosen. A Reflective Framework for Fine-Grained Adaptation of Aspect-Oriented Compositions. In *Proceedings of the 7th International Symposium on Software Composition*, volume 4954 of *Lecture Notes in Computer Science*, pages 215–230. Springer-Verlag, 2008.

328. V. Gradinescu, C. Gorgorin, and L. Iftode. Adaptive Traffic Lights Using Car-to-Car Communication. In *International Vehicular Technology Conference*, Dublin, Ireland, April 2007.

329. B. Greenstein, D. Estrin, R. Govindan, S. Ratnasamy, and S. Shenker. DIFS: A Distributed Index for Features in Sensor Networks. In *First IEEE International Workshop on Sensor Network Protocols and Applications (SNPA'03)*, May 2003.

330. B. Greenstein, E. Kohler, and D. Estrin. A Sensor Network Application Construction Kit (SNACK). In *Proceedings of the 2nd International Conference on Embedded Sensor Systems (SENSYS'04)*, pages 69–80, 2004.

331. A. Grilo and M. Nunes. Link Adaptation and Transmit Power Control for Unitcast and Multicast in IEEE 802.11a/h/e WLANs. In *Proceedings of the IEEE International Conferences on Local Computer Networks*, pages 334–345, 2003.

332. R. Grimes. Preventing Buffer Overflows in C++. *Dr Dobb's Journal: Software Tools for the Professional Programmer*, 29(1):49–52, January 2004.

333. M. Grossglauser and D. Tse. Mobility Increases the Capacity of Ad-hoc Wireless Networks. *Proceedings of the 20th IEEE Annual Joint Conference of the IEEE Computer and Communications Societies (IEEE INFOCOM 2001)*, 2001.

334. M. Grossglauser and M. Vetterli. Locating Nodes with EASE: Last Encounter Routing in Ad Hoc Networks Through Mobility Diffusion. *Proceedings of the 22nd IEEE Annual Joint Conference of the IEEE Computer and Communications Societies (IEEE INFOCOM 2003)*, 2003.

335. R. Gruber, B. Krishnamurthy, and E. Panagosf. The Architecture of the READY Event Notification Service. In *Proceedings of The International Conference on Distributed Computing Systems, Workshop on Middleware*, 1999.

336. Gryphon Web Site. http://www.research.ibm.com/gryphon.

337. T. Gu, H. Pung, and D. Zhang. A Middleware for Building Context-Aware Mobile Services. In *Proceedings of the Vehicular Technology Conference (VTC 2004)*. IEEE Computer Society, May 2004.

338. R. Guerraoui and E. Ruppert. Even Small Birds are Unique: Population Protocols with Identifiers. Technical Report CSE-2007-04, Department of Computer Science and Engineering, York University, 2007.

339. R. Guerraoui, S. Handurukande, K. Huguenin, A.-M. Kermarrec, F. Le Fessant, and E. Riviere. GosSkip, an Efficient, Fault-Tolerant and Self Organizing Overlay Using Gossip-based Construction and Skip-Lists Principles. In *Proceedings of the 6th IEEE International Conference on Peer-to-Peer Computing*, 2006.

340. M. Guimarães and L. Rodrigues. A Genetic Algorithm for Multicast Mapping in Publish-Subscribe Systems. In *Proceedings of the 2nd IEEE International Symposium on Network Computing and Applications*, pages 67–74, 2003.

341. E. Guizzo. Network of Traffic Spies Built Into Cars in Atlanta. *IEEE Spectrum*, April 2004. http://spectrum.ieee.org/apr04/3816.

342. R. Gummadi, O. Gnawali, and R. Govindan. Macro-programming Wireless Sensor Networks using Kairos. In *Proceedings of the International Conference on Distributed Computing in Sensor Systems (DCOSS'05)*, June 2005.

343. J. Guo and G. Xing. Using Mobile Agent-Based Middleware to Support Distributed Coordination for Vehicle Telematics. In *International Symposium on Ubiquitous Computing and Intelligence*, pages 374–379, Niagara Falls, Canada, May 2007.

344. P. Gupta and P. Kumar. The Capacity of Wireless Networks. *IEEE Transactions on Information Theory*, 46(2):388–404, March 2000.

345. A. Gupta, O. Sahin, D. Agrawal, and A. El-Abbadi. Meghdoot: Content-Based Publish:Subscribe over P2P Networks. In *Proceedings of the ACM/IFIP/USENIX 5th International Middleware Conference (Middleware'04)*, pages 254–273, 2004.

346. E. Guttman. Attribute List Extension for the Service Location Protocol. Request for Comments 3059, Internet Engineering Task Force, February 2001.

347. E. Guttman. Service Location Protocol Modifications for IPv6. Request for Comments 3111, The Internet Engineering Task Force, May 2001.

348. E. Guttman. Vendor Extensions for Service Location Protocol, Version 2. Request for Comments 3224, Internet Engineering Task Force, January 2002.

349. E. Guttman and J. Kempf. Automatic Discovery of Thin Servers: SLP, Jini and the SLP-JiniBridge. In *Proceedings of the 25th Annual Conference IEEE Industrial Electronics Society (IECON'99)*, volume 2, pages 722–727. IEEE Computer Society, 1999.

350. E. Guttman, C. Perkins, J. Veizades, and M. Day. Service Location Protocol, Version 2. Request for Comments 2608, Internet Engineering Task Force, June 1999.
351. M. Haahr, R. Cunningham, and V. Cahill. Supporting Corba Applications in a Mobile Environment. In *Proceedings of the 5th Annual ACM/IEEE International Conference on Mobile Computing and Networking (MobiCom'99)*, pages 36–47. The ACM Press, 1999.
352. Z. Haas. A New Routing Protocol for the Reconfigurable Wireless Networks. In *Proceedings of the IEEE 6th International Conference on Universal Personal Communications*, volume 2, pages 562–566, October 1997.
353. Z. Haas, J. Halpern, and L. Li. Gossip-Based Ad Hoc Routing. In *Proceedings of the 21st Joint Conference of the IEEE Computer and Communications Societies (INFOCOM 2002)*, pages 1707–1716, June 2002.
354. A. Haeberlen, E. Flannery, A. Ladd, A. Rudys, D. Wallach, and L. Kavraki. Practical Robust Localization over Large-Scale 802.11 Wireless Networks. In *Proceedings of the 10th Annual International Conference on Mobile Computing and Networking (MOBICOM '04)*, pages 70–84. The ACM Press, 2004.
355. C. Hall, A. Carzaniga, J. Rose, and A. Wolf. A Content-Based Networking Protocol For Sensor Networks. Technical Report CU-CS-979-04, Department of Computer Science, University of Colorado, 2004.
356. S. Hallé, J. Laumonier, and B. Chaib-Draa. A Decentralized Approach to Collaborative Driving Coordination. In *International Conference on Intelligent Transportation Systems*, Washington, D.C., USA, October 2004.
357. C. Han, R. Kumar, R. Shea, E. Kohler, and M. Srivastava. A Dynamic Operating System for Sensor Nodes. In *Proceedings of the 3rd International Conference on Mobile Systems, Applications, and Services (MobiSys'05)*, pages 163–176, New York, NY, USA, 2005. The ACM Press.
358. R. Handorean and G.-C. Roman. Service Provision in Ad Hoc Networks. In *Proceedings of the 5th International Conference on Coordination Models and Languages (COORDINATION)*, LNCS 2315. Springer-Verlag, 2002.
359. K. Harras, K. Almeroth, and E. Belding-Royer. Delay Tolerant Mobile Networks (DTMNs): Controlled Flooding Schemes in Sparse Mobile Networks. In *IFIP Networking 2005*, pages 1180–1192, May 2005.
360. H. Hartenstein and K. Laberteaux. A Tutorial Survey on Vehicular Ad Hoc Networks. *IEEE Communications Magazine*, June 2008.
361. M. Hazas, J. Scott, and J. Krumm. Location—Aware Computing Comes of Age. *Computer*, 37(2):95–97, February 2004.
362. W. Heinzelman, J. Kulik, and H. Balakrishnan. Adaptive Protocols for Information Dissemination in Wireless Sensor Networks. In *Proceedings of the 5th Annual ACM/IEEE International Conference on Mobile Computing and Networking (MOBICOM)*, pages 174–185, Seattle, USA, 1999.
363. S. Helal. Standards For Service Discovery And Delivery. *IEEE Pervasive Computing*, 1:95–100, 2002.
364. S. Helal, N. Desai, V. Verma, and C. Lee. Konark - A Service Discovery and Delivery Protocol for Ad-hoc Networks. In *Proceedings of the Third IEEE Conference on Wireless Communication Networks (WCNC'03)*, March 2003.
365. D. Helder and S. Jamin. Banana Tree Protocol, an End-host Multicast Protocol. Technical Report, University of Michigan, 2002.
366. L. Hempel and E. Topfer. CCTV in Europe. http://www.urbaneye.net, August 2004.
367. T. Henderson, D. Kotz, and I. Abyzov. The Changing Usage of a Mature Campus-Wide Wireless Network. In *Proceedings of MobiCom'04*, pages 187–201, New York, NY, USA, 2004. The ACM Press.
368. K. Hermann. Modeling the Sociological Aspect of Mobility in Ad Hoc Networks. In *Proceedings of the 6th ACM International Symposium on Modeling, Analysis and Simulation of Wireless and Mobile Systems (MSWiM'03)*, pages 128–129, San Diego, California, USA, September 2003.

369. G. Higman. Ordering by Divisibility in Abstract Algebras. *Proceedings of the London Mathematical Society*, 3(2):326–336, 1952.

370. J. Hill, R. Szewczyk, A. Woo, S. Hollar, D. Culler, and K. Pister. System Architecture Directions for Networked Sensors. In *Proceedings of 9th International Conference on Architectural Support for Programming Languages and Operating Systems (ASPLOS)*, pages 93–104, Cambridge, USA, November 2000.

371. I. Ho, K. Leung, J. Polak, and R. Mangharam. Node Connectivity in Vehicular Ad Hoc Networks with Structured Mobility. In *Proceedings of the 32nd IEEE Conference on Local Computer Networks (LCN'07)*, pages 635–642. IEEE Press, 2007.

372. J. Hoebeke, I. Moerman, B. Dhoedt, and P. Demeester. Analysis of Decentralized Resources and Service Discovery Mechanisms in Wireless Multi-Hop Networks. *Computer Communications*, 29:2710–2720, August 2006.

373. T. Hofer, W. Schwinger, M. Pichler, G. Leonhartsberger, J. Altmann, and W. Retschitzegger. Context-Awareness on Mobile Devices—the Hydrogen Approach. In *Proceedings of the 36th Annual Hawaii International Conference on System Sciences (HICSS'03)*, Washington, DC, USA, 2003. IEEE Computer Society.

374. G. Holland ad N. Vaidya. Link Failure and Congestion: Analysis of TCP Performance over Mobile Ad Hoc Networks. In *Proceedings of the ACM Annual International Conference on Mobile Computing and Networking (MOBICOM)*, pages 219–230, 1999.

375. K. Honda and M. Tokoro. An Object Calculus for Asynchronous Communication. In *Proceedings of the European Conference on Object-Oriented Programming (ECOOP'91)*, number 512 in LNCS, pages 133–147. Springer-Verlag, 1991.

376. C.-S. Hong, H.-S. Kim, J. Cho, H. Kyu Cho, and H.-C. Lee. Context Modeling and Reasoning Approach in Context-Aware Middleware for URC System. *International Journal of Mathematical, Physical and Engineering Sciences*, 1(4):208–212, 2007.

377. X. Hong, M. Gerla, G. Pei, and C.-C. Chiang. A Group Mobility Model for Ad Hoc Networks. In *Proceedings of the 2nd ACM International Symposium on Modeling, Analysis and Simulation of Wireless and Mobile Systems (MSWiM'99)*, pages 53–60, 1999.

378. W. Hoschek. The Web Service Discovery Architecture. In *Supercomputing '02: Proceedings of the 2002 ACM/IEEE Conference on Supercomputing*, pages 1–15. IEEE Computer Society Press, 2002.

379. W. Hoschek. Peer-to-Peer Grid Databases for Web Service Discovery. In *Grid Computing—Making the Global Infrastructure a Reality, Berman F. et al. (Ed.)*, pages 491–539. John Wiley & Sons, Ltd., 2003.

380. W. Hsu, K. Merchant, H. Shu, C. Hsu, and A. Helmy. Weighted Waypoint Mobility Model and Its Impact on Ad Hoc Networks. *ACM Mobile Computer Communications Review (MC2R)*, pages 59–63, January 2005.

381. W. Hsu, T. Spyropoulos, K. Psounis, and A. Helmy. Modeling Time-Variant User Mobility in Wireless Mobile Networks. *Proceedings of INFOCOM'07*, pages 758–766, May 2007.

382. W. Hu, J. Hiser, D. Williams, A. Filipi, J. Davidson, D. Evans, J. Knight, A. Nguyen-Tuong, and J. Rowanhill. Secure and Practical Defense Against Code-Injection Attacks Using Software Dynamic Translation. In *Proceedings of the 2nd International Conference on Virtual Execution Environments*, pages 2–12, Ottawa, Ontario, Canada, 2006. The ACM Press.

383. Y.-C. Hu and D. Johnson. Exploiting Congestion Information in Network and Higher Layer Protocols in Multihop Wireless Ad Hoc Networks. In *Proceedings of the 24th International Conference on Distributed Computing Systems (ICDCS'04)*, pages 301–310, March 2004.

384. Z. Hu and B. Li. On the Fundamental Capacity and Lifetime Limits of Energy-Constrained Wireless Sensor Networks. In *Proceedings of the 10th IEEE Real-Time and Embedded Technology and Applications Symposium (RTAS'04)*, pages 38–47, Toronto, Canada, 2004.

385. Q. Huang, C. Lu, and G.-C. Roman. Spatiotemporal Multicast in Sensor Networks. In *1st International Conference on Embedded Networked Sensor Systems (SENSYS '03)*, pages 205–217. The ACM Press, 2003.

386. Y. Huang and H. Garcia-Molina. Publish/Subscribe Tree Construction in Wireless Ad-Hoc Networks. In *4th International Conference on Mobile Data Management (MDM 2003)*, pages 122–140, 2003.

387. Y. Huang and H. Garcia-Molina. Publish/Subscribe in a Mobile Environment. *Wireless Networks*, 10(6):643–652, 2004.
388. J. Hui and D. Culler. The Dynamic Behavior of a Data Dissemination Protocol for Network Programming at Scale. In *Proceedings of the 2nd International Conference on Embedded Networked Sensor Systems (ENSS'04)*, pages 81–94. The ACM Press, 2004.
389. P. Hui and J. Crowcroft. Bubble Rap: Forwarding in Small World DTNs in Every Decreasing Circles. Technical Report UCAM-CL-TR684, Univ. of Cambridge, 2007.
390. P. Hui and J. Crowcroft. How Small Labels create Big Improvements. In *Proceedings IEEE ICMAN*, March 2007.
391. P. Hui, A. Chaintreau, J. Scott, R. Gass, J. Crowcroft, and C. Diot. Pockets Switched Networks and Human Mobility in Conference Environments. In *Proceedings of ACM SIGCOMM'05 Workshops*, pages 244–251, August 2005.
392. P. Hui, J. Crowcroft, and E. Yoneki. BUBBLE Rap: Social-Based Forwarding in Delay Tolerant Networks. In *MobiHoc '08: Proceedings of the 9th ACM International Symposium on Mobile Ad Hoc Networking & Computing*, May 2008.
393. B. Hull, V. Bychkovsky, Y. Zhang, K. Chen, M. Goraczko, A. Miu, E. Shih, H. Balakrishnan, and S. Madden. CarTel: A Distributed Mobile Sensor Computing System. In *Conference on Embedded Networked Sensor Systems*, Boulder, Colorado, USA, November 2006.
394. G. Hunt and M. Scott. The Coign Automatic Distributed Partitioning System. In *Proceedings of the 3rd Symposium on Operating Systems Design and Implementation (OSDI'99)*, pages 187–200, Berkeley, CA, USA, 1999. USENIX Association.
395. L. Iannone, K. Kabassanov, and S. Fdida. The meshDVNet Wireless Mesh Network Testbed. In *Proceedings of the 1st International Workshop on Wireless Network Testbeds, Experimental Evaluation & Characterization (WiNTECH '06)*, pages 107–108. The ACM Press, 2006.
396. IEEE Computer Society LAN MAN Standards Committee. Wireless LAN Medium Access Control (MAC) and Physical Layer (PHY) Specifications. Technical Report 802.11, 1999 Edition, The Institute of Electrical and Electronics Engineers, Inc., 1999.
397. IEEE 802 Standard Working Group. Wireless LAN Medium Access Control (MAC) and Physical (PHY) Specifications Amendment: Enhancements for Higher Throughput MA, 1999.
398. IEEE 802 Standard Working Group. Wireless LAN Medium Access Control (MAC) and Physical (PHY) Specifications: High-Speed Physical Layer in the 5 GHz Band, 1999.
399. IEEE 802 Standard Working Group. Wireless LAN Medium Access Control (MAC) and Physical (PHY) Specifications: Further Higher Data Rate Extension in 2.4 GHz Band, 2003.
400. IEEE P802.11. Status of Project IEEE 802.11 Task Group p. http://grouper.ieee.org/groups/802/11/Reports. Accessed February 2008.
401. IEEE 802 Standard Working Group. Working Group for Wireless Personal Area Networks (WPANs), 2003.
402. IEEE Standard. IEEE 1609 Family of Standards for Wireless Access in Vehicular Environments (WAVE).
403. D. Ingalls, T. Kaehler, J. Maloney, S. Wallace, and A. Kay. Back to the Future: The Story of Squeak, a Practical Smalltalk Written in Iself. In *Proceeding of ACM SIGPLAN Conference on Object-Oriented Programming Systems, Languages, and Applications (OOPSLA'97)*. The ACM Press, October 1997.
404. T. Ingerson and R. Buvel. Structure in Asynchronous Cellular Automata. *Physica D*, 10:59–68, 1984.
405. C. Intanagonwiwat, R. Govindan, D. Estrin, J. Heidemann, and F. Silva. Directed Diffusion for Wireless Sensor Networking. *IEEE/ACM Transactions on Networking*, 11(1):2–16, February 2003.
406. J. Ishmael, S. Bury, D. Pezaros, and N.J. Race. Deploying Rural Community Wireless Mesh Networks. *IEEE Internet Computing*, 2008.
407. G. Itkis, L. Levin. Fast and Lean Self-Stabilizing Asynchronous Protocols. In *Proceedings of the 35th Annual Symposium on Foundations of Computer Science*, pages 226–239, 1994.

408. ITU. The Internet of Things. Technical Report, International Telecommunication Union, 2005.

409. R. Jain, A. Puri, and R. Sengupta. Geographical Routing Using Partial Information for Wireless Ad Hoc Networks. Technical Report UCB/ERL M99/69, EECS Department, University of California, Berkeley, 1999.

410. R. Jain, D. Lelescu, and M. Balakrishnan. Model T: An Empirical Model for User Registration Patterns in a Campus Wireless LAN. In *Proceedings of MobiCom'05*, pages 170–184. The ACM Press, 2005.

411. S. Jain, K. Fall, and R. Patra. Routing in a Delay Tolerant Network. In *Proceedings of the 2004 ACM Conference on Applications, Technologies, Architectures, and Protocols for Computer Communications (SIGCOMM 2004)*, pages 145–158. The ACM Press, 2004.

412. J. Jannotti, D. Gifford, K. Johnson, F. Kaashoek, and J. Jr. O'Toole. Overcast: Reliable Multicasting with on Overlay Network. In *OSDI'00: Proceedings of the 4th Conference on Symposium on Operating System Design Implementation*, Berkeley, CA, USA, 2000. USENIX Association.

413. A. Jardosh, E. Belding-Royer, K. Almeroth, and S. Suri. Real World Environment Models for Mobile Ad Hoc Networks. *IEEE Journal on Special Areas in Communications—Special Issue on Wireless Ad Hoc Networks*, 23(3), March 2005.

414. JBoss AOP homepage. http://labs.jboss.com/jbossaop, 2008.

415. M. Jelasity, A. Montresor, and O. Babaoglu. Gossip-based Aggregation in Large Dynamic Networks. *ACM Transactions on Computer Systems*, 23(3):219–252, August 2005.

416. M. Jelasity, S. Voulgaris, R. Guerraoui, A.-M. Kermarrec, and M. van Steen. Gossip-based Peer Sampling. *ACM Transactions on Computer Systems*, 25(3), August 2007.

417. J. Jeong, S. Kim, and A. Broad. Network Reprogramming. Technical Report, University of California at Berkeley, August 2003.

418. J. Jeong, J. Park, and H. Kim. DNS Service for Mobile Ad Hoc Networks. Internet Draft (Work in Progress), Internet Engineering Task Force, February 2004.

419. J. Jeong, J. Park, H. Kim, and D. Kim. Ad Hoc IP Address Autoconfiguration for AODV. Internet Draft (Work in Progress), Internet Engineering Task Force, February 2004.

420. M. Jerbi, S.-M. Senouci, R. Meraihi, and Y. Ghamri-Doudane. An Improved Vehicular Ad Hoc Routing Protocol for City Environments. In *International Conference on Communications*, Glasgow, Scotland, June 2007.

421. M. Jerbi, S.-M. Senouci, T. Rasheed, and Y. Ghamri-Doudane. An Infrastructure-Free Traffic Information System for Vehicular Networks. In *Vehicular Technology Conference (VTC'07)*, Baltimore, MD, USA, October 2007.

422. H. Jiang. *Distributed Systems of Simple Interacting Agents*. PhD thesis. Yale University, 2007.

423. X. Jiang, and T. Camp. A Review of Geocasting Protocols for a Mobile Ad Hoc Network. In *Proceedings of the Grace Hopper Celebration (GHC 2002)*, 2002.

424. T. Jim, G. Morrisett, D. Grossman, M. Hicks, J. Cheney, and Y. Wang. Cyclone: A Safe Dialect of C. In *USENIX Annual Technical Conference*, pages 275–288, Monterey, California, U.S.A., June 2002. USENIX Association.

425. A. Jindal and K. Psounis. Contention-Aware Analysis of Routing Schemes for Mobile Opportunistic Networks. *Proceedings of the 1st International ACM MobiSys Workshop on Mobile Opportunistic Networking (MobiOpp 2007)*, pages 1–8, 2007.

426. J. Jodra, M. Vara, J. Cabero, and J. Bagazgoitia. Service Discovery Mechanism over OLSR for Mobile Ad-Hoc Networks. In *Proceedings of 20th International Conference on Advanced Information Networking and Applications (AINA)*, 2006.

427. P. Johansson, T. Larsson, N. Hedman, B. Mielczarek, and M. Degermark. Scenario-Based Performance Analysis of Routing Protocols for Mobile Ad-Hoc Networks. In *Proceedings of the 5th Annual ACM/IEEE International Conference on Mobile Computing and Networking (MOBICOM '99)*, pages 195–206. The ACM Press, 1999.

428. D. Johnson and D. Maltz. Protocols for Adaptive Wireless and Mobile Networking. *IEEE Personal Communications*, 3(1):34–42, February 1996.

429. D. Johnson and D. Maltz. Dynamic Source Routing in Ad Hoc Wireless Networks. In T. Imielinski and H. Korth, editors, *Mobile Computing*, volume 353, chapter 5, pages 153–181. Kluwer Academic Publishers, 1996.

430. D. Johnson, D. Maltz, and Y.-C. Hu. The Dynamic Source Routing Protocol for Mobile Ad Hoc Networks (DSR), July 2004.

431. W. Jones. Keeping Cars from Crashing. *IEEE Spectrum*, 38(9):40–45, September 2001.

432. E. Jones, L. Li, and P. Ward. Practical Routing in Delay-Tolerant Networks. In *Proceedings WDTN*, 2005.

433. A. Joseph, J. Tauber, and M. Kaashoek. Mobile Computing with the Rover Toolkit. *IEEE Transactions on Computers*, 46(3):337–352, 1997.

434. C. Julien and G.-C. Roman. Egocentric Context-aware Programming in Ad Hoc Mobile Environments. In *Proceedings of the 10th International Symposium on the Foundations of Software Engineering (SIGSOFT'02)*, pages 21–30. The ACM Press, 2002.

435. C. Julien and G.-C. Roman. EgoSpaces: Facilitating Rapid Development of Context-Aware Mobile Applications. *IEEE Trans. Softw. Eng.*, 32(5), 2006.

436. L. Juszczyk, J. Lazowski, and S. Dustdar. Web Service Discovery, Replication, and Synchronization in Ad-Hoc Networks. In *Proceedings of the First International Conference on Availability, Reliability and Security*, pages 847–854, 2006.

437. S. Kale, E. Hazen, F. Cao, and J. Singh. Analysis and Algorithms for Content-based Event Matching. In *Proceedings of the 4th International Workshop on Distributed Event-Based Systems (DEBS '05)*, 2005.

438. T. Kallonen and J. Porras. Use of Distributed Resources in Mobile Environment. In *Proceedings of the International Conference on Software in Telecommunications and Computer Networks (SoftCOM'06)*, September-October 2006.

439. L. Källström, S. Leggio, S. Suoranta, J. Manner, T. Mikkonen, K. Raatikainen, and A. Ylä-Jääski. A Framework for Seamless Service Interworking in Ad-Hoc Networks. *Computer Communications*, 29:3277–3294, October 2006.

440. A. Kamath, R. Motwani, K. Palem, and P. Spirakis. Tail Bounds for Occupancy and the Satisfiability Threshold Conjecture. *Random Structures and Algorithms*, 7:59–80, 1995.

441. P. Kang, C. Borcea, G. Xu, A. Saxena, U. Kremer, and L. Iftode. Smart Messages: A Distributed Computing Platform for Networks of Embedded Systems. *The Computer Journal, Special Focus-Mobile and Pervasive Computing*, pages 475–494, 2004. The British Computer Society. Oxford University Press.

442. T. Karagiannis, J.-Y. Le Boudec, and M. Vojnović. Power Law and Exponential Decay of Inter Contact Times Between Mobile Devices. In *Proceedings of MobiCom'07*, pages 183–194. The ACM Press, 2007.

443. B. Karp. Geographic Routing for Wireless Networks. PhD Thesis, Harvard University, 2000.

444. B. Karps and H. Kung. GPRS: Greedy Perimeter Stateless Routing for Wireless Networks. In *Proceedings of the 6th ACM/IEEE International Conference on Mobile Computing and Networking (MOBICOM '00)*, 2000.

445. S. Kato, S. Tsugawa, K. Tokuda, T. Matsui, and H. Fujii. Vehicle Control Algorithms for Cooperative Driving with Automated Vehicles and Intervehicle Communications. *IEEE Transactions on Intelligent Transportation Systems*, 3(3):155–161, September 2002.

446. G. Kc, A. Keromytis, and V. Prevelakis. Countering Code-Injection Attacks With Instruction-Set Randomization. In *Proceedings of the 10th ACM Conference on Computer and Communications Security (CCS2003)*, pages 272–280, Washington, District of Columbia, U.S.A., October 2003. The ACM Press.

447. J. Kempf and J. Goldschmidt. Notification and Subscription for SLP. Request for Comments 3082, Internet Engineering Task Force, March 2001.

448. J. Kempf and E. Guttman. An API for Service Location. Request for Comments 2614, Internet Engineering Task Force, June 1999.

449. J. Kempf, R. Moats, and P. St. Pierre. Conversion of LDAP Schemas to and from SLP Templates. Request for Comments 2926, Internet Engineering Task Force, September 2000.

450. J. Kephart and D. Chess. The Vision of Autonomic Computing. *IEEE Computer*, 36(1):41–50, January 2003.

451. A.-M. Kermarrec and M. van Steen. Gossiping in Distributed Systems. *ACM Operating System Review*, 41(5), October 2007.

452. J.-M. Kettunen, J. Manner, and A. Ylä-Jääski. Distributed Service Location and Session Management for Ad-hoc Networks. In *Proceedings of the IEEE WoWMoM*, pages 1–6, 2007.

453. G. Kiczales, J. des Rivieres, D. Bobrow. *The Art of Metaobject Protocol*. MIT Press, Cambridge, MA, USA, 1991.

454. G. Kiczales, J. Lamping, A. Menhdhekar, C. Maeda, C. Lopes, J. Loingtier, and J. Irwin. Aspect-Oriented Programming. In *Proceedings European Conference on Object-Oriented Programming*, volume 1241, pages 220–242. Springer-Verlag, 1997.

455. G. Kiczales, E. Hilsdale, J. Hugunin, M. Kersten, J. Palm, and W. Griswold. An Overview of AspectJ. In *Proceedings of the 15th European Conference on Object-Oriented Programming (ECOOP '01)*, pages 327–353, London, UK, 2001. Springer-Verlag.

456. D. Kim, H. Bae, and C. Toh. Improving TCP-Vegas Performance Over MANET Routing Protocols. *IEEE Transactions on Vehicular Technology*, 56(1), January 2007.

457. J. Kim and N. Bambos. Power Efficient MAC Scheme Using Channel Proving in Multirate Wireless Ad Hoc Networks. In *Proceedings of the IEEE Vehicular Technology Conference*, pages 2380–2384, 2002.

458. M. Kim and D. Kotz. Periodic Properties of User Mobility and Access-Point Popularity. *Journal of Personal and Ubiquitous Computing*, 11(6), August 2007.

459. M. Kim, D. Kotz, and S. Kim. Extracting a Mobility Model from Real User Traces. In *Proceedings of IEEE INFOCOM'06*, April 2006.

460. Y.-J. Kim, R. Govindan, B. Karp, and S. Shenker. On the Pitfalls of Geographic Face Routing. In *ACM Joint Workshop on Foundations of Mobile Computing (DIALM-POMC '05)*, pages 34–43. The ACM Press, 2005.

461. Y.-J. Kim, R. Govindan, B. Karp, and S. Shenker. Geographic Routing Made Practical. In *Proceedings of the 2nd Symposium on Networked Systems Design and Implementation (NSDI '05)*, May 2005.

462. T. Kindberg and J. Barton. A Web-Based Nomadic Computing System. *Computer Networks Journal*, 35(4):443–456, March 2001.

463. J. Kingman. Markov Population Processes. *Journal of Applied Probability*, 6:1–18, 1969.

464. K. Kjaer. A Survey of Context-Aware Middleware. In *Proceedings of the 25th Conference on IASTED International Multi-Conference (SE'07)*, pages 148–155, Anaheim, CA, USA, 2007. ACTA Press.

465. M. Klein and B. König-Ries. Multi-Layer Clusters in Ad-hoc Networks—An Approach to Service Discovery. In *Proceedings of the NETWORKING 2002 Workshops on Web Engineering and Peer-to-Peer Computing*, pages 187–210, 2002.

466. B.-J. Ko, V. Misra, J. Padhye, and D. Rubenstein. Distributed Channel Assignment in Multi-Radio 802.11 Mesh Networks. In *Proceedings of the Wireless Communications and Networking Conference (WCNC 2007)*, pages 3978–3983. IEEE Press, 2007.

467. J. Kolodko and L. Vlacic. Cooperative Autonomous Driving at the Intelligent Control Systems Laboratory. *IEEE Intelligent Systems*, 18(4), July/August 2003.

468. F. Kon, M. Roman, P. Liu, J. Mao, T. Yamane, C. Magalha, and R. Campbell. Monitoring, Security, and Dynamic Configuration with the DynamicTAO Reflective ORB. In *IFIP/ACM International Conference on Distributed Systems Platforms (Middleware '00)*, pages 121–143, Secaucus, NJ, USA, 2000. Springer-Verlag, New York.

469. F. Kon, F. Costa, G. Blair, and R. Campbell. The Case for Reflective Middleware. *Communications of the ACM*, 45(6):33–38, 2002.

470. S. Kopparty, S. Krishnamurthy, M. Faloutsos, and S. Tripathi. Split-TCP for Mobile Ad Hoc Networks. In *IEEE Global Telecommunications Conference (GLOBECOM'02)*, Taipei, Taiwan, November 2002.

471. D. Kostic, A. Rodriguez, J. Albrecht, A. Bhirud, and A. Vahdat. Using Random Subsets to Build Scalable Network Services. In *Proceedings of USITS, March 2003*, March 2003.

472. D. Kotz and T. Henderson. CRAWDAD: A Community Resource for Archiving Wireless Data at Dartmouth. *IEEE Pervasive Computing*, 4(4):12–14, October-December 2005.

473. D. Kotz, T. Henderson, and I. Abyzov. CRAWDAD Trace Dartmouth/Campus/Movement/01_04 (v. 2005-03-08). Available at http://crawdad.cs.dartmouth.edu, March 2005.
474. S. Kowshik, D. Dhurjati, and V. Adve. Ensuring Code Safety Without Runtime Checks for Real-Time Control Systems. In *Proceedings of the International Conference on Compilers Architecture and Synthesis for Embedded Systems*, pages 288–297, Grenoble, France, October 2002.
475. U. Kozat and L. Tassiulas. Network Layer Support for Service Discovery in Mobile Ad Hoc Networks. In *Proceedings of IEEE INFOCOM*, pages 1965–1975, 2003.
476. J. Kramer and J. Magee. The Evolving Philosophers Problem: Dynamic Change Management. *IEEE Transactions on Software Engineering*, 16(11):1293–1306, November 1990.
477. E. Kranakis, H. Singh, and J. Urrutia. Compass Routing on Geometric Networks. In *Proceedings 11th Canadian Conference on Computational Geometry*, pages 51–54, Vancouver, August 1999.
478. B. Krishanamachari, D. Estrin, and S. Wicker. The Impact of Data Aggregation in Wireless Sensor Networks. In *Proceedings of the 22nd International Conference on Distributed Computing Systems (ICDCS'02)*, pages 575–578, Vienna, Austria, July 2002.
479. L. Krishnamurthy. Making Radios More Like Human-Ears: Alternative MAC Techniques and Innovative Platforms to Enable Large-Scale Meshes. In *Proceedings of the Microsoft Mesh Networking Summit*, 2004.
480. M. Kristensen. Enabling Cyber Foraging for Mobile Devices. In *Proceedings of the 5th MiNEMA Workshop: Middleware for Network Eccentric and Mobile Applications*, pages 32–36, Magdeburg, Germany, September 2007.
481. F. Kuhn, R. Wattenhofer, and A. Zollinger. Asymptotically Optimal Geometric Mobile Ad-Hoc Routing. In *Proceedings of the 6th International Workshop on Discrete Algorithms and Methods for Mobile Computing and Communications (DIALM '02)*, 2002.
482. F. Kuhn, R. Wattenhofer, Y. Zhang, and A. Zollinger. Geometric Ad-Hoc Routing: Of Theory and Practice. In *Proceedings of the 22nd ACM Symposium on the Principles of Distributed Computing (PODC '03)*, July 2003.
483. J. Kulik, W. Heinzelman, and H. Balakrishnan. Negotiation-based Protocols for Disseminating Information in Wireless Sensor Networks. *Wireless Networks*, 8(2/3):169–185, March-May 2002.
484. H. Kurihata, T. Takahashi, I. Ide, Y. Mekada, H. Murase, Y. Tamatsu, and T. Miyahara. Rainy Weather Recognition from In-vehicle Camera Images for Driver Assistance. In *IEEE Intelligent Vehicles Symposium (IV'05)*, pages 205–210, Las Vegas, NV, USA, June 2005.
485. S. Kurkowski, T. Camp, and M. Colagrosso. MANET Simulation Studies: The Incredibles. *ACM SIGMOBILE Mobile Computing and Communication Review*, 9(4):50–61, 2005.
486. M. Kwon and S. Fahmy. Path-Aware Overlay Multicast. *Computer Networks*, 47(1):23–45, January 2005.
487. B. Lagaisse and W. Joosen. True and Transparent Distributed Composition of Aspect-Components. In *Proceedings of the ACM/Usenix International Middleware Conference*, volume 4290 of *Lecture Notes in Computer Science*, pages 42–61. Springer-Verlag, November 2006.
488. K.-C. Lan, Z. Wang, R. Berriman, T. Moors, M. Hassan, L. Libman, M. Ott, B. Landfeldt, Z. Zaidi, A. Seneviratne, and D. Quail. Implementation of a Wireless Mesh Network Testbed for Traffic Control. In *Proceedings of the 1st International Workshop on Wireless Mesh and Ad Hoc Networks*, pages 1095–2055, 2007.
489. L. Lan and H. Wen-Jing. Localized Delaunay Triangulation for Topological Construction and Routing on MANETs. In *Proceedings of the 2nd ACM Workshop on Principles of Mobile Computing (POMC '02)*, 2002.
490. L. Lao, J. Cui, and M. Gerla. TOMA: A Viable Solution for Large-Scale Multicast Service Support. In *NETWORKING*, pages 906–917, 2005.
491. J.-Y. Le Boudec. Understanding the Simulation of Mobility Models with Palm Calculus. *Performance Evaluation*, 64(2):126–147, 2007.
492. J.-Y. Le Boudec and M. Vojnovic. Perfect Simulation and Stationarity of a Class of Mobility Models. In *Proceedings of IEEE INFOCOM'05*, pages 72–79, March 2005.

493. J.-Y. Le Boudec and M. Vojnovic. The Random Trip Model: Stability, Stationary Regime, and Perfect Simulation. *IEEE/ACM Transactions on Networking*, 14(6):1153–1166, 2006.

494. J. LeBrun, C.-N. Chuah, D. Ghosal, and M. Zhang. Knowledge-Based Opportunistic Forwarding in Vehicular Wireless Ad Hoc Networks. In *IEEE International Vehicular Technology Conference (VTC'05)*, volume 5, Stockholm, Sweden, June 2005.

495. M. Leclercq, V. Quema, J. Stefani. DREAM: A Component Framework for the Construction of Resource-Aware, Reconfigurable Moms. In *Proceedings of the 3rd Workshop on Adaptive and Reflective Middleware (ARM '04)*, pages 250–255, New York, NY, USA, 2004. The ACM Press.

496. J.-K. Lee and J. Hou. Modeling Steady-State and Transient Behaviors of User Mobility: Formulation, Analysis, and Application. In *Proceeding of MobiHoc'06*, pages 85–96, New York, NY, USA, 2006. The ACM Press.

497. S.-B. Lee, J. Cho, and A. Campbell. A Hotspot Mitigation Protocol for Ad Hoc Networks. *Ad Hoc Networks*, 1(1):87–106, July 2003.

498. S.-J. Lee and M. Gerla. Dynamic Load-Aware Routing in Ad Hoc Networks. In *Proceedings of the IEEE International Conference on Communications (ICC 2001)*, volume 10, pages 3206–3210, June 2001.

499. S. Lee, M. Gerla, and C. Chiang. On-Demand Multicast Routing Protocol (ODMRP). In *Proceedings of IEEE WCNC '99*, pages 1298–1302, New Orleans, LA, September 1999.

500. S.-J. Lee, M. Gerla, and C.-K. Toh. A Simulation Study of Table-Driven and On-Demand Routing Protocols for Mobile Ad Hoc Networks. *IEEE Network*, 13(4):48–54, July 1999.

501. S. Lee, W. Su, and M. Gerla. On-Demand Multicast Routing Protocol in Multihop Wireless Mobile Networks. *Mobile Networks and Applications*, 7(6):441–453, 2002.

502. U. Lee, E. Magistretti, B. Zhou, M. Gerla, P. Bellavista, and A. Corradi. Efficient Data Harvesting in Mobile Sensor Platforms. In *Second IEEE International Workshop on Sensor Networks and Systems for Pervasive Computing (PerSeNS'06)*, Pisa, Italy, March 2006.

503. U. Lee, J.-S. Park, E. Amir, and M. Gerla. FleaNet: A Virtual Market Place on Vehicular Networks. In *International Conference on Mobile and Ubiquitous Systems: Networking & Services*, San Jose, California, July 2006.

504. J. Leguay, T. Friedman, and V. Conan. Evaluating Mobility Pattern Space Routing for DTNs. *Proceedings of the 25th IEEE Annual Joint Conference of the IEEE Computer and Communications Societies (INFOCOM 2006)*, pages 1–10, 2006.

505. D. Lelescu, U. Kozat, R. Jain, and M. Balakrishnan. Model T++:: an Empirical Joint Space-Time Registration Model. In *Proceedings of MobiHoc'06*, pages 61–72. The ACM Press, 2006.

506. V. Lenders, M. May, and B. Plattner. Service Discovery in Mobile Ad Hoc Networks: A Field Theoretic Approach. In *Proceedings of IEEE WoWMoM*, pages 120–130, 2005.

507. I. Leontiadis. Publish/Subscribe Notification Middleware for Vehicular Networks. In *Middleware Doctoral Symposium (MDS'07)*, Newport Beach, CA, November 2007.

508. P. Levis and D. Culler. Maté: A Tiny Virtual Machine for Sensor Networks. In *Proceedings of the International Conference on Architectural Support for Programming Languages and Operating Systems (ASPLOS X)*, pages 85–95. The ACM Press, 2002.

509. P. Levis, D. Gay, and D. Culler. Bridging the Gap: Programming Sensor Networks with Application Specific Virtual Machines. Technical Report UCB//CSD-04-1343, University of California at Berkeley, August 2004.

510. P. Levis, N. Patel, D. Culler, and S. Shenker. Trickle: A Self-Regulating Algorithm for Code Propagation and Maintenance in Wireless Sensor Networks. In *First Symposium on Network Systems Design and Implementation (NSDI'04)*, 2004.

511. G. Li, A. Cheung, Sh. Hou, S. Hu, V. Muthusamy, R. Sherafat, A. Wun, H.-A. Jacobsen, and S. Manovski. Historic Data Access in Publish/Subscribe. In *Proceedings of the 1ˢᵗ Int. Conf. on Distributed Event-based Systems (DEBS)*, 2007.

512. J. Li, J. Jannotti, D. De Couto, D. Karger, and R. Morris. A Scalable Location Service for Geographic Ad-Hoc Routing. In *Proceedings of the 6th ACM International Conference on Mobile Computing and Networking (MOBICOM '00)*, pages 120–130, August 2000.

513. L. Li and L. Lamont. A Lightweight Service Discovery Mechanism for Mobile Ad Hoc Pervasive Environment using Cross-Layer Design. In *Proceedings of the 3rd IEEE Workshop on Pervasive Computing and Communications (PerCom)*, pages 55–59, 2005.

514. Q. Li, J. Aslam, and D. Rus. Online Power-Aware Routing in Wireless Ad-Hoc Networks. In *Proceedings of the 7th Annual International Conference on Mobile Computing and Networking (MobiCom '01)*, pages 97–107. The ACM Press, 2001.

515. X.-Y. Li, G. Calinescu, and P.-J. Wan. Distributed Construction of a Planar Spanner and Routing for Ad Hoc Wireless Networks. In *Proceedings of the the 21st Annual Joint Conference of the IEEE Computer and Communications Societies (INFOCOM '02)*, 2002.

516. B. Liang and Z. Haas. Predictive Distance-Based Mobility Management for Multidimensional PCS Networks. *IEEE/ACM Transactions on Networking*, 11(5):718–732, 2003.

517. D. Liben-Nowell, H. Balakrishnan, and D. Karger. Analysis of the Evolution of Peer-to-Peer Systems. In *Proceedings of the Twenty-First Annual ACM Symposium on Principles of Distributed Computing (PODC 2002)*, pages 233–242, 2002.

518. J. Liebeherr and M. Nahas. Application-Layer Multicast with Delaunay Triangulations. Technical Report, University of Virginia, Department of Computer Science, November 2001.

519. J. Lifton, D. Seetharam, M. Broxton, and J. Paradiso. Pushpin Computing System Overview: A Platform for Distributed. In *Proceedings of the Pervasive Computing Conference (Pervasive'02)*, 2002.

520. M. Lim, A. Greenhalgh, J. Chesterfield, and J. Crowcroft. Hybrid Routing: A Pragmatic Approach to Mitigating Position Uncertainty in Geo-Routing. Technical Report UCAM-CL-TR-629, University of Cambridge, April 2005.

521. A. Lindgren, A. Doria, and O. Schelen. Probabilistic Routing in Intermittently Connected Networks. *ACM Mobile Computing and Communications Review*, 7(3):19–20, 2003.

522. D. Litchfield. Defeating the Stack Based Buffer Overflow Prevention Mechanism of Microsoft Windows 2003 Server. Available at http://www.nextgenss.com/papers/defeating-w2k3-stack-protection.pdf, September 2003.

523. C. Liu and J. Kaiser. A Survey of Mobile Ad Hoc Network Routing Protocols. Technical Report 8, University of Ulm, Germany, 2003.

524. F. Liu, X. Lu, Y. Peng, and J. Huang. An Efficient Distributed Algorithm for Constructing Delay and Degree-Bounded Application-Level Multicast Tree. In *ISPAN '05: Proceedings of the 8th International Symposium on Parallel Architectures, Algorithms and Networks*, Washington, DC, USA, 2005. IEEE Computer Society.

525. J. Liu and V. Issarny. Enhanced Reputation Mechanism for Mobile Ad Hoc Networks. In *Proceedings of the Second International Conference on Trust Management (iTrust2004)*, volume 2995 of *Lecture Notes in Computer Science*, pages 48–62. Springer-Verlag, March 2004.

526. J. Liu and S. Singh. ATCP: TCP for Mobile Ad Hoc Networks. *IEEE Journal on Selected Areas in Communications*, 19(7), July 2001.

527. T. Liu, C. Sadler, P. Zhang, and M. Martonosi. Implementing Software on Resource-Constrained Mobile Sensors: Experiences with Impala and ZebraNet. In *Proceedings of the 2nd International Conference on Mobile Systems, Applications and Services (MobiSys'04)*, June 2004.

528. Y. Liu and B. Plale. Survey of Publish/Subscribe Event Systems. In *Indiana University Computer Science Technical Report TR-574*, 2003.

529. M. Livny and R. Raman. High-Throughput Resource Management. In *The Grid—Blueprint for a New Computing INfrastructure, Foster I. and Kesselman C. (Ed.)*, pages 312–337. Morgan Kaufmann Publishers Inc., 1999.

530. C. Lochert, H. Hartenstein, J. Tian, H. Fler, D. Hermann, and M. Mauve. A Routing Strategy for Vehicular Ad Hoc Networks in City Environments. In *IEEE Intelligent Vehicles Symposium (IV'03)*, Columbus, OH, USA, June 2003.

531. C. Lochert, M. Mauve, H. Fler, and H. Hartenstein. Geographic Routing in City Scenarios. *ACM Mobile Computing and Communications Review (MC2R)*, 9(1), January 2005.

532. L. Lopes, F. Martins, M. Silva, and J. Barros. A Process Calculus Approach to Sensor Network Programming. In *Proceedings of the International Conference on Sensor Technologies and Applications (SENSORCOMM'07)*. IEEE Press, 2007.

533. Ltd. Matsushita Electric Industrial Co. Electronic toll collection system. US Patent, May 2004.

534. E. Lua, J. Crowcroft, M. Pias, R. Sharma, and S. Lim. A Survey and Comparison of Peer-to-Peer Overlay Network Schemes. *IEEE Communications Surveys and Tutorials*, 7:72–93, 2005.

535. H. Lundgren, K. Ramachandran, E. Belding-Royer, K. Almeroth, M. Benny, A. Hewatt, A. Touma, and A. Jardosh. Experiences from the Design, Deployment, and Usage of the UCSB MeshNet Testbed. *IEEE Wireless Communications*, 13(2):18–29, 2006.

536. J. Luo, P. Eugster, and J.-P. Hubaux. Route Driven Gossip: Probabilistic Reliable Multicast in Ad Hoc Networks. In *Proceedings of the 22nd Annual Joint Conference of the IEEE Computer and Communications Societies (INFOCOM 2003)*, volume 3, pages 2229–2239. IEEE Press, March 2003.

537. M. Machado, O. Goussevskaia, R. Mini, A. Loureiro, G. Mateus, and J. Nogueira. Data Dissemination in Autonomic Wireless Sensor Networks. *IEEE Journal on Selected Areas in Communications*, 23(12):2305–2319, December 2005.

538. S. Madden, M. Franklin, J. Hellerstein, and W. Hong. TAG: A Tiny Aggregation Service for Ad-Hoc Sensor Networks. In *Proceedings of the 5th Symposium on Operating Systems Design and Implementation (OSDI'02)*, pages 131–146, December 2002.

539. S. Madden, M. Franklin, J. Hellerstein, and W. Hong. TinyDB: An Acquisitional Query Processing System for Sensor Networks. *ACM Transactions on Database Systems*, 2005.

540. K. Maeda, K. Sato, K. Konishi, A. Yamasaki, A. Uchiyama, H. Yamaguchi, K. Yasumo-toy, and T. Higashino. Getting Urban Pedestrian Flow from Simple Observation: Realistic Mobility Generation in Wireless Network Simulation. In *Proceedings of the 8th ACM International Symposium on Modeling, Analysis and Simulation of Wireless and Mobile Systems (MSWiM'05)*, pages 151–158, September 2005.

541. P. Maes. Concepts and Experiments in Computational Reflection. In *Proceedings of the 2nd International Conference on Object-Oriented Programming Systems, Languages, and Applications (OOPSLA'87)*, volume 22, pages 147–155, Orlando, Florida, October 1987. ACM SIGPLAN Notices.

542. R. Mahajan, M. Rodrig, D. Wetherall, and J. Zahorjan. Analyzing the MAC-Level Behavior of Wireless Networks in the Wild. *SIGCOMM Comput. Commun. Rev.*, 36:75–86, 2006.

543. Q. Mahmoud. J2ME and Location-Based Services. http://developers.sun.com, 2004.

544. C. Maihoefer. A Survey on Geocast Routing Protocols. *IEEE Communications Survey and Tutorials*, 6(2):32–42, 2004.

545. C. Maihofer and M. Bechler. Design Alternatives for IP in Vehicles. In *International Vehicular Technology Conference (VTC'03)*, Jeju Island, Korea, April 2003.

546. C. Maihofer and R. Eberhardt. Time-Stable Geocast for Ad Hoc Networks and Its Application with Virtual Warning Signs. *Computer Communications*, 27(11), July 2004.

547. G. Mainland, M. Welsh, and G. Morrisett. Flask: A Language for Data-driven Sensor Network Programs. Technical Report TR-13-06, Harvard University, May 2006.

548. G. Malkin. RIP version 2. Request For Comments 2453, November 1998.

549. T. Malone and K. Crowston. The Interdisciplinary Study of Coordination. *ACM Computing Surveys*, 26(1):87–119, March 1994.

550. M. Mamei and F. Zambonelli. Programming Pervasive and Mobile Computing Applications with the TOTA Middleware. In *Proceedings of the 2nd IEEE International Conference on Pervasive Computing and Communications (PerCom)*, 2004.

551. R. Mangharam, D. Weller, D. Stancil, R. Rajkumar and J. Parikh. GrooveSim: A Topography-Accurate Simulator for Geographic Routing in Vehicular Networks. In *Proceedings of VANET'05*, pages 59–68, New York, NY, USA, 2005. The ACM Press.

552. J. Manner, S. Leggio, T. Mikkonen, J. Saarinen, P. Vuorela, and A. Ylä-Jääski. Seamless Service Interworking of Ad-Hoc Networks and the Internet. *Computer Communications*, 31:2293–2307, 2008.

553. R.S. Marin-Perianu, P.H. Hartel, and J. Scholten. A Classification of Service Discovery Protocols. Technical Report TR-CTIT-05-25, University of Twente, Enschede, June 2005.

554. F. Martins, L. Lopes, M. Silva, and J. Barros. Robust Programming for Sensor Networks. Technical Report DCC-2008-01, Department of Computer Science, Faculty of Sciences, University of Porto, 2008.

555. C. Mascolo, L. Capra, and W. Emmerich. Mobile Computing Middleware. In *Advanced Lectures on Networking*, pages 20–58. Springer-Verlag, 2002.

556. C. Mascolo, L. Capra, S. Zachariadis, and W. Emmerich. XMIDDLE: A Data-Sharing Middleware for Mobile Computing. *Wireless Personal Communications*, 21(1):77–103, 2002.

557. J. Mathew, S. Sarker, and U. Varshney. M-Commerce Services: Promises and Challenges. *Communications of the AIS (CAIS)*, Vol. 14 (Article 25), 2004.

558. L. Mathy, R. Canonico, and D. Hutchison. An Overlay Tree Building Control Protocol. In *NGC '01: Proceedings of the Third International COST264 Workshop on Networked Group Communication*, London, UK, 2001. Springer-Verlag.

559. M. McNett and G. Voelker. Access and Mobility of Wireless PDA User. *Mobile Computing Communications Review*, 9(2):40–55, April 2005.

560. R. Meier and V. Cahill. STEAM: Event-Based Middleware for Wireless Ad Hoc Network. In *Proceedings of the 1st International Workshop on Distributed Event-Based Systems (DEBS '02)*, Vienna, Austria, 2002. IEEE Computer Society.

561. R. Meier and V. Cahill. Taxonomy of Distributed Event-Based Programming Systems. *The Computer Journal*, 48(5):602–626, June 2005.

562. R. Meier, B. Hughes, R. Cunningham, and V. Cahill. Towards Real-Time Middleware for Applications of Vehicular Ad Hoc Networks. In *Proceedings of the 1st ACM International Workshop on Vehicular Ad Hoc Networks (VANET '04:)*, pages 95–96, Philadelphia, PA, USA, 2004. The ACM Press.

563. T. Meindl, W. Moniaci, E. Pasero, and M. Riccardi. An Embedded Hardware-Software System to Detect and Foresee Road Ice Formation. In *International Joint Conference on Neural Networks*, Vancouver, Canada, July 2006.

564. F. Michaud, P. Lepage, P. Frenette, D. Letourneau, and N. Gaubert. Coordinated Maneuvering of Automated Vehicles in Platoons. *IEEE Transactions on Intelligent Transportation Systems*, 7(4):437–447, December 2006.

565. P. Michiardi and R. Molva. Core: A Collaborative Reputation Mechanism to Enforce Node Cooperation in Mobile Ad Hoc Networks. In *Proceedings of the IFIP TC6/TC11 Sixth Joint Working Conference on Communications and Multimedia Security*, pages 107–121. Kluwer, B.V., 2002.

566. MeshDynamics. Available at http://www.meshdynamics.com.

567. MeshnetNetworks. Available at http://www.motorola.com.

568. Microsoft. *Web Services Dynamic Discovery Specification*. February 2004.

569. Microsoft .Net Framework. http://www.microsoft.com/net. Accessed February 2008.

570. Microsoft Mesh Research. Available at http://research.microsoft.com/mesh.

571. Microsoft. Windows Embedded CE. www.microsoft.com, 2008.

572. Microsoft. Vault: A Programming Language for Reliable Systems. Available at http://research.microsoft.com/vault.

573. R. Miller and Q. Huang. An adaptive peer-to-peer collision warning system. In *IEEE International Vehicular Technology Conference (VTC'02)*, volume 1, pages 317–321, Birmingham, AL, USA, May 2002.

574. R. Milner, J. Parrow, and D. Walker. A Calculus of Mobile Processes, (Parts I and II). *Information and Computation*, 100(1):1–77, 1992.

575. R. Mini, A. Loureiro, and B. Nath. The Distinctive Design Characteristics of a Wireless Sensor Network: The Energy Map. *Computer Communications*, 27(10):935–945, June 2004.

576. R. Mini, A. Loureiro, and B. Nath. The Best Energy Map of a Wireless Sensor Network. In *Proceedings of the 7th ACM International Symposium on Modeling, Analysis and Simulation of Wireless and Mobile Systems (MSWiM '04)*, pages 165–169, Venice, Italy, October 2004.

577. R. Mini, A. Loureiro, and B. Nath. A State-based Energy Dissipation Model for Wireless Sensor Nodes. In *Proceedings of the 10th IEEE International Conference on Emerging Technologies and Factory Automation (ETFA)*, pages 1–8, Catania, Italy, September 2005.

578. M. Minsky. *Computation: Finite and Infinite Machines*. Prentice-Hall, Inc., 1967.

579. H. Miranda. *Gossip-Based Data Distribution in Mobile Ad Hoc Networks*. PhD thesis, Universidade de Lisboa, October 2007.

580. H. Miranda and L. Rodrigues. Using a Fairness Monitoring Service to Improve Load-Balancing in DSR. In *Proceedings of the 1st International Workshop on Services and Infrastructure for the Ubiquitous and Mobile Internet (SIUMI'05), in conjunction with ICDCS'05*, pages 314–320, June 2005.

581. H. Miranda, A. Pinto, and L. Rodrigues. Appia: A Flexible Protocol Kernel Supporting Multiple Coordinated Channels. In *Proceedings of the The 21st International Conference on Distributed Computing Systems (ICDCS '01)*, pages 707–710, Los Alamitos, CA, USA, 2001. IEEE Computer Society.

582. H. Miranda, S. Leggio, L. Rodrigues, and K. Raatikainen. A Stateless Neighbour-Aware Cooperative Caching Protocol for Ad-Hoc Networks. DI/FCUL TR 05–23, Department of Informatics, University of Lisbon, December 2005. Also as Technical Report Number C–2005–76. Computer Science Department, University of Helsinki.

583. H. Miranda, S. Leggio, L. Rodrigues, and K. Raatikainen. A Power-Aware Broadcasting Algorithm. In *Proceedings of the 17th Annual IEEE International Symposium on Personal, Indoor and Mobile Radio Communications (PIMRC)*, 2006.

584. H. Miranda, S. Leggio, L. Rodrigues, and K. Raatikainen. An Algorithm for Dissemination and Retrieval of Information in Wireless Ad Hoc Networks. In A.-M. Kermarrec, L. Bougé, and T. Priol, editors, *Proceedings of the 13th International Euro-Par Conference, Euro-Par 2007*, volume 4641 of *LNCS*, pages 891–900. Springer-Verlag, August 2007.

585. A. Misra, S. Das, A. Mcauley, and S. Das. Autoconfiguration, Registration and Mobility Management for Pervasive Computing. *IEEE Personal Communications Systems Magazine (Special Issue of Pervasive Computing)*, 8:24–31, August 2001.

586. N. Mitra. SOAP version 1.2 part 0: Primer. http://www.w3.org/TR/soap12-part0, June 2003. Accessed February 2008.

587. J. Montgomery. A real-time traffic and weather reporting system for motorists. In *Second IEEE Consumer Communications and Networking Conference (CCNC'05)*, pages 580–581, Las Vegas, NV, USA, January 2005.

588. H.D. Moore. Cracking the iPhone. Available at http://blog.metasploit.com/2007/10/cracking-iphone-part-1.html.

589. L. Mottola, G. Cugola, and G. Picco. Tree Overlays for Publish-Subscribe in Mobile Ad Hoc Networks. In *Technical Report, Politecnico di Milano*, 2005.

590. L. Mottola, A. Pathak, A. Bakshi, G. Picco, and V. Prasanna. Enabling Scope-Based Interactions in Sensor Network Macroprogramming. In *Proceedings of the 4th Int. Conf. on Mobile Ad-Hoc and Sensor Systems (MASS)*, 2007.

591. J. Moy. OSPF Version 2. Technical Report, Proteon, Inc., March 1994.

592. S. Mueller and D. Ghosal. Multipath Routing in Mobile Adhoc Networks: Issues and Challenges. In *M.C.o Calzarossa, E. Gelenbe (eds.) Lecture Notes in Computer Science*, 2004.

593. G. Muhl. Generic Constraints for Content-Based Publish/Subscribe. In *Proceedings of the 6th International Conference on Cooperative Information Systems (CoopIS)*, 2001.

594. G. Muhl. *Large-Scale Content-Based Publish/Subscribe Systems*. PhD thesis, Department of Computer Science, Darmstadt University of Technology, 2002.

595. G. Mühl, L. Fiege, F. Gärtner, and A. Buchmann. Evaluating Advanced Routing Algorithms for Content-Based Publish/Subscribe Systems. In *The 10th IEEE/ACM International Symposium on Modeling, Analysis and Simulation of Computer and Telecommunication Systems (MASCOTS 2002)*, pages 167–176, 2002.

596. L. Mui, M. Mohtashemi, and A. Halberstadt. A Computational Model of Trust and Reputation. In *Proceedings of the 35th Annual Hawaii International Conference on System Sciences (HICSS'02)*, pages 2431–2439, January 2002.

597. R. Murch and K. Letaief. Antenna Systems for Broadband Wireless Access. *IEEE Communications Magazine* 40(4): 76–83, 2002.

598. A. Murphy and G. Picco. Using Coordination Middleware for Location-Aware Computing: A LIME Case Study. In *Proceedings of the 6th Int. Conf. on Coordination Models and Languages (COORD04)*, LNCS 2949, pages 263–278. Springer-Verlag, February 2004.

599. A. Murphy and G. Picco. Using LIME to Support Replication for Availability in Mobile Ad Hoc Networks. In *Proceedings of the 8th International Conference on Coordination Models and Languages (COORD06)*, Bologna (Italy), June 2006.

600. A. Murphy, G. Picco, and G.-C. Roman. Lime: A Middleware for Physical and Logical Mobility. In *Proceedings of the 21st International Conference on Distributed Computing Systems*, pages 524–536. IEEE Computer Society, 2001.

601. A. Murphy, G. Picco, and G.-C. Roman. LIME: A Coordination Model and Middleware Supporting Mobility of Hosts and Agents. *ACM Trans. on Software Engineering and Methodology (TOSEM)*, 15(3), 2006.

602. M. Musolesi and C. Mascolo. A Community Based Mobility Model for Ad Hoc Network Research. In *Proceedings of the 2nd ACM/SIGMOBILE International Workshop on Multihop Ad Hoc Networks: From Theory to Reality (REALMAN'06)*. The ACM Press, May 2006.

603. M. Musolesi and C. Mascolo. Designing Mobility Models Based on Social Network Theory. *ACM SIGMOBILE Mobile Computing and Communication Review*, 11(3), July 2007.

604. M. Musolesi, H. Stephen, and C. Mascolo. An Ad Hoc Mobility Model Founded on Social Network Theory. In *Proceedings of the 7th ACM International Symposium on Modeling, Analysis and Simulation of Wireless and Mobile Systems*. The ACM Press, October 2004.

605. M. Musolesi, S. Hailes, and C. Mascolo. Adaptive Routing for Intermittently Connected Mobile Ad Hoc Networks. *Proceedings of the IEEE International Symposium on a World of Wireless, Mobile and Multimedia Networks (WoWMoM 2005)*, pages 183–189, 2005.

606. V. Muthusamy, M. Petrovic, and H.-A. Jacobsen. Effects of Routing Computations in Content-based Routing Networks with Mobile Data Sources. In *Proceedings of the 11th Annual International Conference on Mobile Computing and Networking (MobiCom '05)*, pages 103–116, 2005.

607. T. Nadeem, S. Dashtinezhad, C. Liao, and L. Iftode. TrafficView: Traffic Data Dissemination Using Car-to-Car Communication. *Mobile Computing and Communications Review (M2CR)*, 8(3):6–19, July 2004.

608. T. Nadeem, P. Shankar, and L. Iftode. A Comparative Study of Data Dissemination Models for VANETs. In *3rd Annual International Conference on Mobile and Ubiquitous Systems (MobiQuitous'06)*, San Jose, California, July 2006.

609. P. Nain, D. Towsley, B. Liu, and Z. Liu. Properties of Random Direction Models. In *Proceedings of INFOCOM'05*, March 2005.

610. E. Nakamura, A. Loureiro, and A. Frery. Information Fusion for Wireless Sensor Networks: Methods, Models, and Classifications. *ACM Computing Surveys*, 39(3):1–55, August 2007.

611. A. Nandan, S. Tewari, S. Das, M. Gerla, and L. Kleinrock. AdTorrent: Delivering Location Cognizant Advertisements to Car Networks. In *Conference on Wireless On Demand Network Systems and Services (WONS'06)*, Les Menuires, France, January 2006.

612. J. Naugle, K. Kasthurirangan, and G. Ledford. TN3270E Service Location and Session Balancing. Request for Comments 3049, Internet Engineering Task Force, January 2001.

613. W. Navidi and T. Camp. Stationary Distributions for the Random Waypoint Mobility Model. *IEEE Transactions on Mobile Computing*, 3(1):99–108, 2004.

614. G. Necula. Proof-Carrying Code. In *ACM SIGPLAN-SIGACT Symposium on Principles of Programming Languages (POPL'97)*, pages 106–119. The ACM Press, January 1997.

615. G. Necula, S. McPeak, and W. Weimer. CCured: Type-safe Retrofitting of Legacy Code. In *Conference Record of POPL 2002: The 29th SIGPLAN-SIGACT Symposium on Principles of Programming Languages*, pages 128–139, Portland, Oregon, U.S.A., January 2002. The ACM Press.

616. M. Nekovee and G. Bilchev. Wireless Networks on the Road: the Promises and Challenges of Vehicular Ad Hoc Networks. In *Networking and Electronic Commerce Research Conference (NAEC'05)*, Riva Del Garda, Italy, October 2005.

617. M. Newman. The Structure of Scientific Collaboration Networks. *Proceedings of the National Academy of Science*, 98:404–409, 2001.

618. M. Newman. The Structure and Function of Complex Networks. *SIAM Review*, 19(1):1–42, 2003.

619. M. Newman. Detecting Community Structure in Networks. *Eur. Phys. J. B*, 38:321–330, 2004.

620. M. Newman. Power Laws, Pareto Distributions and Zipf's Law. *Contemporary Physics*, 46:323, 2005.

621. M. Newman and M. Girvan. Finding and Evaluating Community Structure in Networks. *Physical Review E*, 69:026113, February 2004.

622. M. Newman and J. Park. Why Social Networks are Different from Other Types of Networks. *Physical Review E*, 68, 2003.

623. R. Newton and M. Welsh. Region Streams: Functional Macroprogramming for Sensor Networks. In *First International Workshop on Data Management for Sensor Networks (DMSN'04), Toronto, Canada*, 2004.

624. R. Newton, Arvind, and M. Welsh. Building up to Macroprogramming: An Intermediate Language for Sensor Networks. In *Proceedings of the 4th International Conference on Information Processing in Sensor Networks (IPSN'05)*, April 2005.

625. H. Nguyen and S. Giordano. PROSAN: Probabilistic Opportunistic Routing in SANETs. *Proceedings of ACM MobiCom/SANETs*, September 2007.

626. H. Nguyen, S. Giordano, and A. Puiatti. Probabilistic Routing Protocol for Intermittently Connected Mobile Ad Hoc Networks (PROPICMAN). *Proceedings of IEEE WoWMoM/AOC—The First IEEE WoWMoM Workshop on Autonomic and Opportunistic Communications*, June 2007.

627. Y. Ni, U. Kremer, A. Stere, and L. Iftode. Programming Ad-Hoc Networks of Mobile and Resource-Constrained Devices. *SIGPLAN Not.*, 40(6):249–260, 2005.

628. A. Nicoara, G. Alonso, and T. Roscoe. Controlled, Systematic, and Efficient Code Replacement for Running Java Programs. In *Proceedings of the 3rd EuroSys Conference*, pages 233–246. The ACM Press, April 2008.

629. D. Niculescu and B. Nath. Trajectory-Based Forwarding and Its Applications. In *9th Annual International Conference on Mobile Computing and Networking (MOBICOM)*, pages 260–272, San Diego, USA, September 2003.

630. D. Niculescu, and B. Nath. VOR Base Stations for Indoor 802.11 Positioning. In *Proceedings of the 10th Annual International Conference on Mobile Computing and Networking (MOBICOM '04)*, pages 58–69. The ACM Press, 2004.

631. M. Nidd. Service Discovery in DEAPspace. *IEEE Personal Communications*, 8:39–45, August 2001.

632. B. Noble, M. Satyanarayanan, D. Narayanan, J.E. Tilton, J. Flinn, and K.R. Walker. Agile Application-Aware Adaptation for Mobility. In *Proceedings of the 16th ACM Symposium on Operating Systems Principles (SOSP'97)*, pages 276–287. The ACM Press, December 1997.

633. E. Nordstrom, P. Gunningberg, and H. Lundgren. A Testbed and Methodology for Experimental Evaluation of Wireless Mobile Ad Hoc Networks. In *Proceedings of the First International Conference on Testbeds and Research Infrastructures for the DEvelopment of NeTworks and COMmunities (TRIDENTCOM'05)*, pages 100–109. IEEE Press, 2005.

634. J. Nykvist and K. Phanse. Modeling Connectivity in Mobile Ad-hoc Network Environments. In *Proceedings of the 6th Scandinavian Workshop on Wireless Ad-hoc Networks (AD-HOC'06)*, 2006.

635. Object Management Group. CORBA Event Service Specification, Version 1.1. OMG Document formal/2000-03-01, 2001.

636. R. Ogier, F. Templin, and M. Lewis. Topology Dissemination Based on Reverse-Path Forwarding (TBRPF). Request for Comments 3684, Internet Engineering Task Force, February 2004.

637. Y. Oiwa, T. Sekiguchi, E. Sumii, and A. Yonezawa. Fail-Safe ANSI-C Compiler: An Approach to Making C Programs Secure: Progress Report. In *Proceedings of International Symposium on Software Security 2002*, pages 133–153, Tokyo, Japan, November 2002.

638. S. Okasha. Altruism, Group Selection and Correlated Interaction. *British Journal for the Philosophy of Science*, 56(4):703–725, December 2005.

639. B. Oki, M. Pfluegel, A. Siegel, and D. Skeen. The Information Bus—An Architecture for Extensive Distributed Systems. In *Proceedings of the 1993 ACM Symposium on Operating Systems Principles*, December 1993.

640. OLSR-NG — FunkFeuer Wiki. Available at http://wiki.funkfeuer.at/index.php/OLSR-NG.

641. OLSR routing protocol (rfc3626). Available at http://hipercom.inria.fr/olsr.

642. OMG. *Common Object Request Broker Architecture (CORBA) Core Specification 3.0.2*, 2003. http://www.omg.org/technology/documents.

643. OOLSR—OLSR Implementation. Available at http://hipercom.inria.fr/OOLSR.

644. OPNET Technologies Inc. OPNET Modeler, 2004.

645. L. Opyrchal, M. Astley, J.S. Auerbach, G. Banavar, R. Strom, and D. Sturman. Exploiting IP Multicast in Content-Based Publish-Subscribe Systems. In *Proceedings of the IFIP/ACM International Conference on Distributed Systems Platforms (Middleware 2000)*, pages 185–207, 2000.

646. T. Ormandy. LibTIFF TiffFetchShortPair Remote Buffer Overflow Vulnerability. Available at http://www.securityfocus.com/bid/19283, Aug 2006.

647. OSGI Alliance. About the OSGi Service Platform. Technical Report, OSGI Alliance, 2004.

648. G. Ou. Painful Lesson in OLPC Mesh Networking for Mongolians. ZDNet Blogs, 30 January 2008. http://blogs.zdnet.com/Ou/?p=981.

649. G. Pandurangan, P. Raghavan, and E. Upfal. Building Low-Diameter P2P Networks. In *IEEE Symposium on Foundations of Computer Science*, pages 492–499, 2001.

650. S. Papanastasiou, M. Ould-Khaoua, and L. Mackenzie. On the Evaluation of TCP in MANETs. In *International Workshop on Wireless Ad Hoc Networking (IWWAN'05)*, London, UK, May 2005.

651. J. Paradiso and T. Starner. Energy Scavenging for Mobile and Wireless Electronics. *Pervasive Computing, IEEE*, 4(1):18–27, Jan.-March 2005.

652. J. Patel, I. Gupta, and N. Contractorn. JetStream: Achieving Predictable Gossip Dissemination by Leveraging Social Network Principles. In *Proceedings of the IEEE International Symposium on Network Computing and Applications, (NCA'06)*, pages 32–39. IEEE Press, July 2006.

653. G. Păun. Computing with Membranes. *Journal of Computer and System Sciences*, 61(1):108–143, August 2000.

654. R. Pawlak, L. Seinturier, L. Duchien, and G. Florin. JAC: A Flexible Solution for Aspect-Oriented Programming in Java. In *Proceedings of the 3rd International Conference on Metalevel Architectures and Separation of Crosscutting Concerns (REFLECTION '01)*, pages 1–24, London, UK, 2001. Springer-Verlag.

655. M. Pearlman, Z. Haas, P. Sholander, and S. Tabrizi. On the Impact of Alternate Path Routing for Load Balancing in Mobile Ad Hoc Networks. In *Proceedings of the 1st ACM International Symposium on Mobile Ad Hoc Networking & Computing*, pages 3–10. IEEE Press, 2000.

656. M. Pease, R. Shostak, and L. Lamport. Reaching Agreements in the Presence of Faults. *Journal of the ACM*, 27(2):228–234, April 1980.

657. L. Pelusi, A. Passarella, and M. Conti. Opportunistic Networking: Data Forwarding in Disconnected Mobile Ad Hoc Networks. *IEEE Communications Magazine*, 44(11), 2006.

658. L. Pelusi, A. Passarella, and M. Conti. *Handbook of Wireless Ad Hoc and Sensor Networks*, chapter Encoding for Efficient Data Distribution in Multi-hop Ad Hoc Networks. Wiley and Sons Publisher, 2007.

659. D. Pendarakis, S. Shi, D. Verma, and M. Waldvogel. ALMI: An Application Level Multicast Infrastructure. In *USITS'01: Proceedings of the 3rd Conference on USENIX Symposium on Internet Technologies and Systems*, Berkeley, CA, USA, 2001. USENIX Association.

660. J. Pereira, F. Fabret, F. Llirbat, and D. Shasha. Efficient Matching for Web-Based Publish/-Subscribe Systems. In *Proceedings of the 7th International Conference on Cooperative Information Systems*, volume 1901 of *LNCS*. Springer-Verlag, 2001.

661. C. Perkins. Service Location Protocol for Mobile Users. In *Ninth IEEE International Symposium on Personal, Indoor and Mobile Radio Communication*, volume 1, pages 141–146. IEEE Press, September 1998.

662. C. Perkins and P. Bhagwat. Highly Dynamic Destination-Sequenced Distance-Vec tor Routing (DSDV) for Mobile Computers. In *Proceedings of the Conference on Communications Architecture, Protocols and Applications (SIGCOMM 94)*, August 1994.

663. C. Perkins and E. Guttman. DHCP Options for Service Location Protocol. Request for Comments 2610, Internet Engineering Task Force, June 1999.

664. C. Perkins, E. Belding-Royer, and S. Das. Ad hoc On-Demand Distance Vector (AODV) Routing. Request for Comments 3561, Internet Engineering Task Force, July 2003.

665. L. Pesonen and J. Bacon. Secure Event Types in Content-based, Multi-Domain Publish/Subscribe Systems. In *Proceedings of SEM 2005*. The ACM Press, September 2005.

666. K. Petersen, M. Spreitzer, D. Terry, M. Theimer, and A. Demers. Flexible Update Propagation for Weakly Consistent Replication. In *Proceedings of the 16th ACM Symposium on Operating Systems Principles (SOSP-16)*, pages 288–301, 1997.

667. M. Petrovic and M. Aboelaze. Performance of TCP/UDP over Ad Hoc IEEE 802.11. In *Proceedings of the International Conference on Telecommunications*, pages 700–708, 2003.

668. G. Picco and M. Buschini. Exploiting Transiently Shared Tuple Spaces for Location Transparent Code Mobility. In *Proceedings of the 5th International Conference on Coordination Models and Languages (COORDINATION)*, LNCS 2315. Springer-Verlag, 2002.

669. G. Picco and P. Costa. Semi-Probabilistic Publish/Subscribe. In *Proceedings of 25th IEEE International Conference on Distributed Computing Systems (ICDCS'05)*, 2005.

670. G. Picco, A. Murphy, and G.-C. Roman. LIME: Linda Meets Mobility. In D. Garlan, editor, *Proceedings of the 21st International Conference on Software Engineering (ICSE'99)*, pages 368–377, Los Angeles, CA, USA, May 1999. The ACM Press.

671. G. Picco, G. Cugola, and A. Murphy. Efficient Content-Based Event Dispatching in the Presence of Topological Reconfiguration. In *Proceeding of the 23rd International Conference on Distributed Computing Systems (ICDCS'03)*, pages 234–243, 2003.

672. T. Pietraszek and C. Berghe. Defending Against Injection Attacks Through Context-Sensitive String Evaluation. In *Proceedings of the 8th International Symposium on Recent Advances in Intrusion Detection*, pages 124–145, Seattle, Washington, U.S.A., 2005. Springer-Verlag.

673. P. Pietzuch and J. Bacon. Hermes: a Distributed Event-Based Middleware Architecture. In *Proceedings of the International Workshop on Distributed Event-Based Systems (DEBS'02)*, 2003.

674. P. Pietzuch and J. Bacon. Peer-to-Peer Overlay Broker Networks in an Event-Based Middleware. In *Proceedings of the 2nd International Workshop on Distributed Event-Based Systems (DEBS'03)*, 2003.

675. M. Piorkowski, N. Sarafijanovic-Djukic, and M. Grossglauser. On Clustering Phenomenon in Partitioned Mobile Networks. In *Proceedings of First ACM SIGMOBILE International Workshop on Mobility Models for Networking Research (MobilityModels'08)*, May 2008.

676. K. Pister, J. Kahn, and B. Boser. Smart Dust: Wireless Networks of Millimeter-Scale Sensor Nodes, 1999.

677. E. Pitt and K. McNiff. *Java.rmi: The Remote Method Invocation Guide*. Addison-Wesley, January 1998.

678. J. Polastre, J. Hill, and D. Culler. Versatile Low Power Media Access for Wireless Sensor Networks. In *2nd International Conference on Embedded Networked Sensor Systems (SenSys)*, pages 95–107, Baltimore, USA, November 2004.

679. A. Popovici, A. Frei, and G. Alonso. A Proactive Middleware Platform for Mobile Computing. In *ACM/IFIP/USENIX International Middleware Conference*, volume 2672 of *Lecture Notes in Computer Science*, pages 455–473. Springer-Verlag, 2003.

680. J. Porras, A. Valtaoja, and P. Hiirsalmi. Peer-to-Peer Communication Approach for a Mobile Environment. In *Proceedings of the 37th Annual Hawaii International Conference on Systems Sciences (HICSS04)*, January 2004.

681. G. Pottie and W. Kaiser. Wireless Integrated Network Sensors. *Communications of the ACM*, 43(5):51–58, May 2000.

682. R. Preotiuc-Pietro, J. Pereira, F. Llirbat, F. Fabret, K. Ross, and D. Shasha. Publish/Subscribe on the Web at Extreme Speed. In *Proceedings of ACM SIGMOD Conf. on Management of Data*, Cairo, Egypt, 2000.

683. M. Presburger. Über die Vollständigkeit eines gewissen Systems der Arithmetik ganzer Zahlen, in welchem die Addition als einzige Operation hervortritt. In *Comptes-Rendus du I Congrès de Mathématiciens des Pays Slaves*, pages 92–101, Warszawa, 1929.

684. D. Qiao, S. Choi, and K. Shin. Goodput Analysis and Link Adaptation for IEEE 802.11a Wireless LANs. *IEEE Transactions on Mobile Computing*, 1(4):278–292, 2002.

685. V. Raghunathan, C. Schurgers, S. Park, and M. Srivastava. Energy-aware Wireless Microsensor Networks. *IEEE Signal Processing Magazine*, 19(2):40–50, March 2002.

686. A. Ranganathan, J. Al-Muhtadi, S. Chetan, R. Campbell, and M. Mickunas. MiddleWhere: A Middleware for Location Awareness in Ubiquitous Computing Applications. In *Proceedings of the 5th ACM/IFIP/USENIX International Conference on Middleware*, pages 397–416, New York, NY, USA, 2004. Springer-Verlag, New York.

687. A. Raniwala and T. Chiueh. Architecture and Algorithms for an IEEE 802.11 Based Multi-Channel Wireless Mesh Network. In *Proceedings of the IEEE Infocom*, 2005.

688. A. Rao, C. Papadimitriou, S. Shenker, and I. Stoica. Geographic Routing without Location Information. In *Proceedings of the 9th Annual International Conference on Mobile Computing and Networking (MOBICOM '03)*, pages 96–108. The ACM Press, 2003.

689. S. Ratnasamy, P. Francis, M. Handley, R. Karp, and S. Schenker. A Scalable Content-Addressable Network. In *SIGCOMM '01: Proceedings of the 2001 Conference on Applications, Technologies, Architectures, and Protocols for Computer Communications*. The ACM Press, 2001.

690. S. Ratnasamy, M. Handley, R. Karp, and S. Shenker. Application-Level Multicast Using Content-Addressable Networks. In *NGC'01: Proceedings of the Third International COST264 Workshop on Networked Group Communication*, volume 2233 of *LNCS*, pages 14–34. Springer-Verlag, 2001.

691. S. Ratnasamy, B. Karp, L. Yin, F. Yu, D. Estrin, R. Govindan, and S. Shenker. GHT: A Geographic Hash Table for Data-Centric Storage in Sensornets. In *Proceedings of the 1st ACM International Workshop on Wireless Sensor Networks and Applications (WSNA '02)*, September 2002.

692. O. Ratsimor, D. Chakraborty, A. Joshi, and T. Finin. Allia: Alliance-Based Service Discovery for Ad-Hoc Environments. In *Proceedings of the 2nd International Workshop on Mobile Commerce (WMC'02)*, pages 1–9, New York, NY, USA, 2002. The ACM Press.

693. P.-G. Raverdy, O. Riva, A. de La Chapelle, R. Chibout, V. Issarny. Efficient Context-aware Service Discovery in Multi-Protocol Pervasive Environments. In *Proceedings of the 7th International Conference on Mobile Data Management (MDM'06)*, page 3. IEEE Computer Society, 2006.

694. N. Ravi, C. Borcea, P. Kang, and L. Iftode. Portable Smart Messages for Ubiquitous Java-Enabled Devices. In *Proceedings of the 1st Annual International Conference on Mobile and Ubiquitous Systems: Computing, Networking and Services (MobiQuitous'04)*, pages 412–421. IEEE Computer Society, 2004.

695. N. Ravi, S. Smaldone, L. Iftode, and M. Gerla. Lane Reservation for Highways (Position Paper). In *10th International Conference on Intelligent Transportation Systems*, Seattle, Washington, September 2007.

696. B. Ray. Symbian signing is no protection from spyware. Available at http://www.theregister.co.uk/2007/05/23/symbian_signed_spyware, 2007.

697. M. Raya, P. Papadimitratos, and J.-P. Hubaux. Securing Vehicular Communications. *IEEE Wiresless Communications*, 13(5), October 2006.

698. RDF—Resource Description Framework. Available at http://www.w3.org/RDF.

699. M. Regenold. From Waving Arms to LED's: A Brief History of Traffic Signals. *Tech Transfer Newsletter*, Fall 2007.

700. N. Reijers and K. Langendoen. Efficient Code Distribution in Wireless Sensor Networks. In *Proceedings of the 2nd ACM International Conference on Wireless Sensor Networks and Applications (WSNA'03)*, pages 60–67. The ACM Press, 2003.

701. D. Reilly and A. Taleb-Bendiab. A Service-Based Architecture for In-Vehicle Telematics Systems. In *International Workshop on Smart Appliances and Wearable Computing (IW-SAWC'02)*, Washington, DC, USA, July 2002.

702. G. Resta and P. Santi. An Analysis of the Node Spatial Distribution of the Random Waypoint Mobility Model for Ad Hoc Networks. In *Proceedings of POMC'02*, pages 44–50. The ACM Press, 2002.

703. G. Resta and P. Santi. The QoS-RWP Mobility and User Behavior Model for Public Area Wireless Networks. In *Proceedings of the 9th ACM International Symposium on Modeling, Analysis and Simulation of Wireless and Mobile Systems (MSWiM'06)*, pages 375–384. The ACM Press, October 2006.

704. P. Reynolds and R. Brangeon. DOLMEN—Service Machine Development for an Open Long-Term Mobile and Fixed Network Environment (Project Report). Available at ftp://ftp.cordis.europa.eu/pub/infowin/docs/fr-036.pdf, 1996.

705. I. Rhee, M. Shin, S. Hong, K. Lee, and S. Chong. On the Lévy-walk Nature of Human Mobility. In *Proceedings of INFOCOM'08*, Arizona, USA, 2008.

706. A. Riabov, Z. Liu, J. Wolf, P. Yu, and L. Zhang. Clustering Algorithms for Content-Based Publication-Subscription Systems. In *Proceedings of the 22nd International Conference on Distributed Computing Systems (ICDCS'02)*, 2002.

707. A. Riabov, Z. Liu, J. Wolf, P. Yu, and L. Zhang. New Algorithms for Content-Based Publication-Subscription Systems. In *Proceedings of the 23rd International Conference on Distributed Computing Systems (ICDCS'03)*, pages 678–686, 2003.

708. G. Richard. Service Advertisement and Discovery: Enabling Universal Device Cooperation. *IEEE Internet Computing*, 4(5):18–26, 2000.

709. O. Riva, T. Nadeem, C. Borcea, and L. Iftode. Context-Aware Migratory Services in Ad Hoc Networks. *IEEE Transactions on Mobile Computing*, 6(12):1313–1328, 2007.

710. W. Rjaibi, K. Dittrich, and D. Jaepel. Event Matching in Symmetric Subscription Systems. In *Proceedings of the 12th Annual IBM Centers for Advanced Studies Conference (CASCON '02)*, 2002.

711. T. Robbins. Libformat. Available at http://www.securityfocus.com/tools/1818, October 2001.

712. V. Rodoplu and T. Meng. Minimum Energy Mobile Wireless Networks. *IEEE Journal on Selected Areas in Communication*, 17(8):1333–1344, August 1999.

713. M. Roman and N. Islam. Dynamically Programmable and Reconfigurable Middleware Services. In *Proceedings of the 5th ACM/IFIP/USENIX International Conference on Middleware (Middleware '04)*, pages 372–396, New York, NY, USA, 2004. Springer-Verlag, New York.

714. G.-C. Roman, A. Murphy, and G. Picco. Coordination and Mobility. In A. Omicini, F. Zambonelli, M. Klusch, and R. Tolksdorf, editors, *Coordination of Internet Agents: Models, Technologies, and Applications*, pages 254–273. Springer-Verlag, 2000. Invited contribution.

715. L. Rosa, A. Lopes, and L. Rodrigues. Policy-Driven Adaptation of Protocol Stacks. In *Proceedings of the International Conference on Autonomic and Autonomous Systems (ICAS '06)*, Washington, DC, USA, 2006. IEEE Computer Society.

716. A. Rowston, A.-M. Kermarrec, M. Castro, and P. Druschel. SCRIBE: The Design of a Large-scale Notification Infrastructure. In *Proceedings of the 3rd International Workshop on Networked Group Communication (NGC2001)*, 2001.

717. A. Rowstron. WCL: A Coordination Language for Geographically Distributed Agents. *World Wide Web Journal*, 1(3):167–179, 1998.

718. A. Rowstron, and P. Druschel. Pastry: Scalable, Decentralized Object Location, and Routing for Large-Scale Peer-to-Peer Systems. In *Proceedings of the IFIP/ACM International Conference on Distributed Systems Platforms (Middleware 2001)*, pages 329–350, 2001.

719. E. Royer and C. Perkins. Multicast Operation of the Ad-Hoc On-demand Distance Vector Routing Protocol. In *Proceedings of the 5th Annual ACM/IEEE International Conference on Mobile Computing and Networking (MOBICOM '99)*, pages 207–218. The ACM Press, 1999.

720. M. Rutherford, A. Carzaniga, and A. Wolf. Simulation-Based Test Adequacy Criteria for Distributed Systems. In *Proceedings of the 14th ACM SIGSOFT International Symposium on Foundations of Software Engineering (FSE-14)*, pages 231–241. The ACM Press, 2006.

721. S. Sadjadi, P. McKinley, and E. Kasten. Architecture and Operation of an Adaptable Communication Substrate. In *Proceedings of the The 9th IEEE Workshop on Future Trends of Distributed Computing Systems (FTDCS'03)*, pages 46–55, Washington, DC, USA, 2003. IEEE Computer Society.

722. A. Saha and D. Johnson. Modeling Mobility for Vehicular Ad-Hoc Networks. In *Proceedings of VANET'04*, pages 91–92. The ACM Press, October 2004.

723. F. Sailhan and V. Issarny. Scalable Service Discovery for MANET. In *Proceedings of the Third IEEE International Conference on Pervasive Computing and Communications (PerCom'05)*, pages 235–244. IEEE Computer Society, 2005.

724. Salutation. Available at http://www.salutation.org.

725. B. Sapkota, D. Roman, S. Kruk, and D. Fensel. Distributed Web Service Discovery Architecture. In *Proceedings of the Advanced International Conference on Telecommunications and International Conference on Internet and Web Applications and Services (AICT/ICIW 2006)*, 2006.

726. T. Sarkar, M. Wicks, M. Salazar-Palma, and R. Bonneau. A Survey of Various Propagation Models for Mobile Communication. *Smart Antennas*, pages 239–307, 2004.

727. Y. Sasson, D. Cavin, and A. Schiper. Probabilistic Broadcast for Flooding in Wireless Mobile Ad hoc Networks. In *Proceedings of the IEEE Wireless Communications and Networking Conference (WCNC)*, March 2003.

728. M. Satyanarayanan. Mobile Information Access. *IEEE Personal Communications*, 3(1):26–33, 1996.

729. M. Satyanarayanan. Pervasive Computing: Vision and Challenges. *Personal Communications, IEEE [see also IEEE Wireless Communications]*, 8(4):10–17, 2001.

730. M. Satyanarayanan, J. Kistler, P. Kumar, M. Okasaki, E.H. Siegel, and D.C. Steere. Coda: A Highly Available File System for a Distributed Workstation Environment. *Transactions on Computers*, 39(4):447–459, 1990.

731. A. Scaglione and S. Servetto. On the Interdependence of Routing and Data Compression in Multi-Hop Sensor Networks. In *Proceedings of the International Conference on Mobile Computing and Networking (MobiCom'02)*, Atlanta, GA, 2002.

732. B. Schilit, N. Adams, and R. Want. Context-Aware Computing Applications. In *IEEE Workshop on Mobile Computing Systems and Applications (WMCSA'94)*, pages 89–101, Santa Cruz, CA, US, 1994. IEEE Computer Society.

733. A. Schill, B. Bellmann, W. Bohmak, and S. Kummel. Infrastructure Support for Cooperative Mobile Environments. In *Proceedings of the 4th Workshop on Enabling Technologies: Infrastructure for Collaborative Enterprises*, pages 171–178, 1995.

734. R. Schmitz, A. Leiggener, A. Festag, L. Eggert, and W. Effelsberg. Analysis of Path Characteristics and Transport Protocol Design in Vehicular Ad Hoc Networks. In *IEEE International Vehicular Technology Conference (VTC'06)*, Melbourne, Australia, May 2006.

735. A. Schönhage. Storage Modification Machines. *SIAM Journal on Computing*, 9(3):490–508, August 1980.

736. J. Scott. *Social Networks Analysis: A Handbook*. Sage Publications, London, United Kingdom, second edition, 2000.

737. J. Scott, R. Gass, J. Crowcroft, P. Hui, C. Diot, and A. Chaintreau. CRAWDAD Data Set cambridge/haggle (v. 2006-01-31). Available at http://crawdad.cs.dartmouth.edu/cambridge/haggle, January 2006.

738. K. Seada, C. Westphal, and C. Perkins. Analyzing Path Accumulation for Route Discovery in Ad Hoc Networks. In *Proceedings of the IEEE Wireless Communications & Networking Conference (WCNC '07)*, March 2007.

739. S3MS. Security of Software and Services for Mobile Systems. Available at http://www.s3ms.org, 2007.

740. M. Seddigh, J. González, and I. Stojmenovic. RNG and Internal Node Based Broadcasting Algorithms for Wireless One-to-One Networks. *ACM SIGMOBILE Mobile Computing and Communication Review*, 5(2):37–44, 2001.

741. B. Segall and D. Arnold. Elvin Has Left the Building: A Publish/Subscribe Notification Service with Quenching. In *Proceedings of the 1997 Australian UNIX and Open Systems Users Group Conference*, 1997.

742. B. Segall, D. Arnold, J. Boot, M. Henderson, and T. Phelps. Content Based Routing with Elvin4. In *Proceedings of AUUG2K, Canberra, Australia*, June 2000.

743. A. Senart, M. Bouroche, and V. Cahill. Modelling an Emergency Vehicle Early-Warning System using Real-time Feedback. *International Journal of Intelligent Information and Database Systems (IJIIDS)*, February-March 2008.

744. SETI@home. Available at http://setiathome.berkeley.edu.

745. H. Shacham, M. Page, B. Pfaff, E.-J. Goh, N. Modadugu, and D. Boneh. On the Effectiveness of Address-Space Randomization. In *Proceedings of the 11th ACM Conference on Computer and Communications Security*, pages 298–307, Washington, District of Columbia, U.S.A., October 2004. The ACM Press.

746. R. Shah, S. Roy, S. Jain, and W. Brunette. Data MULEs: Modeling a Three-Tier Architecture for Sparse Sensor Networks. *IEEE Workshop on Sensor Network Protocols and Applications (SNPA)*, 2003.

747. G. Sharma and R. Mazumdar. Scaling Laws for Capacity and Delay in Wireless Ad Hoc Networks with Random Mobility. In *IEEE International Conference on Communications (ICC'04)*, pages 3869– 3873, June 2004.

748. Z. Shelby, C. Pomalaza-Ráez, H. Karvonen, and J. Haapola. Energy Optimization in Multi-hop Wireless Embedded and Sensor Networks. *International Journal of Wireless Information Networks*, 12(1):11–21, January 2005.

749. W. Sheng, Q. Yang, and Yi. Guo. Experimental Testbed and Distributed Algorithm for Cooperative Driving in VII Simulation. In *IEEE Conference on Intelligent Transportation Systems (ITSC'06)*, Toronto, ON, Canada, September 2006.

750. W. Sheng, Q. Yang, and Y. Guo. Cooperative Driving Based on Inter-vehicle Communications: Experimental Platform and Algorithm. In *International Conference on Intelligent Robots and Systems (IROS'06)*, Beijing, China, October 2006.

751. Y. Shi. *Design of Overlay Networks for Internet Multicast*. PhD thesis, Washington University, 2002.

752. K. Shin, S. Tsugawa, T. Kiyohito, M. Takeshi, and F. Haruki. Vehicle Control Algorithms for Cooperative Driving with Automated Vehicles and Intervehicle Communications. *IEEE Transactions on Intelligent Transportation Systems*, 3(3):155–161, September 2002.

753. Shockfish. Spotme. discover people. www.shockfish.com, 2006.

754. SIG working group. *Bluetooth Specification Version 1.2*. SIG working group, 1.2 edition, November 2003. SDP.

755. M. Sihvonen. *Adaptive Personal Service Environment*. PhD thesis, University of Oulu, December 2007.

756. R. Silva Filho and D. Redmiles. A Survey of Versatility for Publish/Subscribe Infrastructures. In *ISR Technical Report UCI-ISR-05-8, Institute for Software Research, University of California, Irvine*, 2005.

757. D. Simon, C. Cifuentes, D. Cleal, J. Daniels, and D. White. Java on the Bare Metal of Wireless Sensor Devices—The Squawk Java Virtual Machine. In *Proceedings of VEE'06*. The ACM Press, June 2006.

758. J. Singh, N. Bambos, and B. Srinivasan. Detlef Clawin. Cross-layer Multi-hop Wireless Routing for Inter-Vehicle Communication. In *2nd International Conference on Testbeds and Research Infrastructures for the Development of Networks and Communities (TRIDENT-COM'06)*, March 2006.

759. M. Singh and M. Huhns. *Service-Oriented Computing: Semantics, Processes, Agents*. John Wiley & Sons, Ltd., January 2005.

760. S. Singh, M. Woo, and C. Raghavendra. Power-Aware Routing in Mobile Ad Hoc Networks. In *Proceedings of the 4th Annual ACM/IEEE International Conference on Mobile Computing and Networking (MOBICOM '98)*, pages 181–190. The ACM Press, 1998.

761. T. Sivaharan, G. Blair, and G. Coulson. GREEN: A Configurable and Re-configurable Publish-Subscribe Middleware for Pervasive Computing. In *Proceedings of DOA 2005*, 2005.

762. T. Sohn, W. Griswold, J. Scott, A. LaMarca, Y. Chawathe, I. Smith, and M. Chen. Experiences with Place Lab: An Open Source Toolkit for Location-Aware Computing. In *Proceedings of ICSE'06*, pages 462–471. The ACM Press, 2006.

763. R. Sombrutzki, A. Zubow, M. Kurth, and J. Redlich. Self-Organization in Community Mesh Networks The Berlin RoofNet. In *Proceedings of the 1st Workshop on Operator-Assisted (Wireless Mesh) Community Networks*, pages 1–11, 2006.

764. C. Sommer and F. Dressler. The DYMO Routing Protocol in VANET Scenarios. In *Proceedings of the 66th IEEE Vehicular Technology Conference (VTC2007-Fall)*, pages 16–20. IEEE Press, September 2007.

765. C. Sorensen, M. Wu, T. Sivaharan, G. Blair, P. Okanda, A. Friday, and Duran-Limon. A Context-aware Middleware for Applications in Mobile Ad Hoc Environments. In *Proceedings of the 2nd Workshop on Middleware for Pervasive and Ad-hoc Computing (MPAC'04)*, pages 107–110. The ACM Press, 2004.

766. A. Spyropoulos and C. Raghavendra. Energy Efficient Communications in Ad Hoc Networks Using Directional Antenna. In *Proceedings of the IEEE Annual Conference on Computer Communications (INFOCOM)*, pages 220–228, 2002.

767. T. Spyropoulos and K. Psounis. Spray and Focus: Efficient Mobility-Assisted Routing for Heterogeneous and Correlated Mobility. *Proceedings of IEEE Percom International Workshop on Intermittently Connected Mobile Ad Hoc Networks*, March 2007.

768. T. Spyropoulos, K. Psounis, and C. Raghavendra. Performance Analysis of Mobility-Assisted Routing. In *Proceedings of MobiHoc'06*, pages 49–60. The ACM Press, 2006.

769. T. Spyropoulos, K. Psounis, and C. Raghavendra. Efficient Routing in Intermittently Connected Mobile Networks: The Multiple-Copy Case. *ACM/IEEE Transactions on Networking*, 16, 2007.

770. M. Srivatsa and L. Liu. Securing Publish-Subscribe Overlay Services with EventGuard. In *CCS '05: Proceedings of the 12th ACM Conference on Computer and Communications Security*, pages 289–298. The ACM Press, 2005.

771. P. Stanley-Marbell, C. Borcea, K. Nagaraja, and L. Iftode. Smart Messages: A System Architecture for Large Networks of Embedded Systems. In *Position summary in the Proceedings of the 8th Workshop on Hot Topics in Operating Systems (HotOS-VIII)*, May 2001.

772. T. Stathopoulos, J. Heidemann, and D. Estrin. A Remote Code Update Mechanism for Wireless Sensor Networks. Technical Report, University of California at Los Angeles, 2003.

773. I. Stoica, R. Morris, D. Karger, M.F. Kaashoek, and H. Balakrishnan. Chord: A Scalable Peer-to-Peer Lookup Service for Internet Applications. In *Proceedings of ACM SIGCOMM*, 2001.

774. I. Stoica, D. Adkins, S. Ratnasamy, S. Shenker, S. Surana, and S. Zhuang. Internet Indirection Infrastructure. In *Proceedings of the First International Workshop on Peer-to-Peer Systems*, 2002.

775. I. Stojmenovic and X. Lin. Power-Aware Localized Routing in Wireless Networks. *IEEE Transactions on Parallel and Distributed Systems*, 12(11):1122–1133, October 2001.

776. I. Stojmenovic and X. Lin. Loop-Free Hybrid Single-Path/Flooding Routing Algorithms with Guaranteed Delivery for Wireless Networks. *IEEE Transactions on Parallel and Distributed Systems*, 12(10), October 2001.

777. J. Su, A. Chin, A. Popivanova, A. Goel, and E. de Lara. User Mobility for Opportunistic Ad-Hoc Networking. In *Proceedings of the Sixth IEEE Workshop on Mobile Computing Systems and Applications (WMCSA'04)*, pages 41–50. IEEE Press, 2004.

778. Y. Su and J. Flinn. Slingshot: Deploying Stateful Services in Wireless Hotspots. In *Proceedings of the 3rd International Conference on Mobile Systems, Applications, and Services (MobiSys'05)*, pages 79–92. The ACM Press, 2005.

779. Q. Sun and H. Garcia-Molina. Using Ad-hoc Inter-Vehicle Network for Regional Alerts. Technical Report, Stanford University, October 2004.

References

780. K. Sundaresan, V. Anantharaman, H.-Y. Hsieh, and R. Sivakumar. ATP: A Reliable Transport Protocol for Ad Hoc Networks. In *4th ACM International Symposium on Mobile Ad Hoc Networking and Computing (MobiHoc'03)*, Annapolis, MA, USA, June 2003.

781. Sun Developer Network (SDN). Java ME at a Glance. http://java.sun.com/javame, 2008.

782. Sun Developer Network (SDN). The Source for Java Developers. http://java.sun.com, 2008.

783. Sun Microsystems Inc. The Connected Limited Device Configuration. Version 1.1. Java Standards Process JSR 139. Available at http://jcp.org/aboutJava/communityprocess/final/jsr139/index.html.

784. Sun Microsystems Inc. The Connected Device Configuration. Version 1.1.2. Java Standards Process JSR 218. Available at http://jcp.org/aboutJava/communityprocess/mrel/jsr218/index.html.

785. Sun Microsystems Inc. Mobile Information Device Profile. Version 2.0. Java Standards Process JSR 118. Available at http://jcp.org/aboutJava/communityprocess/final/jsr118/index.html.

786. Sun Microsystems. *Jini Specification Version 2.0*, 2003. Available at http://java.sun.com/products/jini.

787. Sun Microsystems. *Jini Architecture Specification 2.0*, June 2003. Available at http://wwws.sun.com/software/jini/specs/jini2_0.pdf.

788. Sun Microsystems. *Jini Technology Core Platform Specification 2.0*, June 2003. Available at http://wwws.sun.com/software/jini/specs/core2_0.pdf.

789. SIENA Web Site. http://www.cs.colorado.edu/users/carzanig/siena.

790. Spring website. http://www.springframework.org, 2008.

791. C. Szyperski. *Component Software: Beyond Object-Oriented Programming*. Addison-Wesley Professional, 1998.

792. H. Takagi and L. Kleinrock. Optimal Transmission Ranges for Randomly Distributed Packet Radio Terminals. *IEEE Transactions on Communications*, 32(3):246–257, March 1984.

793. A. Takano, H. Okada, and K. Mase. Performance Comparison of a Position-Based Routing Protocol for VANET. In *IEEE International Conference on Mobile Adhoc and Sensor Systems (MASS'07)*, Pisa, Italy, October 2007.

794. D. Tang and M. Baker. Analysis of a Local-Area Wireless Network. In *Proceedings of MobiCom'00*, pages 1–10. The ACM Press, 2000.

795. F. Tango, A. Saroldi, M. Alonso, and A. Oyaide. Towards a New Approach in Supporting Drivers Function: Specifications of the SASPENCE System. In *Intelligent Transportation Systems Conference (ITSC'05)*, pages 614–620, Vienna, Austria, September 2005.

796. M. Tariq, M. Ammar, and E. Zegura. Message Ferry Route Design for Sparse Ad Hoc Networks with Mobile Nodes. *Proceedings of ACM Mobihoc*, May 2006.

797. S. Tarkoma. *Efficient Content-based Routing, Mobility-aware Topologies, and Temporal Subspace Matching*. PhD thesis, University of Helsinki, Department of Computer Science, Helsinki, Finland, April 2006.

798. S. Tarkoma and J. Kangasharju. On the Cost and Safety of Handoffs in Content-based Routing Systems. *Computer Networks*, 51(6), April 2007.

799. D. Terdiman. Making Wireless Roaming Fun. Wired News Online, April 2003.

800. K. Terfloth, G. Wittenburg, and J. Schiller. FACTS—A Rule-based Middleware Architecture for Wireless Sensor Networks. In *Proceedings of the 1ˢᵗ Int. Conf. on Communication System Software and Middleware (COMSWARE)*, 2006.

801. W. Terpstra, S. Behnel, L. Fiege, A. Zeidler, and A. Buchmann. A Peer-to-Peer Approach to Content-Based Publish/Subscribe. In *Proceedings of the 2nd International Workshop on Distributed Event-Based Systems (DEBS'03)*, 2003.

802. D. Terry, K. Petersen, M. Spreitzer, and M. Theimer. The Case for Non-Transparent Replication: Examples from Bayou. *IEEE Data Engineering Bulletin*, 21(4):12–20, 1998.

803. The Network Simulator—ns-2. Available at http://www.isi.edu/nsnam/ns, 2008.

804. The ns-3 Project. Available at http://www.nsnam.org.

805. The TinyOS Documentation Project. Available at http://www.tinyos.org.

806. The ZigBee Alliance. Available at http://www.zig-bee.org.

807. S. Tilak, N. Abu-Ghazaleh, and W. Heinzelman. A Taxonomy of Wireless Micro-Sensor Network Models. *Mobile Computing and Communication Review*, 6(2), April 2002.

808. S. Tilak, A. Murphy, and W. Heinzelman. Non-uniform Information Dissemination for Sensor Networks. In *Proceedings of the 11th IEEE Conference on Network Protocols (ICNP'03)*, pages 295–304, November 4–7 2003.

809. C.-K. Toh. Associativity-Based Routing for Ad Hoc Mobile Networks. *Wireless Personal Communications*, 4(2):103–139, March 1997.

810. O. Tonguz, N. Wisitpongphan, J. Parikh, F. Bai, P. Mudalige, and V. Sadekar. On the Broadcast Storm Problem in Ad Hoc Wireless Networks. In *3rd International Conference on Broadband Communications, Networks and Systems (BROADNETS'06)*, pages 1–11, San Jose, California, October 2006.

811. A. Tønnesen. Impementing and Extending the Optimized Link State Routing Protocol. Master's thesis, UniK University Graduate Center, University of Oslo, August 2004.

812. P. Triantafillou and A. Economides. Subscription Summarization: A New Paradigm for Efficient Publish/Subscribe Systems. In *Proceedings of the 24th International Conference on Distributed Computing Systems (ICDCS'04)*, pages 562–571, 2004.

813. Tropos. Available at http://www.tropos.com.

814. E. Truyen. *Dynamic and Context-Sensitive Composition in Distributed Systems*. PhD thesis, K.U. Leuven, Belgium, November 2004.

815. E. Truyen, N. Janssens, F. Sanen, and W. Joosen. Support for Distributed Adaptations in Aspect-Oriented Middleware. In *Proceedings of the 7th International Conference on Aspect-Oriented Software Development (AOSD '08)*, pages 120–131, New York, NY, USA, 2008. The ACM Press.

816. C. Tschudin. Lightweight Underlay Network Ad Hoc Routing LUNAR Protocol. Internet draft (work in progress), Internet Engineering Task Force, March 2004.

817. Y. Tseng, S. Ni, and E. Shih. Adaptive Approaches to Relieving Broadcast Storms in Wireless Multihop Mobile Ad Hoc Networks. In *Proceedings of the 21st International Conference on Distributed Computing Systems (ICDCS'01)*, pages 481–488, 2001.

818. Y. Tseng, S. Ni, Y. Chen, and J. Sheu. The Broadcast Storm Problem in a Mobile Ad Hoc Network. *Wireless Networks*, 8(2/3):153–167, 2002.

819. TSpaces: Intelligent Connectionware. Available at http://www.almaden.ibm.com/cs/TSpaces.

820. C. Tuduce and T. Gross. A Mobility Model Based on WLAN Traces and its Validation. In *Proceedings of INFOCOM'05*, pages 19–24, March 2005.

821. P. Turchin. *Measuring and Modeling Population Redistribution in Animals and Plants*. Sinauer Associates, 1998.

822. UDDI—Universal Description, Discovery and Intergation. Available at http://www. oasis-open.org/committees/uddi-spec.

823. D. Ungar, A. Spitz, and A. Ausch. Constructing a Metacircular Virtual Machine in an Exploratory Programming Environment. In *Proceeding of ACM SIGPLAN Conference on Object-Oriented Programming Systems, Languages, and Applications (OOPSLA'05)*, pages 11–20. The ACM Press, October 2005.

824. UPnP—Universal Plug and Play Forum. Available at http://www.upnp.org.

825. M. Uruena and D. Larrabeiti. eXtensible Service Registration Protocol (XSRP). Internet-draft, Internet Engineering Task Force, March 2004.

826. M. Uruena and D. Larrabeiti. eXtensible Service Subscription Protocol (XSSP). Internet-draft, Internet Engineering Task Force, March 2004.

827. M. Uruena and D. Larrabeiti. Overview of the eXtensible Service Discovery Framework (XSDF). Internet draft, Internet Engineering Task Force, March 2004.

828. M. Uruena and D. Larrabeiti. eXtensible Service Transfer Protocol (XSTP). Internet-draft, Internet Engineering Task Force, March 2004.

829. M. Uruena and D. Larrabeiti. eXtensible Service Discovery Framework (XSDF): Common Elements and Procedures. Internet draft, Internet Engineering Task Force, March 2004.

830. M. Uruena and D. Larrabeiti. eXtensible Service Location Protocol (XSLP). Internet-draft, Internet Engineering Task Force, March 2004.

831. U.S. Census Bureau. TIGER, TIGER/Line and TIGER-Related Products. Available at http://www.census.gov/geo/www/tiger.

832. US National Transportation Safety Board and the Federal Highway Administration. Traffic Signal Preemption for Emergency Vehicles—A Cross Cutting Study, January 2006. www.itsdocs.fhwa.dot.gov/jpodocs/repts_te/14097_files/14097.pdf.

833. A. Vahdat and D. Becker. Epidemic Routing for Partially Connected Ad Hoc Networks. Technical Report CS-2000-06, CS. Dept. Duke Univ., 2000.

834. T. van Dam and K. Langendoen. An Adaptive Energy-Efficient MAC Protocol for Wireless Sensor Networks. In *1st International Conference on Embedded Networked Sensor Systems (SenSys)*, pages 171–180, Los Angeles, USA, November 2003.

835. E. Van de Velde, C. Blondia, and L. Campelli. REACT: Routing Protocol for Emergency Applications in Car-to-Car Networks Using Trajectories. In *Annual Mediterranean Ad Hoc Networking Workshop*, Lipari, Italy, June 2006.

836. P. van Emde Boas. Space Measures for Storage Modification Machines. *Information Processing Letters*, 30(2):103–110, January 1989.

837. R. van Renesse, K. Birman, M. Hayden, A. Vaysburd, and D. Karr. Adaptive Systems Using Ensemble. *Software Practice and Experience*, 28(9):963–979, August 1998.

838. R. van Renesse, K. Birman, and W. Vogels. Astrolabe: A Robust and Scalable Technology for Distributed Systems Monitoring, Management, and Data Mining. *ACM Transactions on Computing Systems*, 21(3), May 2003.

839. A. Varga. The OMNeT++ Discrete Event Simulation System. In *Proceedings of the European Simulation Multiconference (ESM'01)*, June 2001.

840. A. Vasquez, R. Pastor-Satorras, and A. Vespignani. Large-Scale Topological and Dynamical Properties of the Internet. *Physical Review E*, 67, 2003.

841. G. Viswanathan, V. Afanasyev, S. Buldyrev, E. Murphy, P. Prince, and H. Stanley. Lévy Flight Search Patterns of Wandering Albatrosses. *Nature*, 381:413–415, May 1996.

842. G. Viswanathan, S. Buldyrev, S. Havlin, M. da Luz, E. Raposo, and H. Stanley. Optimizing the Success of Random Searches. *Nature*, 401(6756):911–914, 1999.

843. Y. Vogiazou, B. Raijmakers, B. Clayton, M. Eisenstadt, E. Geelhoed, J. Linney, K. Quick, J. Reid, and P. Scott. You Got Tagged!: The City as a Playground. In *Second International Conference on Appliance Design (2AD'04)*, Bristol, UK, May 2004.

844. J. von Neumann and A. Burks. *Theory of Self-Reproducing Automata*. University of Illinois Press, Urbana, Illinois, 1966.

845. S. Voulgaris, E. Riviere, A.-M. Kermarrec, and M. van Steen. SUB-2-SUB: Self-Organizing Content-Based Publish and Subscribe for Dynamic and Large Scale Collaborative Networks. In *International Workshop on Peer-To-Peer Systems*, 2006.

846. R. Wahbe, S. Lucco, T. Anderson, and S. Graham. Efficient Software-Based Fault Isolation. In *Proceedings of the 14th ACM Symposium on Operating System Principles*, pages 203–216, Asheville, North Carolina, U.S.A., December 1993. The ACM Press.

847. R. Wakikawa, J. Malinen, C. Perkins, A. Nilsson, and A. Tuominen. Internet Connectivity for Mobile Ad Hoc Networks. Internet draft (work in progress), Internet Engineering Task Force, November 2002.

848. C. Wang, A. Carzaniga, D. Evans, and A. Wolf. Security Issues and Requirements for Internet-Scale Publish-Subscribe Systems. In *The Proceedings of the Hawaii International Conference on System Sciences*, January 2002.

849. K. Wang and B. Li. Group Mobility and Partition Prediction in Wireless Ad-Hoc Networks. *Proceedings of ICC'02*, 2:1017–1021, 2002.

850. Y. Wang and X.-Y. Li. Geometric Spanners for Wireless Ad Hoc Networks. In *Proceedings of the 22nd IEEE International Conference on Distributed Computing Systems (ICDCS '02)*, 2002.

851. Y. Wang, L. Qiu, D. Achlioptas, G. Das, P. Larson, and H. Wang. Subscription Partitioning and Routing in Content-Based Publish/Subscribe Networks. In *Proceeding of the 16th International Symposium on DIStributed Computing (DISC'02)*, October 2002.

852. Y.-M. Wang, L. Qiu, C. Verbowski, D. Achlioptas, G. Das, and P. Larson. Summary-Based Routing for Content-Based Event Distribution Networks. *SIGCOMM Comput. Commun. Rev.*, 34(5):59–74, 2004.

853. Z. Wang, S. Elbaum, and D. Rosenblum. Automated Generation of Context-Aware Tests. In *Proceedings of ICSE'07*. The ACM Press, 2007.

854. R. Want, K. Farkas, and C. Narayanaswami. Guest Editors' Introduction: Energy Harvesting and Conservation. *IEEE Pervasive Computing*, 4(1):14–17, January/March 2005.

855. S. Wasserman, K. Faust, and D. Iacobucci. *Social Network Analysis: Methods and Applications (Structural Analysis in the Social Sciences)*. Cambridge University Press, November 1994.

856. M. Wattenhofer, R. Wattenhofer, and P. Widmayer. Geometric Routing without Geometry. In *Proceedings of the 12th Colloquium on Structural Information and Communication Complexity*, LNCS. Springer-Verlag, May 2005.

857. D. Watts. *Small Worlds The Dynamics of Networks between Order and Randomness*. Princeton Studies on Complexity. Princeton University Press, 1999.

858. M. Weiser. The Computer for the Twenty-First Century. *Scientific American*, 265(3):94–104, September 1991.

859. Y. Weissn and E. Barrantes. Known/Chosen Key Attacks Against Software Instruction Set Randomization. In *22nd Annual Computer Security Applications Conference*, Miami Beach, Florida, U.S.A., December 2006. IEEE Press.

860. E. Weiß, G. Gehlen, S. Lukas, C.-H. Rokitansky, and B. Walke. MYCAREVENT—Vehicular Communication Gateway for Car Maintenance and Remote Diagnosis. In *11th IEEE Symposium on Computers and Communications (ISCC'06)*, Pula-Cagliari, Sardinia, Italy, June 2006.

861. M. Welsh and G. Mainland. Programming Sensor Networks Using Abstract Regions. In *Proceedings of 1st Symp. on Networked Systems Design and Implementation (NSDI)*, 2004.

862. M. Welsh and G. Mainland. Programming Sensor Networks Using Abstract Regions. In *Proceedings of the First USENIX/ACM Symposium on Networked Systems Design and Implementation (NSDI '04)*, March 2004.

863. K. Weniger. Passive Duplicate Address Detection in Mobile Ad Hoc Networks. In *Proceedings of the IEEE Wireless Communications and Networking Conference (WCNC)*, volume 3, pages 1504–1509, 2003.

864. K. Whitehouse, C. Sharp, E. Brewer, and D. Culler. Hood: A Neighborhood Abstraction for Sensor Networks. In *Proceedings of the 2nd International Conference on Mobile Systems, Applications, and Services (MOBISYS)*, 2004.

865. J. Widmer and J. Le Boudec. Network Coding for Efficient Communication in Extreme Networks. *Proceedings of the ACM SIGCOMM 2005 Workshop on Delay Tolerant Networking (WDTN 2005)*, pages 284–291, 2005.

866. A. Wigley, M. Sutton, S. Wheelwright, R. Burbidge, and R. Mcloud. Microsoft .Net. Compact Framework: Core Reference. Microsoft Press, 2002.

867. L. Wischhof, A. Ebner, H. Rohling, M. Lott, and R. Halfmann. SOTIS—A Self-Organizing Traffic Information System. In *57th IEEE Vehicular Technology Conference (VTC'03)*, pages 2442–2446, Jeju, South Korea, April 2003.

868. G. Wittenburg, K. Terfloth, F. Villafuerte, T. Naumowicz, H. Ritter, and J. Schiller. Fence Monitoring—Experimental Evaluation of a Use Case for Wireless Sensor Networks. In *Proceedings of the 4th European Conf. on Wireless Sensor Networks (EWSN)*, 2007.

869. R. Wojtczuk. Defeating Solar Designer's Non-executable Stack Patch. Available at http://www.insecure.org/sploits/non-executable.stack.problems.html, 1998.

870. Wolfram MathWorld: The Web's Most Extensive Mathematics Resource. Available at http://mathworld.wolfram.com.

871. WSDL—Web Service Description Language. Available at http://www.w3.org/TR/wsdl.

872. D. Wu, D. Gupta, S. Liese, and P. Mohapatra. QuRiNet: Quail Ridge Natural Reserve Wireless Mesh Network. In *Proceedings of the 1st International Workshop on Wireless Network Testbeds, Experimental Evaluation & Characterization (WiNTECH '06)*, pages 109–110. The ACM Press, 2006.

873. K. Xu, X. Hong, and M. Gerla. Landmark Routing in Ad Hoc Networks with Mobile Backbones. *Journal of Parallel Distributed Computing*, 63:110–122, 2003.

874. Q. Xu, K. Hedrick, R. Sengupta, and J. VanderWerf. Effects of Vehicle-Vehicle / Roadside-Vehicle Communication on Adaptive Cruise Controlled Highway Systems. In *56th IEEE International Vehicular Technology Conference (VTC'02)*, volume 2, pages 1249–1253, Vancouver, BC, September 2002.

875. W. Xu, S. Bhatkar, and R. Sekar. Taint-Enhanced Policy Enforcement: A Practical Approach to Defeat a Wide Range of Attacks. In *Proceedings of the 15th USENIX Security Symposium*, Vancouver, British Columbia, Canada, August 2006. USENIX Association.

876. G. Xylomenos, G. Polyzos, P. Mahonen, and M. Saaranen. TCP Performance Issues Over Wireless Links. *IEEE Communications Magazine*, 39:52–58, 2001.

877. T. Yan and H. Garcia-Molina. Index Structures for Selective Dissemination of Information Under the Boolean Model. *ACM Trans. Database Syst.*, 19(2):332–364, 1994.

878. T. Yan and H. Garcia-Molina. The SIFT Information Dissemination System. *ACM Trans. Database Syst.*, 24(4):529–565, 1999.

879. Y. Yang, H. Hassanein, and A. Mawji. A New Approach to Service Discovery in Wireless Mobile Ad Hoc Networks. In *Proceedings of the IEEE ICC*, pages 3838–3843, 2006.

880. J. Yap. Implementing Road and Congestion Pricing—Lessons from Singapore. In *Workshop on Implementing Suistainable Urban Travel Policies in Japan and other Asia-Pacific Countries*, Tokyo, Japan, March 2005.

881. W. Ye, J. Heidemann, and D. Estrin. An Energy-Efficient MAC Protocol for Wireless Sensor Networks. In *21st Annual Joint Conference of the IEEE Computer and Communications Societies (INFOCOM)*, volume 3, pages 1567–1576, New York, June 2002.

882. W. Ye, F. Silva, and J. Heidemann. Ultra-Low Duty Cycle MAC with Scheduled Channel Polling. In *4th International Conference on Embedded Networked Sensor Systems (SenSys)*, pages 321–334, Boulder, USA, November 2006.

883. C. Yeo, B. Lee, and M. Er. A Framework for Multicast Video Streaming over IP Networks. *J. Network and Computer Applications*, 26(3):273–289, 2003.

884. Y. Yokote. The Apertos Reflective Operating System: The Concept and its Implementation. In *Proceedings of the 7th International Conference on Object-Oriented Programming Systems, Languages, and Applications (OOPSLA'92)*, pages 414–434, British Columbia, Canada, October 1992.

885. E. Yoneki and J. Bacon. Content Based Routing with On-Demand Multicast. In *Proceedings of the 24th IEEE International Conference on Distributed Computing Systems, Workshop on Wireless Ad Hoc Networking (ICDCS - WWAN 2004)*, pages 788–793, 2004.

886. E. Yoneki, P. Hui, S. Chan, and J. Crowcroft. A Socio-Aware Overlay for Publish-Subscribe Communication in Delay Tolerant Networks. In *10th ACM-IEEE International Symposium on Modeling, Analysis and Simulation of Wireless Mobile Systems*, 2007.

887. H.-J. Yoon, E.-J. Lee, H. Jeong, and J.-S. Kim. Proximity-Based Overlay Routing for Service Discovery in Mobile Ad Hoc Networks. In *Proceedings of ISCIS*, pages 176–186, 2004.

888. J. Yoon, M. Liu, and B. Noble. Random Waypoint Considered Harmful. In *Proceedings of INFOCOM'03*, 2003.

889. J. Yoon, M. Liu, and B. Noble. Sound Mobility Models. In *Proceedings of MobiCom'03*, pages 205–216. The ACM Press, 2003.

890. J. Yoon, B. Noble, M. Liu, and M. Kim. Building Realistic Mobility Models from Coarse-Grained Traces. In *Proceedings of MobiSys'06*, pages 177–190. The ACM Press, 2006.

891. Y. Younan, W. Joosen, and F. Piessens. Code Injection in C and C++: A Survey of Vulnerabilities and Countermeasures. Technical Report CW386, Departement Computerwetenschappen, Katholieke Universiteit Leuven, July 2004.

892. Y. Younan, W. Joosen, and F. Piessens. A Methodology for Designing Countermeasures against Current and Future Code Injection Attacks. In *Proceedings of the Third IEEE International Information Assurance Workshop 2005 (IWIA2005)*, College Park, Maryland, U.S.A., March 2005. IEEE Press.

893. Y. Younan, W. Joosen, and F. Piessens. Efficient Protection Against Heap-Based Buffer Overflows without Resorting to Magic. In *Proceedings of the International Conference on Information and Communication Security (ICICS 2006)*, Raleigh, North Carolina, U.S.A., December 2006.

894. Y. Younan, W. Joosen, and F. Piessens. Extended Protection Against Stack Smashing Attacks without Performance Loss. In *Proceedings of the Twenty-Second Annual Computer Security Applications Conference*, Miami, Florida, U.S.A., December 2006. IEEE Press.

895. S. Zachariadis, L. Capra, C. Mascolo, and W. Emmerich. Xmiddle: Information Sharing Middleware for a Mobile Environment. In *Proceedings of the 24th International Conference on Software Engineering (ICSE'02)*. The ACM Press, 2002.

896. X. Zeng, R. Bagrodia, and M. Gerla. GloMoSim: A Library for Parallel Simulation of Large-Scale Wireless Networks. In *Proceedings of the 12th Workshop on Parallel and Distributed Simulation (PADS'98)*, pages 154–161. IEEE Press, 1998.

897. B. Zhang, S. Jamin, and L. Zhang. Host Multicast: A Framework for Delivering Multicast To End Users. In *INFOCOM*, 2002.

898. C. Zhang and H. Jacobsen. Refactoring Middleware with Aspects. *IEEE Transactions on Parallel and Distributed Systems*, 14(11):1058–1073, 2003.

899. J. Zhang, L. Liu, L. Ramaswam, and C. Pu. PeerCast: Churn-Resilient End System Multicast on Heterogeneous Overlay Networks. *Journal of Network and Computer Applications (JNCA)*, 2007.

900. L. Zhang, F. Wang, M. Han, and R. Mahajan. A General Model of Wireless Interference. In *Proceedings of the 13th Annual ACM International Conference on Mobile Computing and Networking (MobiCom '07)*, pages 171–182. The ACM Press, 2007.

901. X. Zhang, J. Kurose, B. Levine, D. Towsley, and H. Zhang. Study of a Bus-Based Disruption-Tolerant Network: Mobility Modeling and Impact on Routing. In *Proceedings of Mobi-Com'07*, pages 195–206. The ACM Press, 2007.

902. Z. Zhang. Routing in Intermittently Connected Mobile Ad Hoc Networks and Delay Tolerant Networks: Overview and Challenges. *IEEE Communications Surveys & Tutorials*, 8(1):24–37, 2006.

903. B. Zhao, J. Kubiatowicz, and A. Joseph. Tapestry: An Infrastructure for Fault-tolerant Wide-area Location and Routing. Technical Report, University of California at Berkeley, Berkeley, CA, USA, April 2001.

904. B. Zhao, L. Huang, J. Stribling, S. Rhea, A. Joseph, and J. Kubiatowicz. Tapestry: A Resilient Global-scale Overlay for Service Deployment. *IEEE Journal on Selected Areas in Communications*, 2003.

905. J. Zhao, Y. Zhang, and G. Cao. Data Pouring and Buffering on the Road: A New Data Dissemination Paradigm for Vehicular Ad Hoc Networks. *IEEE Transactions on Vehicular Technology*, 56(6):3266–3277, November 2007.

906. W. Zhao, H. Schulzrinne, E. Guttman, C. Bisdikian, and W. Jerome. Select and Sort Extensions for the Service Location Protocol (SLP). Request for Comments 3421, Internet Engineering Task Force, November 2002.

907. W. Zhao, H. Schulzrinne, and E. Guttman. Mesh-enhanced Service Location Protocol (mSLP). Technical Report 3528, Internet Engineering Task Force, April 2003.

908. Y. Zhao, R. Govindan, and D. Estrin. Residual Energy Scan for Monitoring Sensor Networks. In *Wireless Communications and Networking Conference (WCNC)*, volume 1, pages 356–362, Orlando, USA, March 2002.

909. D. Zhu and M. Mutka. Promoting Cooperation Among Strangers to Access Internet Services from an Ad Hoc Network. In *Proceedings of the Second IEEE International Conference on Pervasive Computing and Communications (PerCom'04)*, pages 229–240. IEEE Computer Society, 2004.

910. F. Zhu, M. Mutka, and L. Ni. Facilitating Secure Ad Hoc Service Discovery in Public Environments. *Journal of Systems and Software*, 76:45–54, April 2005.

911. S. Zhuang, B. Zhao, A. Joseph, R. Katz, and J. Kubiatowicz. Bayeux: An Architecture for Scalable and Fault-Tolerant Wide-Area Data Dissemination. In *Proceedings of the 11th International Workshop on Network and Operating Systems Support for Digital Audio and Video (NOSSDAV '01)*, pages 11–20, New York, NY, USA, 2001. The ACM Press.

912. H. Zimmermann. OSI Reference Model The ISO Model of Architecture for Open Systems Interconnection. *IEEE Transactions on Communications*, 28(4):425–432, April 1980.

Author List

Mouna Allani, University of Lausanne, Switzerland, 191–218

Filipe Araujo, University of Coimbra, Portugal, 63–93

James Aspnes, Yale University, USA, 97–120

Roberto Baldoni, Sapienza Università di Roma, Italy, 219–244

João Barros, Instituto de Telecomunicações, Universidade do Porto, Portugal, 25–41

Mélanie Bouroche, Department of Computer Science, Trinity College Dublin, Ireland, 369–381

Vinny Cahill, Department of Computer Science, Trinity College Dublin, Ireland, 369–381

Marco Conti, IIT-CNR, Italy, 121–147

Paolo Costa, Vrije Universiteit, Amsterdam, The Netherlands, 245–264

Jon Crowcroft, Univ. of Cambridge, Computer Lab, UK, 121–147

Patrick Th. Eugster, Purdue University, USA, 305–322

Roy Friedman, Technion, Israel, 169–190

Benoît Garbinato, University of Lausanne, Switzerland, 191–218, 305–322

Silvia Giordano, SUPSI-DTI, Switzerland, 121–147

Tom Goovaerts, DistriNet, Katholieke Universiteit Leuven, Belgium, 265–284

Paul Grace, Computing Department, Lancaster University, UK, 285–302

Adrian Holzer, University of Lausanne, Switzerland, 305–322

Pan Hui, Univ. of Cambridge, Computer Lab, UK, 121–147

Johnathan Ishmael, Lancaster University, UK, 149–166

Wouter Joosen, DistriNet, Katholieke Universiteit Leuven, Belgium, 265–284

Anne-Marie Kermarrec, INRIA, France, 169–190

Mads Darø Kristensen, University of Aarhus, Denmark, 349–368

Luís Lopes, CRACS/DCC-FCUP, Universidade do Porto, Portugal, 25–41

Antonio A.F. Loureiro, Federal University of Minas Gerais, Brazil, 3–24

Jukka Manner, Helsinki University of Technology, Finland, 323–347

Francisco Martins, LASIGE/DI-FCUL, Universidade de Lisboa, Portugal, 25–41

Cecilia Mascolo, University of Cambridge, UK, 43–62

Raquel A.F. Mini, Pontifical Catholic University of Minas Gerais, Brazil, 3–24

Hugo Miranda, FCUL, Portugal, 63–93, 169–190

Luca Mottola, Politecnico di Milano, Italy, 245–264

Amy L. Murphy, BK-IRST, Italy, 245–264

Mirco Musolesi, Dartmouth College, USA, 43–62

Hoang Anh Nguyen, SUPSI-DTI, Switzerland, 121–147

Andrea Passarella, IIT-CNR, Italy, 121–147

Fernando Pedone, University of Lugano, Switzerland, 191–218

Pieter Philippaerts, DistriNet, Katholieke Universiteit Leuven, Belgium, 265–284

Gian Pietro Picco, University of Trento, Italy, 245–264

Frank Piessens, DistriNet, Katholieke Universiteit Leuven, Belgium, 265–284

Jari Porras, Lappeenranta University of Technology, Finland, 349–368

Leonardo Querzoni, Sapienza Università di Roma, Italy, 219–244

Nicholas Race, Lancaster University, UK, 149–166

Oriana Riva, ETH Zürich, Switzerland, 349–368

Luís Rodrigues, INESC-ID/Instituto Superior Técnico, Portugal, 169–190

Eric Ruppert, York University, Canada, 97–120

Aline Senart, Department of Computer Science, Trinity College Dublin, Ireland, 369–381

Sasu Tarkoma, Helsinki University of Technology and Nokia NRC, Finland, 219–244

Antonino Virgillito, Sapienza Università di Roma, Italy, 219–244

Stefan Weber, Department of Computer Science, Trinity College Dublin, Ireland, 369–381

Bart De Win, DistriNet, Katholieke Universiteit Leuven, Belgium, 265–284

Yves Younan, DistriNet, Katholieke Universiteit Leuven, Belgium, 265–284

Reviewer List

Filipe Araujo, University of Coimbra, Portugal

Vinny Cahill, Department of Computer Science, Trinity College Dublin, Ireland

Marco Conti, IIT-CNR, Italy

Paolo Costa, Vrije Universiteit, Amsterdam, The Netherlands

Patrick Th. Eugster, Purdue University, USA

Silvia Giordano, SUPSI-DTI, Switzerland

Paul Grace, Computing Department, Lancaster University, UK

Adrian Holzer, University of Lausanne, Switzerland

Johnathan Ishmael, Lancaster University, UK

Anne-Marie Kermarrec, INRIA, France

Francisco Martins, LASIGE/DI-FCUL, Universidade de Lisboa, Portugal

José Mocito, INESC-ID, University of Lisbon, Portugal

Mirco Musolesi, Dartmouth College, USA

Fernando Pedone, University of Lugano, Switzerland

Jari Porras, Lappeenranta University of Technology, Finland

Nicholas Race, Lancaster University, UK

Oriana Riva, ETH Zürich, Switzerland

Eric Ruppert, York University, Canada

Antonino Virgillito, Sapienza Università di Roma, Italy

Bart De Win, DistriNet, Katholieke Universiteit Leuven, Belgium

Index